Laser and Fiber Optic Gas Absorption Spectroscopy

An invaluable text for the teaching, design and development of gas sensor technology. This excellent resource synthesises the fundamental principles of spectroscopy, laser physics, and photonics technology and engineering to enable the reader to fully understand the key issues and apply them in the design of optical gas absorption sensors. It provides a straightforward introduction to low-cost and highly versatile near-IR systems, as well as an extensive review of mid-IR systems. Fibre laser systems for spectroscopy are also examined in detail, especially the emerging technique of frequency comb spectroscopy. Featuring many examples of real-world application and performance, as well as MATLAB computer programs for modelling and simulation, this exceptional work is ideal for postgraduate students, researchers and professional engineers seeking to gain an in-depth understanding of the principles and applications of fibre optic and laser-based gas sensors.

George Stewart is Research Professor at Strathclyde University.

Laser and Fiber Optic Gas Absorption Spectroscopy

GEORGE STEWART
University of Strathclyde

CAMBRIDGE
UNIVERSITY PRESS

University Printing House, Cambridge CB2 8BS, United Kingdom

One Liberty Plaza, 20th Floor, New York, NY 10006, USA

477 Williamstown Road, Port Melbourne, VIC 3207, Australia

314–321, 3rd Floor, Plot 3, Splendor Forum, Jasola District Centre, New Delhi – 110025, India

79 Anson Road, #06–04/06, Singapore 079906

Cambridge University Press is part of the University of Cambridge.

It furthers the University's mission by disseminating knowledge in the pursuit of
education, learning, and research at the highest international levels of excellence.

www.cambridge.org
Information on this title: www.cambridge.org/9781107174092
DOI: 10.1017/9781316795637

First published 2021

Printed in the United Kingdom by TJ Books Limited

A catalogue record for this publication is available from the British Library.

ISBN 978-1-107-17409-2 Hardback

Additional resources for this publication at www.cambridge.org/stewart.

This book is dedicated to my mother who sadly passed away before it was completed. She never really understood the work I did but was always immensely proud of it and was always there to support me throughout her life. She will be greatly missed.

Contents

Preface

This book has been born from the author's teaching and research experience of more than 40 years in the field of photonics, including 20 years in laser spectroscopy. It is instructive to look back over these years to identify important contributions that combine photonics with the science of spectroscopy and to bring them together in this book under the single topic of laser and fibre optic absorption spectroscopy for gas sensing applications.

In considering this topic, it becomes evident that knowledge of a range of disciplines is necessary in order to gain a full understanding of the technology and to be in a position to capitalise on future opportunities. It is therefore a primary objective of the book to bring together in a single volume the relevant fundamental physics, laser theory, physical chemistry, photonics and engineering design principles to enable the reader to have both a thorough understanding of the background theory and the ability to apply the results in the design of optical gas sensors for practical applications. Of course it is recognised that some readers may already have sufficient or greater knowledge of certain topics that are included in this book and hence the contents are presented in such a way that a reader may skip a chapter that he is already familiar, for example, the first chapter on the basics of absorption spectroscopy which will be well known to those conversant with gas spectroscopy. Some readers may be familiar with the content of later chapters on the fundamental theory of diode and fibre lasers, but it should be noted that, while there are a large number of textbooks on this topic, the theoretical treatment given here is oriented toward spectroscopy applications which have rather different constraints and requirements from those pertaining to the common application of diode lasers in data communication systems.

A further important point is worthy of mention here. Most of the content relates to the near-IR spectral region for absorption spectroscopy of gases where, in this context, the near-IR means the use of light and laser wavelengths at less than approximately two microns, within the transmission window of standard silica optical fibres. Of course this is not usually the most sensitive spectral region for gas spectroscopy, but there are several reasons for this approach. First, sensors operating in the near-IR provide the simplest, cheapest, most versatile and most mature solutions and should always be considered first for meeting the requirements of a particular situation. Second, the principles and techniques of spectroscopy discussed in detail in the book, such as wavelength modulation spectroscopy, photoacoustic spectroscopy, evanescent-wave spectroscopy, cavity-enhanced spectroscopy, etc., are all equally applicable to the mid-IR region and this is highlighted, as appropriate, throughout the text. However, mid-IR sources and sensing applications are growing rapidly in maturity and, with this in mind, the final chapter provides an important overview of mid-IR gas absorption sensors.

There are also a number of useful on-line resources available with this book. Several MATLAB programs allow the reader to simulate the signal outputs from practical gas sensors under different conditions of operation from those illustrated in the book figures. PowerPoint presentations highlight the key points of each chapter, including all the illustrations and figures and may be used directly for teaching purposes. These resources are located at: www.cambridge.org/stewart.

I would like to express my gratitude to a number of colleagues and researchers without whose inputs it would have been impossible to write this book. First, I would like to acknowledge the invaluable support provided by Professors Brian Culshaw and Walter Johnstone who have been my close colleagues for more than 30 years and whose research leadership, theoretical insights, acute experimental skills, along with countless discussions over the years, have been instrumental in realising so much of the original contributions reported here. In addition their work in forming the spin-out company OptoSci Ltd., along with the other company directors, Drs Iain Mauchline, Douglas Walsh, Micky Pery and Professor Julian Jones, has resulted in a number of the ideas being transformed into commercial products. I am also deeply indebted to Professor Deepak Uttamchandani, who has been a close colleague for more than 30 years, for many stimulating discussions on related topics in photonics and to more recent colleagues, Dr Michael Lengden and Dr Ralf Bauer for their contributions in diode laser and photoacoustic spectroscopy. Most important, however, have been the considerable number of research students and post-doctoral researchers who have done most of the hard labour in the laboratories over many years in testing and exploring new ideas. In this regard, I would like to acknowledge the pioneering research of Wei Jin, currently Professor of Photonic Instrumentation at the Hong Kong Polytechnic University, and Dr Wayne Philp who were the first to explore evanescent-wave and fibre optic gas sensing in our research laboratories at Strathclyde University. Of particular note is the pioneering PhD research work performed by many research students at the time, notably Dr Kevin Duffin whose perseverance in the face of our limited understanding eventually opened up new insights on the importance of phase in the measurement of harmonic signals from diode lasers, leading to further original contributions from Drs Andrew McGettrick, Keith Ruxton, James Bain and Arup L. Chakraborty. On advancing the understanding and application of fibre lasers and fibre optic systems in spectroscopy, I am particularly grateful for the painstaking work of Drs Kathryn Atherton, Miha Završnik, Gillian Whitenett, Joanna Bhullar Marshall, Zhou Meng, M. Aleem Mirza, Norhana Arsad, Li Dawn Li and Min Li.

Finally, and most important, I would like to express my deepest gratitude to those nearest and dearest to me, my father and mother who sadly have passed away, my brother Willie and to my dearest friend and companion over the past year, Sari.

1 Absorption Spectroscopy of Gases

1.1 Introduction

The science of spectroscopy dates back to the seventeenth century with Newton's demonstration of the dispersion of sunlight into a spectrum of colours, but it was not until the nineteenth century that it was recognised that elements and molecules possess a characteristic emission or absorption spectrum. Observation of emission or absorption line spectra became an important method for the identification of chemical compounds and the only method for determining the composition of stars and other objects in our universe. The advent of the laser in the second half of the twentieth century marked the era of modern spectroscopy, opening up new techniques based on the laser's unique properties of high intensity, spectral purity and coherence. Additionally, over the past 30 years the widespread deployment of fibre optic data communication has led to the ready availability of low cost and versatile laser sources in the near-IR region of the spectrum, as well as low-loss fibres and a variety of fibre components. Today, laser absorption spectroscopy in the near-IR and mid-IR spectral regions is a well-established technique for the identification of gases and for the measurement of gas parameters such as concentration, pressure and temperature. The primary focus of this book is on near-IR gas absorption spectroscopy to take advantage of the maturity of near-IR lasers and photonics, but it should be noted that the principles and methods discussed throughout are equally applicable to the longer wavelength mid-IR region, which will be considered in detail in Chapter 8. In this chapter we will review the fundamental principles and parameters used to describe the absorption spectroscopy of gases, covering topics such as the Beer–Lambert law, lineshapes and line-broadening mechanisms, fundamental and overtone absorption lines and the measurement of gas parameters from lineshape information.

1.2 Fundamentals of Optical Absorption

1.2.1 Definition of Parameters

There are several parameters used to describe the absorption of light by gas molecules. We start with the definition of the absorption cross-section for a single gas molecule, where the molecule is viewed as having an effective area, σ, for the capture or absorption of a photon from a beam of intensity, $I(x,y)$, as shown in Figure 1.1.

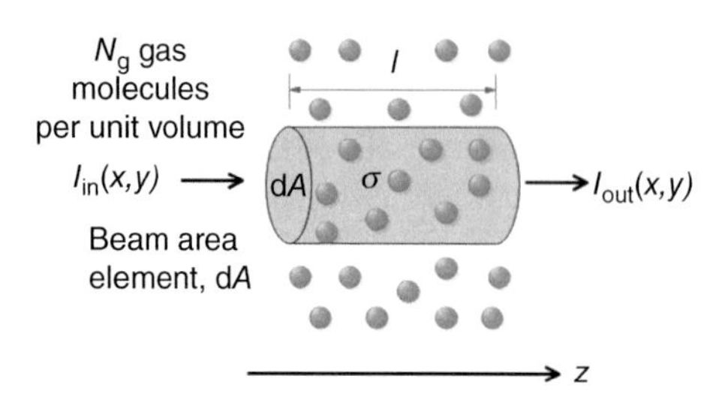

Figure 1.1 Description of the parameters used in the Beer–Lambert law.

Considering a cross-sectional area of $dA = dxdy$, the probability of photon capture or absorption by the molecule is σ/dA. Over a short length, dz, in the propagation direction, the beam will encounter $N_g dAdz$ gas molecules and the rate of photon absorption or change in beam intensity will therefore be:

$$dI(x,y) = -I(x,y)\left(\frac{\sigma}{dA}\right)\left(N_g dAdz\right) = -I(x,y)\sigma N_g dz \tag{1.1}$$

where N_g is the number density of the target gas molecules (number per unit volume) and we assume that this density is uniform in the (x,y) transverse plane and along beam direction, z.

The power in the beam is $P = \iint I(x,y)dxdy$ and hence integration of (1.1) leads to the Beer–Lambert law for the transmitted power, P_{out}, after a path length, l, over which the beam and gas interact:

$$P_{out} = P_{in}e^{-\sigma N_g l} \tag{1.2}$$

where P_{in} is the incident beam power.

As shown in Figure 1.2, the absorption cross-section, σ, is dependent on the optical frequency, ν, gas pressure, p, and temperature, T, and may be written as $\sigma(\nu,p,T) = S(T)\phi(\nu,p,T)$, where $\phi(\nu,p,T)$ describes the lineshape function and S is the line strength per molecule defined by:

$$S(T) = \int_{-\infty}^{\infty} \sigma(\nu,p,T)d\nu \tag{1.3}$$

Note that with the above definitions, $\phi(\nu,p,T)$ is normalised so that $\int_{-\infty}^{\infty}\phi(\nu,p,T)d\nu = 1$.

The number density, N_g, in (1.2) may be expressed in terms of the mole fraction or partial pressure of the target gas using the ideal gas law and Dalton's law:

$$p = \left(\frac{n}{V}\right)RT = \left(\frac{n_1 + n_2 + \cdots + n_i}{V}\right)RT = p_1 + p_2 + \cdots + p_i \tag{1.4}$$

where n is the total number of moles and n_i and p_i are the number of moles and the partial pressure of each of the constituent gases, respectively.

Hence we can write for the number density of the target gas:

$$N_g = N_A\left(\frac{n_g}{V}\right) = \frac{pX_g}{k_B T} = \frac{p_g}{k_B T} \tag{1.5}$$

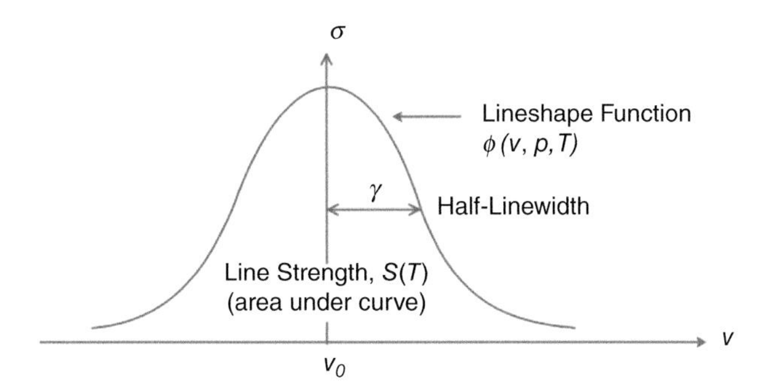

Figure 1.2 Typical absorption cross-section of a gas molecule and the associated parameters.

where N_A is Avogadro's number, $k_B = R/N_A$ is the Boltzmann constant, p is the total gas pressure, n_g is the number of moles, p_g the partial pressure and $X_g = n_g/n$ is the mole fraction of the target gas.

Hence the Beer–Lambert law may be written as:

$$P_{\text{out}} = P_{\text{in}} \exp\left[-S'(T)\phi(\nu, p, T)pX_g l\right] = P_{\text{in}} \exp\left[-S'(T)\phi(\nu, p, T)p_g l\right] \tag{1.6}$$

where $S'(T) = S(T)/(k_B T)$.

Another commonly used parameter is the absorption coefficient, α, where the absorption cross-section or line strength for a single molecule is multiplied by N_0 as follows:

$$\alpha(\nu, p, T) = N_0\sigma(\nu, p, T) = N_0 S(T)\phi(\nu, p, T) \tag{1.7}$$

where $N_0 = p_0/(k_B T_0) \cong 2.5 \times 10^{25}\text{m}^{-3}$ is the molecular density at normal temperature and pressure (NTP, $T_0 = 293.15$ K, $p_0 = 1$ atm pressure or 101.325 kPa).

The Beer–Lambert law is then:

$$P_{\text{out}} = P_{\text{in}} \exp\left[-\alpha C l\right] \tag{1.8}$$

where the target gas concentration is given by $C = N_g/N_0 = (T_0/T)(pX_g/p_0) = (T_0/T)(p_g/p_0)$, with $T_0 = 293.15$ K and $p_0 = 1$atm pressure.

Equations (1.2), (1.6) and (1.8) represent useful equivalent forms of the Beer–Lambert law for calculating the gas absorption.

1.2.2 Absorption Lineshape Functions

The absorption lineshape function, $\phi(\nu, p, T)$, for gases is determined by three line-broadening mechanisms, namely, natural (or lifetime) broadening, Doppler broadening and collisional (or pressure) broadening [1, 2].

1.2.2.1 Natural Broadening

Photon absorption occurs when an electron, atom or molecule is raised from a lower to a higher energy level. However, in accordance with the uncertainty principle, $\Delta E \Delta t \geq h$, the levels have an uncertainty in their energy giving rise to a small range of optical frequencies over which the absorption occurs. Similar to the exponential decay in spontaneous emission of light from downward transitions, which is characterised by a Lorentzian function in the frequency domain, the absorption lineshape from natural broadening is also described by a normalised Lorentzian function with a natural half-linewidth, γ_n. Natural broadening represents the limit in the sharpness of a spectral line, but will not be considered further here since for most practical applications the line-width is dominated by Doppler or pressure-broadening effects, as discussed below.

1.2.2.2 Doppler Broadening

The random thermal motion of gas molecules means that the optical frequency 'seen' by a molecule is shifted due to the Doppler effect. The frequency shift will be positive or negative depending on whether the *z*-component of the molecular velocity, v_z, is in the opposite or in the same direction as the beam propagation along the *z*-axis (the velocity component perpendicular to the beam will have no effect). The frequency shift is given by:

$$\frac{\nu - \nu_0}{\nu_0} = v_z/c \tag{1.9}$$

where ν_0 is the centre frequency of the transition.

The number of gas molecules with a velocity component between v_z and $v_z + \mathrm{d}v_z$ is given by the one-dimensional Maxwell–Boltzmann distribution:

$$N(v_z)\mathrm{d}v_z = N_T\left(\frac{m}{2\pi k_B T}\right)^{\frac{1}{2}} \exp\left\{-\frac{mv_z^2}{2k_B T}\right\}\mathrm{d}v_z \tag{1.10}$$

where N_T is the total number of molecules and m is the molecular mass.

Substituting $v_z = (c/\nu_0)(\nu - \nu_0)$ and $dv_z = (c/\nu_0)d\nu$ from (1.9) into (1.10) and, since the absorbed power in the range ν to $\nu + \mathrm{d}\nu$ will be proportional to the number of molecules in the range v_z to $v_z + \mathrm{d}v_z$, we obtain a normalised Gaussian lineshape for Doppler broadening as:

$$\phi_G(\nu, T) = \frac{1}{\gamma_D\sqrt{\pi}} \exp\left\{-\left(\frac{\nu - \nu_0}{\gamma_D}\right)^2\right\} \tag{1.11}$$

with the $1/e$ half-linewidth,

$$\gamma_D = \nu_0\sqrt{\frac{2k_B T}{mc^2}} \cong \frac{1.29 \times 10^2}{\lambda_0}\sqrt{\frac{T}{M}} \tag{1.12}$$

where M is the molecular weight in atomic mass units. The half-width-half-maximum linewidth is $\gamma_{HWHM} = \gamma_D\sqrt{\ln(2)}$. Typically, for CO_2 at an absorption wavelength of ~2 μm, with $M = 44$ and T ~300 K, γ_{HWHM} ~ 0.14 GHz.

1.2.2.3 Collisional Broadening

Gas molecules are subject to random collisions and, as the molecules approach each other during collisions, their energy levels will be perturbed, each level by differing amounts.

Hence the frequency of photon absorption during a collision will differ from the unperturbed situation, resulting in collisional broadening of the absorption line. From a simple analysis, we would expect the line broadening to be dependent on the collision rate or inversely on the mean time, τ_c, between collisions. In turn, the collision rate will be proportional to the collision cross-section, σ_c, the total number density of molecules present, $N_D = p/k_BT$ (from the ideal gas law) and the average molecular speed, $\bar{v} = \sqrt{8k_BT/\pi m}$ (from the Maxwell–Boltzmann distribution). Hence:

$$\frac{1}{\tau_c} \sim \sigma_c N_D \bar{v} = \sigma_c p \sqrt{\frac{8}{\pi m k_B T}} \tag{1.13}$$

Collisional (or pressure) broadening produces a normalised Lorentzian lineshape of the form:

$$\phi_L(\nu) = \left(\frac{1}{\pi\gamma_c}\right)\left\{1 + \left(\frac{\nu - \nu_0}{\gamma_c}\right)^2\right\}^{-1} \tag{1.14}$$

where, from (1.13), the half-linewidth will have a pressure and temperature dependence of the form:

$$\gamma_c = \gamma_0 \frac{p}{p_0}\left(\frac{T_0}{T}\right)^n \tag{1.15}$$

where γ_0 is the half-linewidth at NTP and the temperature index, n, is ½ in the simple model outlined above, but in practice may differ from this value.

Broadening will, however, depend on whether collisions occur between molecules of the same species (self-broadening by the target gas) or cross-broadening from different species. The overall broadening may be approximated as the weighted sum of self-broadening and cross-broadening effects. For example, for the target gas in air:

$$\gamma_c \cong \frac{\{p_g\gamma_{0s} + (p - p_g)\gamma_{0a}\}}{p_0}\left(\frac{T_0}{T}\right)^n \tag{1.16}$$

where γ_{0s} and γ_{0a} are the self- and air-broadened half-widths at NTP and p_g is the partial pressure of the target gas.

Note from the above discussion that the Doppler linewidth is dependent only on temperature, whereas pressure broadening depends on pressure and inversely on temperature. At NTP, pressure-broadened linewidths are typically of the order of several GHz and dominate over Doppler broadening. However, at lower pressures and/or higher temperatures where the pressure and Doppler-broadened linewidths are comparable, the lineshape may be described by a convolution of the normalised Gaussian and Lorentzian lineshapes:

$$\phi_V(\nu - \nu_0) = \left\{\left(\frac{1}{\gamma_D\sqrt{\pi}}\right)\exp\left[-\left(\frac{\nu - \nu_0}{\gamma_D}\right)^2\right]\right\} \otimes \left\{\left(\frac{1}{\pi\gamma_c}\right)\left[1 + \left(\frac{\nu - \nu_0}{\gamma_c}\right)^2\right]^{-1}\right\} \tag{1.17}$$

This convolution function is also normalised and may be written as:

$$\phi_V(\nu - \nu_0) = \frac{1}{\gamma_D \sqrt{\pi}} V(x, a) \tag{1.18}$$

where $V(x, a)$ is the Voigt function [3] defined by the integral:

$$V(x, a) = \frac{a}{\pi} \int_{-\infty}^{\infty} \frac{e^{-u^2}}{a^2 + (x - u)^2} \mathrm{d}u \tag{1.19}$$

with $x = (\nu - \nu_0)/\gamma_D$, $a = \gamma_c/\gamma_D$.

The Voigt function cannot be evaluated analytically, but various approximations are given in the literature [4–7]. Approximations to the Voigt linewidth are discussed by Olivero and Longbothum [5].

1.3 Extraction of Gas Parameters from Absorption Line Measurements

Measurement of the half-width, γ, and line centre depth, d_c, of an absorption line yields information on the gas concentration, pressure or temperature. In situations where Doppler broadening is dominant, the target gas temperature may be determined uniquely from the linewidth using (1.12). The normalised depth of a Doppler-broadened line from (1.6) and (1.11) is:

$$d_c = \left(\frac{I_{in} - I_{out}}{I_{in}} \right)_{\nu = \nu_0} = 1 - \exp\left[-S'(T) \frac{p_g l}{\gamma_D \sqrt{\pi}} \right] \cong S'(T) \frac{p_g l}{\gamma_D \sqrt{\pi}} \tag{1.20}$$

which gives $p_g \cong (\sqrt{\pi}\, \gamma_D d_c)/\{S'(T) l\}$ and the approximation applies for small absorbance.

Hence knowing the line strength and gas cell length, the partial pressure (or target gas concentration) may be determined from the measured line depth and width, but no information can be gained on the total gas pressure if other gases are present.

Where collisional broadening is dominant, the normalised depth of the Lorentzian profile from (1.6) and (1.14) is:

$$d_c = \left(\frac{I_{in} - I_{out}}{I_{in}} \right)_{\nu = \nu_0} = 1 - \exp\left[-S'(T) \frac{p_g l}{\pi \gamma_c} \right] \cong S'(T) \frac{p_g l}{\pi \gamma_c} \tag{1.21}$$

which gives $p_g \cong (\pi \gamma_c d_c)/\{S'(T) l\}$.

So again the partial pressure (or target gas concentration) may be determined from the measured line depth and width, knowing the line strength and gas cell length.

However, the linewidth depends on both pressure and temperature for Lorentzian and Voigt profiles, as shown by (1.16), and hence additional information is required to determine both pressure and temperature. For this, we can make use of the fact that the line strength, defined by (1.3) as the area under the absorption cross-section distribution, is a function of temperature only and is not dependent on the linewidth. This is because the broadening mechanisms spread the absorption probability distribution function over a wider range of frequencies, but the total integrated probability remains the same for the same number density of target gas.

Hence if we simultaneously measure the widths (γ_1 and γ_2) and normalised depths (d_{c1} and d_{c2}) of two nearby absorption lines of the same species, then, since the gas concentration and cell length are the same for both lines, the ratio of line strengths may be calculated from (1.20) or (1.21) for small absorbance as:

$$\frac{S'_1(T)}{S'_2(T)} = \frac{S_1(T)}{S_2(T)} = \left(\frac{\gamma_1}{\gamma_2}\right)\left(\frac{d_{c1}}{d_{c2}}\right) \tag{1. 22}$$

This ratio of line strengths, which is purely a function of temperature, may be calibrated against temperature for two particular absorption lines and used for gas temperature measurements. (Alternatively for large absorbance, this ratio may be computed from the ratio of areas under the two lines.) For this procedure it is clearly advantageous to choose two lines where the line strength variation with temperature is significantly different; see Arroyo [8] for more information.

Finally, we note that high-speed gas flows can be monitored from the Doppler effect on the central position of an absorption line. If the absorption line is monitored by two beams, one parallel and one orthogonal to the direction of gas flow, then the shift, $\Delta\nu_0$, in the line centre position between the two measurements is simply related to the gas flow velocity, V_g, by: $\Delta\nu_0 = (V_g/c)\nu_0$.

1.4 Absorption Spectra of Gases

Here we briefly review the origin and nature of the absorption lines of gases, with a particular interest in the near-IR region. A more detailed account may be found in references [9–12]. In general, absorption lines arise from the excitation of rotational and vibrational states of gas molecules and from electron transitions between orbitals of the atoms involved. As in the case of electronic transitions, the rotational and vibrational states are quantised, giving rise to a discrete set of energy levels and absorption lines. In broad terms, the energy separations of rotational levels correspond to the microwave region, vibrational transitions to the mid-IR region and electronic transitions to the <1 μm wavelength region. However, as we shall see, both rotational and vibrational transitions have an important bearing on the near-IR absorption spectrum. Diatomic molecules possessing two atoms of the same type, such as O_2 or H_2 undergo no changes in electric dipole moment under rotation or vibration so have no rotational or vibrational spectra and their spectral features of interest arise from electronic transitions.

1.4.1 Rotational Lines of Gases

A molecule may be excited into rotational states through the interaction of the oscillatory electric field of the radiation with the electric dipole moment of the molecule. Hence the existence of a pure rotational line spectrum depends on whether the molecule has a permanent dipole moment. Symmetric linear molecules such as CO_2 and highly symmetric spherical molecules such as CH_4 have no permanent electric dipole moment and hence no pure rotational line spectrum in the microwave region. However, as discussed below, rotational line spectra for these molecules may appear in combination with vibrational states.

In describing molecular rotation, it is convenient to consider three principal axes of rotation through the centre of gravity of the molecule. A key factor in determining the rotational energy states and hence the rotational line spectrum is the moment of inertia associated with each axis, which in turn depends on the shape and symmetry of the molecule. For diatomic gases such as CO and linear molecules such as CO_2 (see Figure 1.3a), the moment of inertia for rotation around the bond axis is approximately zero, while rotation around the other two orthogonal axes has the same moment of inertia so these molecules possess only one series of rotational absorption lines. Highly symmetric spherical molecules such as CH_4 have, to a first approximation, the same moment of inertia for all three principal axes so likewise have a single series of rotational lines (although fine

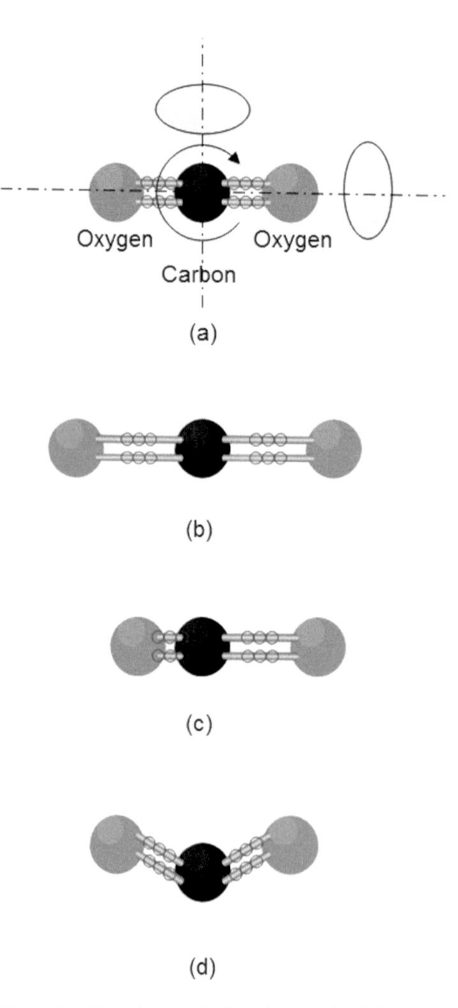

Figure 1.3 Rotation and vibration of the CO_2 molecule (a) the three axes of rotation, (b) symmetric stretch mode, (c) anti-symmetric stretch mode, (d) bending mode.

structure exists and may be observed at low pressure). Most molecules have three different moments of inertia, e.g. H_2O, SO_2, NO_2, etc., so possess three basic series of rotational line spectra and combination series from rotation about more than one axis.

1.4.2 Vibrational Lines of Gases

In a simple model, the bond between the atoms of a molecule may be viewed as a spring obeying Hooke's law, and the vibrations of a molecule as a mass–spring system forming a simple harmonic oscillator. According to the rules of quantum mechanics, the allowed energy states of the oscillator are quantised according to $E_n = (v + \frac{1}{2})hf_0$ where f_0 is the classical oscillation frequency, $v = 0, 1, 2, \ldots$, and changes in the vibrational state are restricted to: $\Delta v = \pm 1$. However, the real behaviour of molecular bonds differs from the simple Hooke's law model and the vibrations are better described as an anharmonic oscillator, as shown in Figure 1.4, with quantised energy states, $E_n = (v + \frac{1}{2}) hf_a\{1 - x_a(v + \frac{1}{2})\}$, where f_a is a characteristic frequency and x_a is the anharmonicity factor. The spacing between the energy states is no longer uniform and the allowed transitions are governed by the selection rule, $\Delta v = \pm 1, \pm 2, \pm 3, \ldots$ However, the probability of large jumps is small and at normal temperatures the populations of states with $v \geq 1$ is small, so the main vibrational transitions of interest are the fundamental, first and second overtones, corresponding to v changing from $0 \rightarrow 1$, $0 \rightarrow 2$, $0 \rightarrow 3$, respectively, with diminishing probability and line strength. The fundamental transition $0 \rightarrow 1$ is typically in the mid-IR region for gases, with the near-IR region corresponding to first and second overtones giving absorption lines which are two or more orders of magnitude weaker than the fundamental. Note that at high temperature, transitions from the $v = 1$ state (known as a vibrational hot-band) may become important.

The above discussion applies in principle to each vibration mode of a molecule. In general, a molecule containing N atoms can undergo $3N - 6$ modes of vibration ($3N - 5$ for

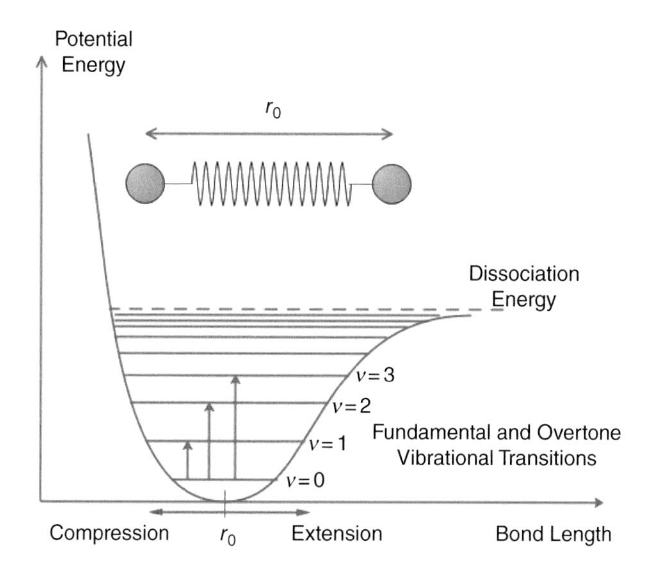

Figure 1.4 The energy levels of the anharmonic oscillator.

linear molecules). This includes $N - 1$ bond-stretching motions, with the others being bending motions. Of course, a vibration mode must produce a change in the dipole moment of the molecule for it to interact with the exciting radiation; for example, the symmetric stretching mode of the linear CO_2 molecule (see Figure 1.3b) does not give rise to vibrational absorption lines, whereas the anti-symmetric stretching mode (Figure 1.3c) does. Animations of the vibration modes of various molecules, such as methane, may be viewed on the web, see, for example, reference [13].

As well as pure vibrational modes, a molecule may be excited into a superposition of vibration modes giving vibrational energy levels corresponding to combination and difference bands. For example if ν_2 represents the (doubly degenerate) vibration mode of CH_4 at ~ 6.5 μm and ν_3 represents the (triply degenerate) asymmetric stretching mode ~3.3 μm, then the first overtone, $2\nu_3$, gives rise to near-IR absorption lines around 1.66 μm and the combination, $\nu_2 + 2\nu_3$, gives lines around 1.33 μm (note that IR-inactive vibration modes may appear in combination with other modes). Table 1.1 provides a list of the vibration modes and absorption regions ($\lambda > 1$ μm) for some common gases.

Table 1.1 Vibrational modes and absorption regions ($\lambda > 1$ μm) for some common gases

Gas molecule	Vibrational mode	Wavelength region of absorption lines (μm)
Carbon monoxide (CO)	Stretching mode, ν_1	4.666
Linear molecule, N = 2, one stretching mode	$2\nu_1$	2.347
only, with P and R branches only	$3\nu_1$	1.575
Carbon dioxide (CO_2)	Symmetric stretch, ν_1	IR inactive
Linear molecule, N = 3, four fundamental vibrational modes (two degenerate bending modes)	2-degenerate bend, ν_2	14.986
	Asymmetric stretch, ν_3	4.257
	$2\nu_2+\nu_3$	2.768
	$\nu_1+\nu_3$	2.692
	$4\nu_2+\nu_3$	2.060
	$\nu_1+2\nu_2+\nu_3$	2.01
	$2\nu_1+\nu_3$	1.961
	$6\nu_2+\nu_3$	1.646
	$\nu_1+4\nu_2+\nu_3$	1.606
	$2\nu_1+2\nu_2+\nu_3$	1.575
	$3\nu_1+\nu_3$	1.538
	$3\nu_3$	1.434
	$2\nu_2+3\nu_3$	1.221
	$\nu_1+3\nu_3$	1.206
Methane (CH_4)	Symmetric stretch, ν_1	IR inactive
Tetrahedral molecule, N = 5, giving nine vibrational modes but several are degenerate	2-degenerate bend, ν_2	6.523
	3-degenerate asymmetric stretch, ν_3	3.312

Table 1.1 (*cont.*)

Gas molecule	Vibrational mode	Wavelength region of absorption lines (μm)
	3-degenerate asymmetric bend, ν_4	7.628
	$2\nu_4$	3.828
	$\nu_2+\nu_4$	3.534
	$\nu_2+2\nu_4$	2.425
	$\nu_1+\nu_4$	2.368
	$\nu_3+\nu_4$	2.304
	$\nu_2+\nu_3$	2.203
	$\nu_3+2\nu_4$	1.791
	$\nu_1+\nu_2+\nu_4$	1.732
	$\nu_2+\nu_3+\nu_4$	1.706
	$2\nu_3$	1.666
	$\nu_2+2\nu_3$	1.331
	$2\nu_1+2\nu_4$	1.187
Water vapour (H_2O) *Non-linear molecule, N = 3, three fundamental vibrational modes*	Symmetric stretch, ν_1	2.734
	Bending mode, ν_2	6.270
	Asymmetric stretch, ν_3	2.662
	$2\nu_2$	3.173
	$3\nu_2$	2.143
	$\nu_1+\nu_2$	1.910
	$\nu_2+\nu_3$	1.876
	$\nu_1+2\nu_2$	1.476
	$2\nu_2+\nu_3$	1.455
	$2\nu_1$	1.389
	$\nu_1+\nu_3$	1.379
	$2\nu_3$	1.343
	$\nu_1+3\nu_2$	1.209
	$3\nu_2+\nu_3$	1.194
	$2\nu_1+\nu_2$	1.141
	$\nu_1+\nu_2+\nu_3$	1.135
	$\nu_2+2\nu_3$	1.111
Hydrogen sulfide (H_2S) *Non-linear molecule of similar shape to water, N = 3, three fundamental vibrational modes*	Symmetric stretch, ν_1	3.726
	Bending mode, ν_2	7.752
	Asymmetric stretch, ν_3	3.830
	$2\nu_2$	4.129
	$\nu_1+\nu_2$	2.646
	$\nu_2+\nu_3$	2.639
	$\nu_1+\nu_3$	1.940
	$2\nu_1+\nu_2$	1.590
	$\nu_1+\nu_2+\nu_3$	1.590
	$\nu_2+2\nu_3$	1.565

1.4.3 Rovibrational Lines

In general, molecules may simultaneously undergo both vibration and rotation, giving rise to rovibrational energy states. This includes molecules such as CO_2, CH_4, etc., which have no permanent dipole moment and hence no pure rotational spectra, but undergo dipole changes during vibration. As noted earlier, the energy separation of the vibrational states of a gas molecule is typically in the mid-IR region, whereas for rotational states the energy separation is two or three orders of magnitude smaller, in the microwave region. The large difference in these energies means that, to a first approximation, the total energy of a combined rovibrational state is simply the sum of the separate rotational and vibrational energies. Associated with each vibrational absorption line there is a series of closely spaced rotational lines. For linear molecules where the vibration mode produces dipole changes parallel to the major axis of rotational symmetry (e.g. the anti-symmetric

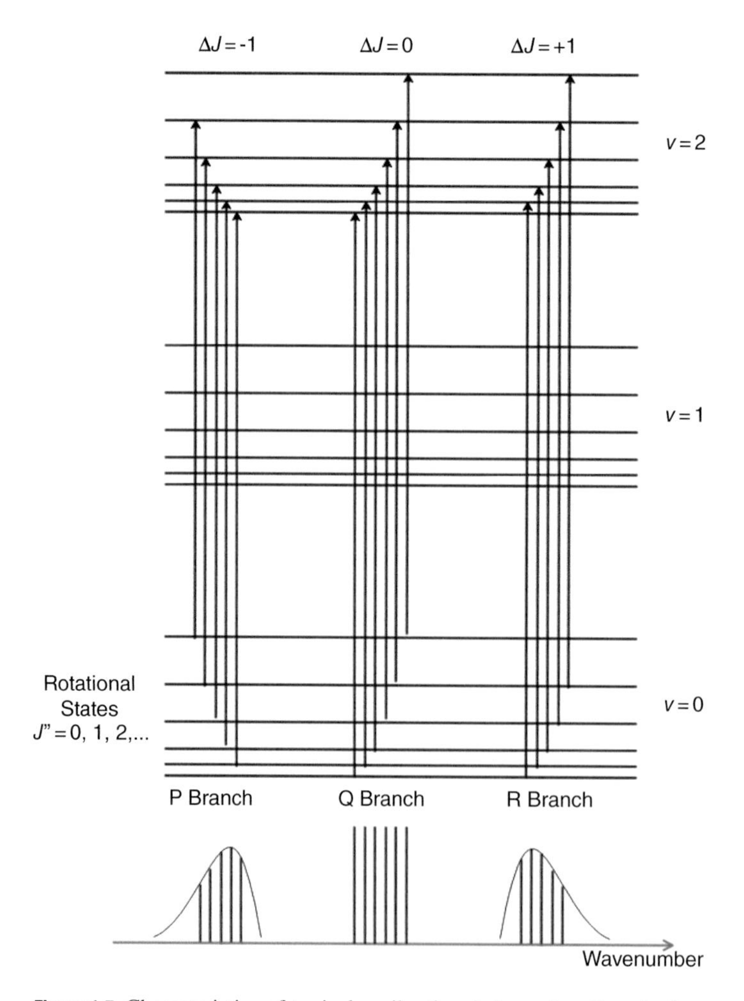

Figure 1.5 Characteristics of typical rovibrational absorption lines in the near-IR showing the P, Q and R branches.

stretching mode of CO_2 in Figure 1.3c), transitions between vibrational states must be accompanied by a change in the rotational state, $\Delta J = \pm 1$, but for a vibration mode producing perpendicular dipole changes (e.g. the bending mode of N_2O), $\Delta J = 0, \pm 1$ is allowed. For the vibration modes of non-linear molecules, the selection rules are more complex, but $\Delta J = 0, \pm 1$ is often allowed (e.g. the asymmetric stretching mode of CH_4). Rovibrational lines corresponding to $\Delta J = -1, 0, +1$ are described as the P, Q, R branches, respectively. Figure 1.5 illustrates the above principles showing the origin of typical rovibrational absorption lines in the near-IR with the P, Q and R branches.

1.4.4 Examples of Gas Absorption Spectra

An extremely useful source of information on spectral line parameters of atmospheric gases and pollutants is the *HITRAN* database [14, 15] which also includes an extensive list of references. The *HITRAN* Application Programming Interface (HAPI) [16, 17] and *HITRAN* on the Web [18] may be used for the modelling of absorption spectra, including plotting absorption coefficients, and absorption and transmission functions for a variety of important gases and mixtures, with various adjustable parameters such as temperature, pressure, lineshape (Voigt, Lorentz or Doppler), path length, wavenumber range, etc. Outputs are provided in graphical or tabular form which may be downloaded for processing by EXCEL or other programmes.

Figures 1.6–1.12 show some of the stronger near-IR absorption lines for several common and important gases plotted using *HITRAN* on the Web [18] with a concentration length product of $Cl = 1$ cm in (1.8). For comparison, the mid-IR absorption lines for CO, CO_2 and CH_4 with a much smaller value of $Cl = 0.01$ cm are shown in Figure 8.1 in Chapter 8, clearly showing the relatively weak strength of the near-IR lines. Note that for our purposes we have plotted the spectra in terms of wavelength

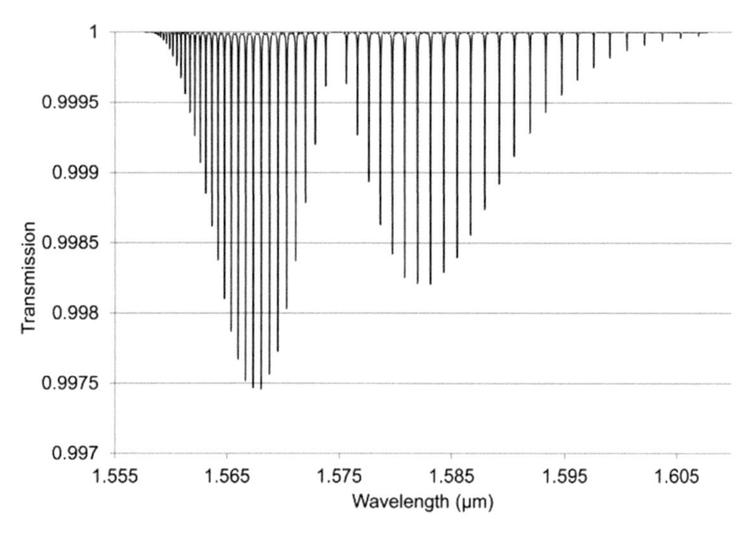

Figure 1.6 Near-IR absorption lines for CO plotted using *HITRAN* on the Web [18].

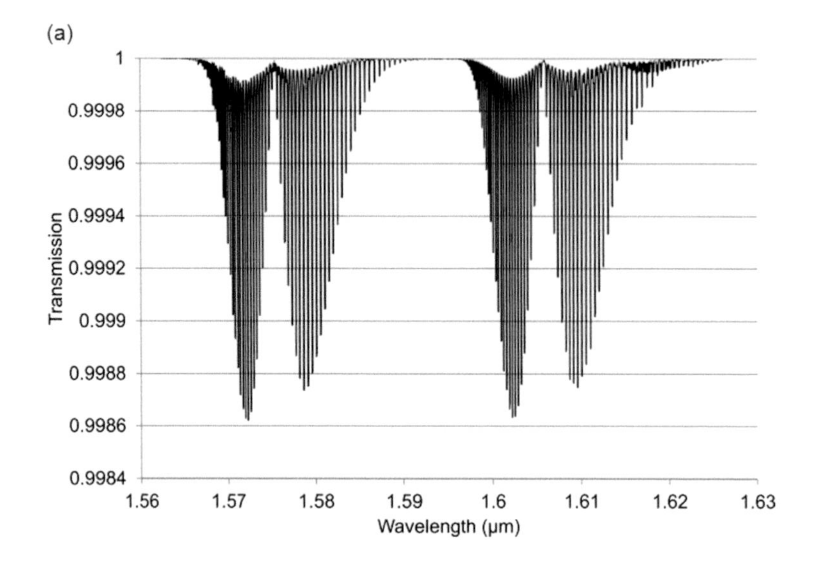

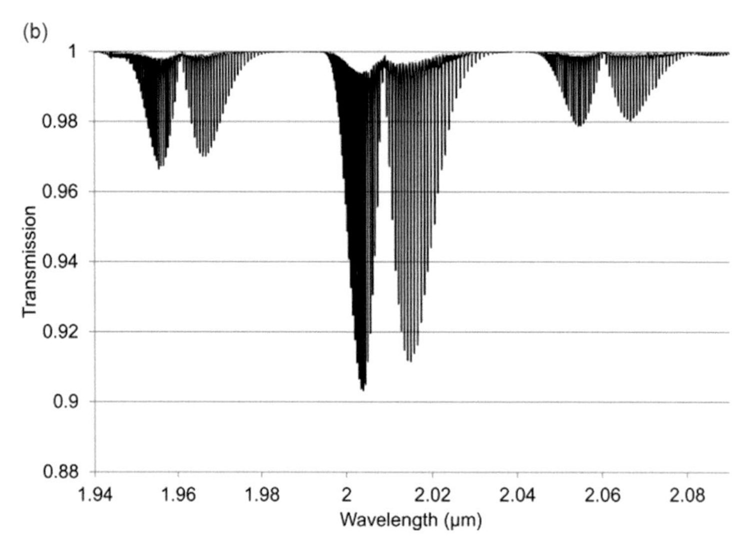

Figure 1.7 Near-IR absorption lines for CO_2: (a) 1.6 μm region (b) 2 μm region plotted using *HITRAN* on the Web [18].

(μm) rather than the commonly used wavenumber (cm^{-1}) and hence the order of the P, Q and R branches is reversed. The transmittance (T) on the vertical axis corresponds to the ratio, P_{out}/P_{in}, as defined by (1.2), (1.6) or (1.8). Note that the vertical scale differs between graphs to clearly show the line depths. All the examples of the near-IR spectra shown in Figures 1.6–1.12 are for the most abundant isotopologue of the pure gas at a pressure of 1 atm, temperature of 296 K and an absorption path length of 1 cm (hence Cl = 1 cm) with the assumption of a Voigt profile for the lineshape function.

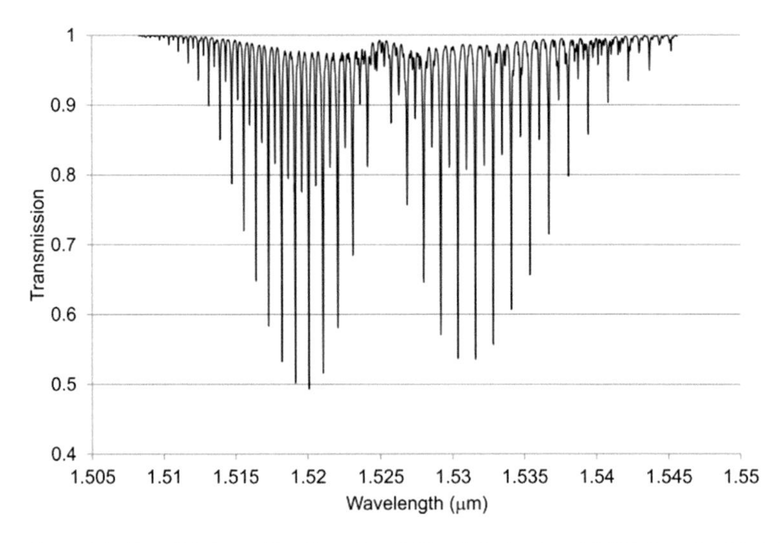

Figure 1.8 Near-IR absorption lines for C_2H_2 plotted using *HITRAN* on the Web [18].

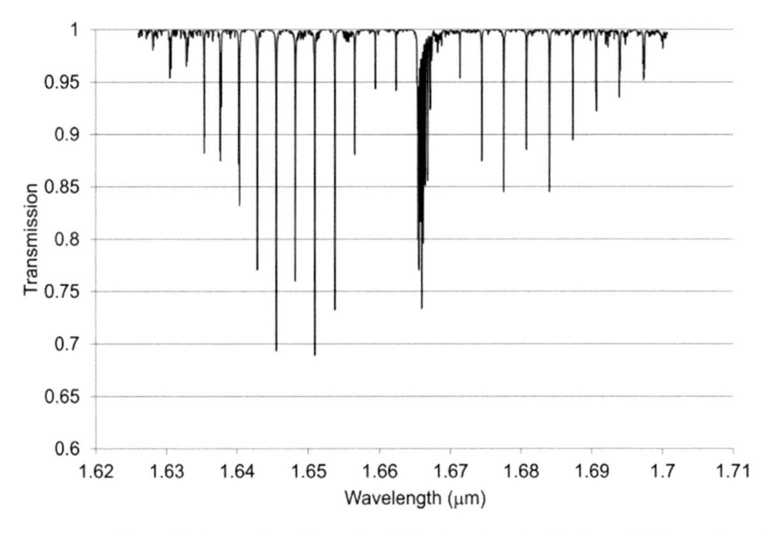

Figure 1.9 Near-IR absorption lines for CH_4 showing the P, Q and R branches plotted using *HITRAN* on the Web [18].

The absorption coefficient defined by (1.8) at line centre (in units of cm^{-1}) for a particular line may be easily obtained from these figures as approximately the line depth when it is small or more accurately as $\alpha_0 = -\ln(T_{min})$ where T_{min} is the line centre transmittance. The optical attenuation at line centre (in dB cm^{-1}) is $4.34\alpha_0$. Table 1.2 provides a list of the line centre absorption coefficient and attenuation for some absorption lines calculated from the figures, which will be useful later in discussing

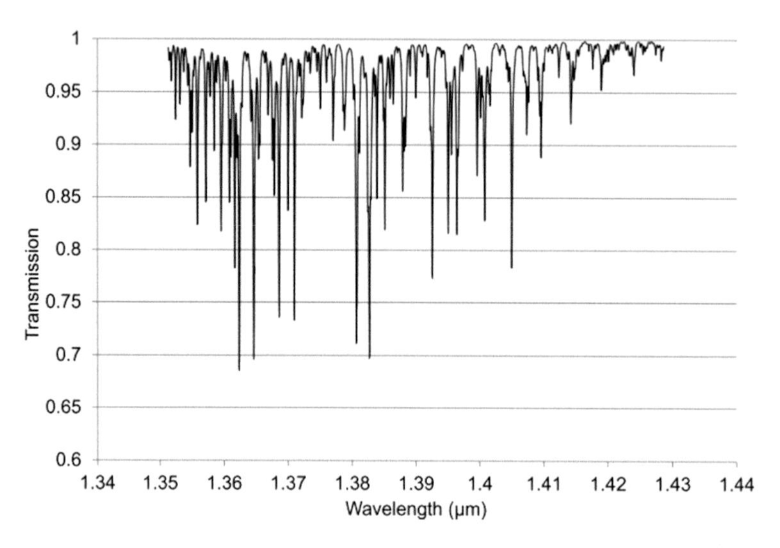

Figure 1.10 Near-IR absorption lines for H_2O in the 1.4μm region plotted using *HITRAN* on the Web [18].

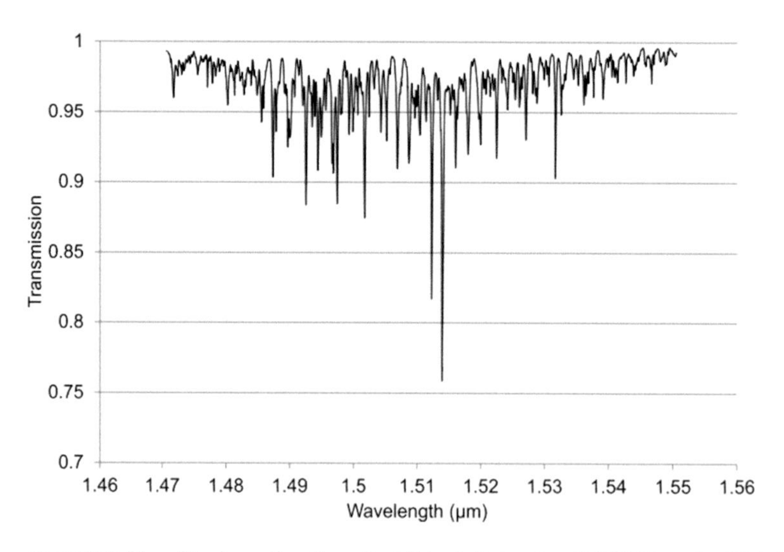

Figure 1.11 Near-IR absorption lines for NH_3 plotted using *HITRAN* on the Web [18].

the sensitivity of near-IR and fibre optic systems. Similar data for a range of other gases may be obtained from *HITRAN*, including gases such as O_2 which have no rotational or vibrational spectra but have electronic bands (O_2 has a spectral band centred around 762 nm).

Note the characteristic shape of the intensity envelope of the P and R branches as clearly shown in Figures 1.6–1.8. This arises from the multiplication of two functions

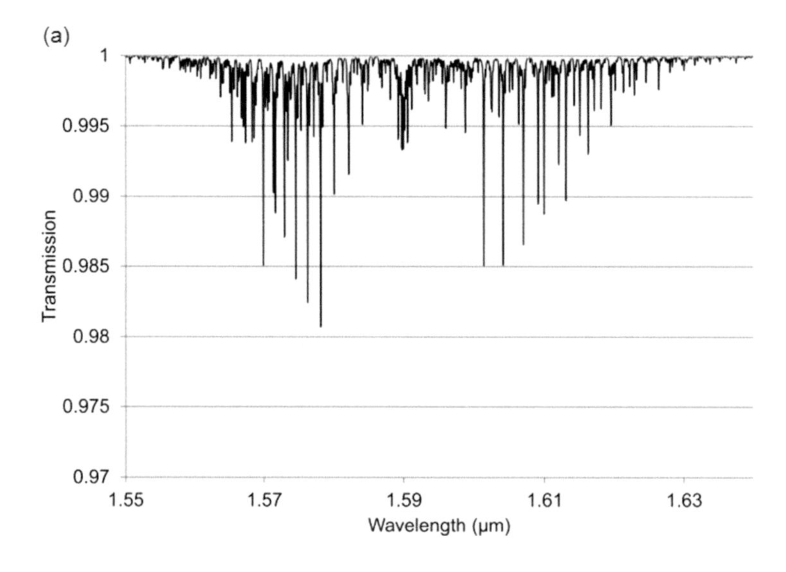

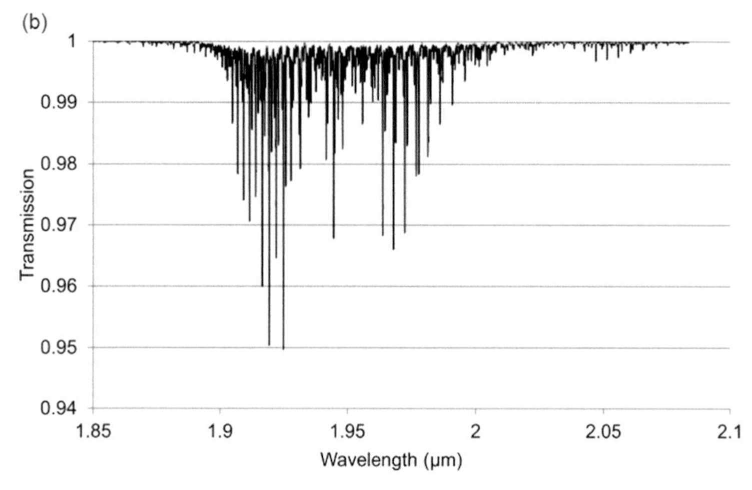

Figure 1.12 Near-IR absorption lines for H_2S: (a) 1.6 μm region, (b) 1.9 μm region plotted using *HITRAN* on the Web [18].

governing the population of the rotational levels, namely, the exponentially decreasing function describing the population of the J-th state following Boltzmann statistics and the linearly increasing degeneracy of the J-th state by the factor $(2J + 1)$. An additional factor determining the population of rotational levels is the influence of nuclear spin [10]. As a result, for linear molecules possessing a centre of symmetry such as CO_2, alternate rotational levels in the P and R branches have zero intensity. For C_2H_2, the rotational lines alternate in strength due to the nuclear spin of hydrogen atoms (see Figure 1.8).

Table 1.2 Approximate values of line centre absorption coefficient and attenuation for common pure gases at atmospheric pressure and a temperature of 296 K

Gas molecule	Wavelength of absorption line (μm)	α_0 (cm^{-1})	Optical attenuation at line centre (dB cm^{-1})
Carbon monoxide (CO)	1.5680	0.0025	0.011
Carbon dioxide (CO_2)	1.5723	0.0014	0.006
	2.0035	0.1	0.434
Acetylene (C_2H_2)	1.5201	0.69	2.99
Methane (CH_4)	1.6510	0.36	1.56
Water vapour (H_2O)	1.3827	0.36	1.56
Ammonia (NH_3)	1.5139	0.27	1.19
Hydrogen sulfide (H_2S)	1.5781	0.019	0.08
	1.9250	0.05	0.22

1.5 Relative Merits of Near-IR and Mid-IR Absorption Spectroscopy

As noted previously, absorption lines in the near-IR are typically two or three orders of magnitude weaker than the fundamental transitions. A question therefore arises as to why one would use near-IR lines when much higher sensitivity is possible in the mid-IR. Certainly if high sensitivity is the main consideration then the fundamental absorption lines in the mid-IR are clearly the best choice. However, for many practical situations, the sensitivities that can be achieved in the near-IR may be perfectly adequate and there are several other issues that should be considered when reviewing the relative merits of the near-IR compared with the mid-IR region of the spectrum:

(i) *Laser sources*: the vast fibre optic data communications market has led to the ready availability of low cost, compact laser sources in the near-IR region, particularly DFB lasers, which can be tailor-made for a wide range of near-IR wavelengths and are extremely flexible in terms of wavelength tuning and modulation through control of the diode current and temperature. There have, of course, been major advances in the availability and performance of mid-IR sources in the last decade, particularly with regard to the development of quantum cascade lasers, but mid-IR sources are currently more expensive and generally less mature than their near-IR counterparts.

(ii) *Optical detectors*: mid-IR detectors can perform as well as near-IR devices but may require cooling at longer IR wavelengths to maintain noise performance and again are generally more expensive.

(iii) *Optical fibres*: near-IR systems offer unprecedented levels of flexibility in the design of spectroscopic systems through use of standard optical fibres, fibre amplifiers and photonic components from the data communications market. Techniques borrowed from data communication systems such as spatial-, time- and wavelength-division multiplexing may be applied in principle for the design of multi-point and multi-gas sensors, with fibre amplifiers to boost signal powers as required. By contrast, while the performance of mid-IR fibres will no doubt improve with further research, they are currently more expensive and are limited to relatively short lengths from fibre losses and may, in some cases, suffer from durability issues.

We shall return to these issues in Chapter 8, where a review of sources, detectors and fibres for mid-IR gas absorption sensors is given.

1.6 Conclusion

This chapter has reviewed the fundamental parameters involved in the design of gas absorption sensors, particularly for deciding appropriate wavelengths of operation, for predicting the expected sensitivity and for the extraction of concentration, pressure and temperature under different conditions of operation. The focus here is on near-IR spectroscopy since it is always prudent to consider this spectral region first to ascertain whether the required sensitivity for an optical gas sensor could be met with low-cost near-IR photonics. Of course certain gases may not possess suitable near-IR lines or their intensity may be too weak, in which case there is no option but to use the mid-IR spectral region, as reviewed in the last chapter of this book. However, near-IR systems will no doubt always have an important role in gas spectroscopy due to the maturity, low cost and versatility of sources, detectors and photonic components in this region of the spectrum.

References

1. W. Demtroder, *Laser Spectroscopy Vol 1: Basic Principles*, Berlin, Springer, 2008.
2. M. Fox, Radiative transitions in atoms, in *Quantum Optics: An Introduction*, Oxford, England, Oxford University Press, ch. 4, 48–71, 2006.
3. W. Voigt, Das Gesetz der Intensitätsverteilung innerhalb der Linien eines Gasspektrums, *München Sitz. Ak. Wiss. Math-Phys. Kl.* 42, 603, 1912.
4. S. R. Drayson, Rapid computation of the Voigt profile, *J. Quant. Spectrosc. Radiat. Transfer*, 16, 611–614, 1976.
5. J. J. Olivero and R. L. Longbothum, Empirical fits to the Voigt line width: a brief review, *J. Quant. Spectrosc. Radiat. Transfer*, 17, 233–236, 1977.
6. A. B. McLean, C. E. J. Mitchell and D. M. Swanston, Implementation of an efficient analytical approximation to the Voigt function for photoemission lineshape analysis, *J. Electron Spectrosc. Relat. Phenom.*, 69, 125–132, 1994.
7. X. Huang and Y. L. Yung, A common misunderstanding about the Voigt line profile, *J. Atmos. Sci.*, 61, 1630–1632, 2004.
8. M. P. Arroyo and R. K. Hanson, Absorption measurements of water-vapor concentration, temperature, and line-shape parameters using a tunable InGaAsP diode laser, *Appl. Opt.*, 32, (30), 6104–6116, 1993.
9. J. Michael Hollas, *Modern Spectroscopy*, 4th edn., Chichester, England, John Wiley & Sons Ltd, 2004.

10. C. N. Banwell and E. M. McCash, *Fundamentals of Molecular Spectroscopy*, 4th edn., New York, McGraw-Hill, 1995.
11. G. Herzberg, *Molecular Spectra and Molecular Structure*, 1–3, New York, Van Nostrand, 1964–1966.
12. G. M. Barrow, *Introduction to Molecular Spectroscopy*, London, McGraw-Hill, 1962.
13. N. Greeves. Vibrations of methane. 2019. [Online]. Available: www.chemtube3d.com/vibrationsCH4.htm (accessed March 2020)
14. I. E. Gordon, L. S. Rothman, C. Hill, et al., The HITRAN2016 molecular spectroscopic database, *J. Quant. Spectrosc. Radiat. Transfer*, 203, 3–69, 2017.
15. L. S. Rothman. HITRAN database. 2019. [Online]. Available: www.hitran.org (accessed March 2020)
16. R. V. Kochanov, I. E. Gordon, L. S. Rothman, et al., HITRAN Application Programming Interface (HAPI): a comprehensive approach to working with spectroscopic data, *J. Quant. Spectrosc. Radiat. Transfer*, 177, 15–30, 2016.
17. R. V. Kochanov, I. E. Gordon, L. S. Rothman, et al. HAPI: The HITRAN Application Programming Interface. 2019. [Online]. Available: www.hitran.org/hapi (accessed March 2020)
18. L. S. Rothman. HITRAN on the web. 2019. [Online]. Available: http://hitran.iao.ru/home (accessed March 2020)

2 DFB Lasers for Near-IR Spectroscopy

2.1 Introduction

In general, a wide variety of laser sources have been deployed for spectroscopy, depending on the wavelength region of interest. Typical sources for near-IR spectroscopy include Fabry–Perot (FP) lasers, vertical cavity surface emitting lasers (VCSELs) [1], distributed feedback (DFB) lasers, erbium-doped fibre lasers (which will be discussed in Chapters 6 and 7) and LED sources [2]. Later in Chapter 8 we review the range of sources available for mid-IR spectroscopy. In this chapter we focus in detail on DFB lasers since they are relatively inexpensive for near-IR spectroscopy, provide a broad coverage of the near-IR region, possess a narrow linewidth and are versatile in terms of operation [3–5]. However, the fundamental concepts discussed apply equally well to diode lasers for other spectral regions, including the mid-IR, as discussed in Chapter 8. Of course, the theory and properties of DFB lasers have been extensively studied in the literature for fibre optic data and communication systems, but here we consider the electronic, thermal and optical properties of DFB lasers particularly relevant to the requirements for spectroscopy, including a detailed analysis of the DC tuning and wavelength modulation characteristics from thermal and carrier effects. We shall also briefly discuss how to measure some of the fundamental parameters of DFB lasers that are important in spectroscopic applications.

2.2 Structure of DFB Lasers

Figure 2.1 shows the general construction of a buried double heterostructure (BDH) laser diode which ensures that there is good carrier and optical confinement to the active layer [6, 7]. For near-IR operation, the laser structure is typically grown on an n-type InP substrate, with various lattice-matched layers formed from $In_{1-x}Ga_xAs_yP_{1-y}$ alloys. The composition of the active layer determines the bandgap and hence the wavelength of laser operation. The composition of the surrounding confinement layers is selected to have a slightly larger bandgap and hence smaller refractive index to enable waveguiding in the active layer. A ridge waveguide structure over the active region may also be used for optical confinement.

In many cases a multiple quantum well (MQW) structure is used for the active layer. Figure 2.2 shows the typical construction of the active layer of a separate confinement heterostructure (SCH) device where the quantum wells and barrier layers are

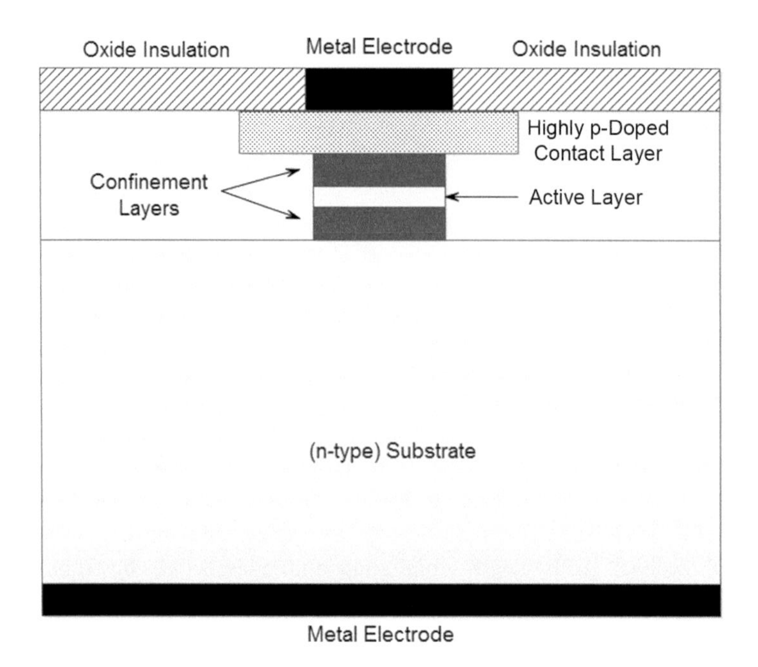

Figure 2.1 Typical cross-section of a buried double heterostructure laser diode.

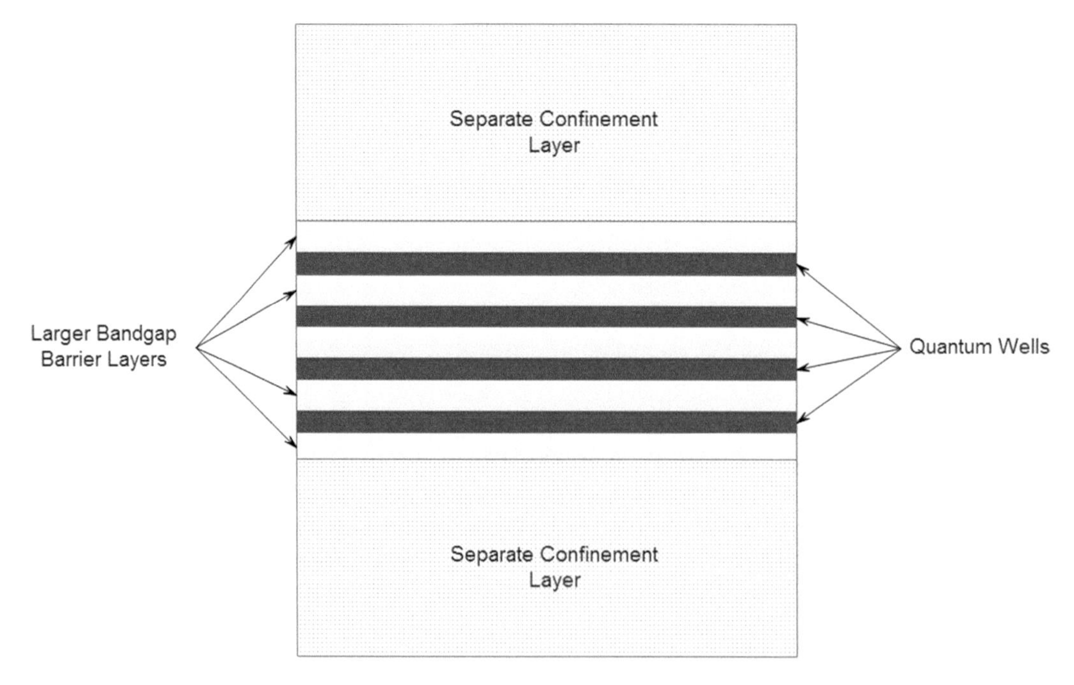

Figure 2.2 Typical construction of the active layer of a separate confinement heterostructure (SCH) device.

sandwiched between separate confinement layers. The dimension of a quantum well is sufficiently small (typically 5–10 nm thick) so that the allowed energy states for electrons and holes are quantised according to a one-dimensional potential energy well. This structure has a number of advantages: (i) the quantised energy levels are dependent on the well dimension and this may be used to adjust the operation wavelength, (ii) quantisation results in the restriction of available states so population inversion is more readily obtained with a consequent lowering of the threshold current and (iii) a narrow gain linewidth is possible since carriers are restricted to the quantised states.

To obtain a narrow linewidth, single-mode operation is required and this is achieved in the DFB laser by use of a diffraction grating. Here the grating is optically coupled to the active region (for example, etched in a waveguiding region adjacent to the active region) so that the partial reflections from the grating corrugations are subject to gain. As a result, the wavelengths for laser oscillation do not exactly correspond to the Bragg wavelength, $\lambda_B = 2n_e\Lambda/k$, but, in a simplified analytical model [8], are symmetrically spaced around λ_B according to:

$$\lambda = \lambda_B \pm \left(m + \frac{1}{2}\right)\frac{\lambda_B^2}{2n_eL_g} \tag{2.1}$$

where n_e is the effective refractive index, Λ is the grating period, k is the diffraction order, L_g is the grating length and m is an integer, $m = 0, 1, 2, \ldots$.

In laser operation, the $m = 0$ modes have the lowest gain threshold so higher-order modes are suppressed. Furthermore any asymmetry in the grating results in one of the $m = 0$ modes being favoured over the other. This is commonly done by introducing a $\Lambda/4$ shift in the grating phase so that single mode operation is obtained. Since $L_g \gg \lambda_B$ the lasing wavelength is close to λ_B. Typically, outputs powers of 1–20 mW may be obtained with CW linewidths of <10 MHz.

2.3 Application of DFB Lasers in Tunable Diode Laser Spectroscopy

As will be discussed in detail in Chapter 3, the general procedure for using DFB lasers in tunable diode laser spectroscopy (TDLS) is as follows (see Figure 2.3):

(i) The operation wavelength of the laser is thermally tuned to the vicinity of the target absorption line using an in-built thermo-electric cooler (TEC) and thermistor along with an external temperature controller unit.

(ii) The laser diode current is adjusted to a suitable DC bias level above threshold and a linear current scan (typically a sawtooth or triangular waveform at 1–10 Hz frequency) is applied to the diode current to scan the laser wavelength through the absorption line.

(iii) For high sensitivity, low noise and accurate recovery of lineshapes, TDLS is frequently combined with wavelength modulation spectroscopy (WMS), where an additional sinusoidal modulation is applied to the diode current, typically with a modulation frequency of 10 kHz–1 MHz, to give a wavelength modulation on the DFB output with an amplitude similar to that of the width of the absorption line.

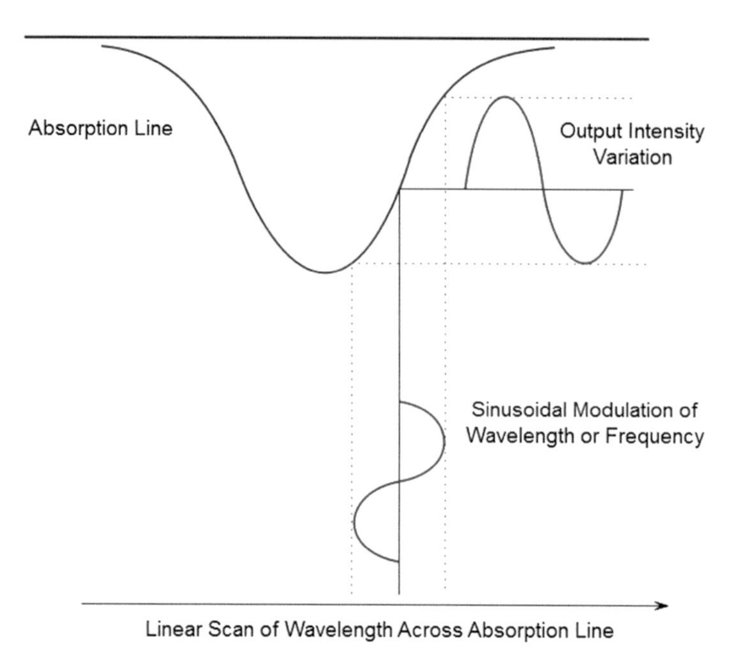

Figure 2.3 Basic principle of wavelength modulation spectroscopy where the wavelength (or frequency) modulation is converted to an intensity variation.

However, it is well known that current modulation of semiconductor lasers gives rise to both intensity and wavelength modulation of the laser output with a phase difference between the two. As will be clear from Chapter 3, the recovered harmonic signals in WMS depend critically on the absolute magnitudes of the wavelength and intensity modulation and their relative magnitude and phase. This is often quantified by the chirp-to-power ratio (CPR) in the context of high-performance fibre optic data communication systems where it is desirable to minimise the CPR [9, 10] to reduce pulse dispersion. In general the CPR depends on the detailed electronic, optical and thermal characteristics of the particular laser used and on the chosen modulation parameters, notably the modulation frequency.

Factors affecting the CPR have been extensively studied for DFB lasers but, in contrast to fibre optic data systems, it is often desirable to maximise the CPR for WMS applications. Furthermore, WMS systems are usually operated at relatively low frequencies (<1 MHz), so thermal effects are very important. In this chapter we present in detail the pertinent thermal and optical properties of DFB lasers within the context of TDLS. Approximate analytical models are used to derive the intensity and wavelength modulation characteristics of DFB lasers to assist in the choice of parameters in the design and operation of TDLS and WMS systems [11]. We first consider the thermal tuning and modulation properties of DFB lasers and then consider the effect of carrier density modulation on the wavelength and intensity modulation. As we shall see, the magnitude and phase of the resultant frequency (or wavelength) modulation in relation to the current modulation is determined by contributions from both slow thermal mechanisms and fast carrier injection effects, with their relative importance dependent on the chosen modulation frequency for WMS.

2.4 Thermal Tuning and Modulation

As noted, thermal modulation plays a very important role in most WMS systems because of the relatively low frequencies that are usually employed. The expansion of the optical path length in the DFB laser grating period with temperature (and hence with drive current) causes an increase in the output wavelength, so thermal wavelength modulation is in phase with the current modulation at low modulation frequencies. However, as the modulation frequency is increased, thermal delay means that the wavelength modulation progressively lags the current modulation and its contribution also diminishes.

The temperature modulation produced within a laser diode from modulation of the drive current can be numerically computed from solution of the heat conduction equation [12] for particular laser structures, taking into account the properties of the various layers and 2-D heat flow [13–20]. However, here we present simple 1-D analytical models which help to identify the important parameters in the operation of DFB lasers for WMS systems.

2.4.1 RC Thermal Model of Laser Diode

The commonly used RC thermal model [21–23] is based on the laser chip having a thermal capacity C_T and a thermal resistance R_T to the heat sink, which conducts all the heat energy away from the chip, as illustrated in Figure 2.4.

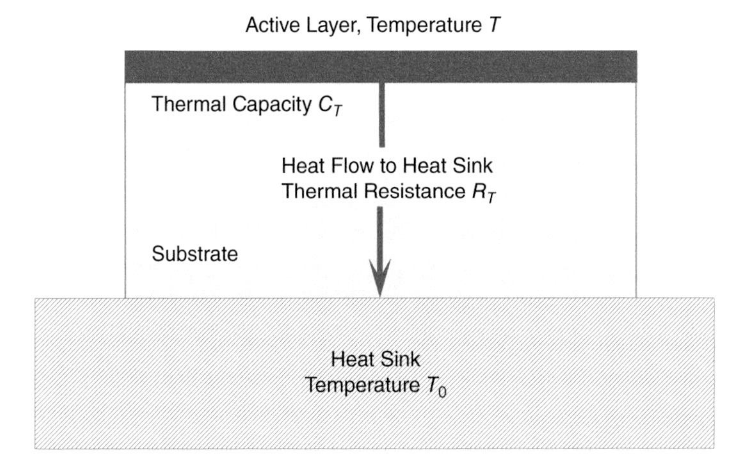

Figure 2.4 Basic RC model for heat flow in a simplified laser structure.

This leads to a simple thermal equation for the laser diode temperature T defined by the simple statement:

'Rate of energy storage + rate of energy flow to the heat sink = rate of energy supplied'

This translates to mathematical form as:

$$C_T \frac{\mathrm{d}T}{\mathrm{d}t} + \frac{T - T_0}{R_T} = P_{el} \tag{2.2}$$

where P_{el} is the electrical power dissipated in the chip and T_0 is the heat sink temperature.

As noted earlier for WMS, the forward injection current, i_f, supplied to the DFB laser takes the form: $i_f = I_b + i_m \cos(\omega_m t)$ where $I_b = I_{\text{bias}} + I_{\text{scan}}$ is the DC bias current and scan current, i_m is the modulation current amplitude and ω_m is the modulation (angular) frequency. We assume that the linear current scan for sweeping the wavelength through the absorption line is at a sufficiently low frequency that its effect is similar to a DC change in current.

The forward diode voltage is approximated by $v_f \cong V_j + i_f r_s$, so the electrical power dissipated in the chip is:

$$P_{el} = v_f i_f - P_{opt} = \left[V_j i_f - P_{opt}\right] + i_f^2 r_s \tag{2.3}$$

where r_s is the series (slope) resistance and V_j is the junction voltage of the laser chip (both of which may be approximately obtained from the *VI* characteristic of the laser diode). P_{opt} is the output optical power given by: $P_{opt} = \eta(I_b - I_{th})$ where I_{th} is the threshold current and η is the slope efficiency (slope of the optical power versus current characteristic). We assume that the laser is above threshold and spontaneous emission power can be neglected.

Substituting the expression for i_f into (2.3), the DC and AC heat generated follows as:

$$P_{el} = P_{DC} + A_1 \cos(\omega_m t) + A_2 \cos(2\omega_m t) \tag{2.4}$$

where P_{DC}, A_1 and A_2 are given by:

$$P_{DC} = \left[V_j I_b - \eta(I_b - I_{th})\right] + \left[r_s I_b^2 + \frac{r_s i_m^2}{2}\right] \tag{2.5}$$

$$A_1 = (V_j - \eta)i_m + 2r_s I_b i_m \tag{2.6}$$

$$A_2 = \frac{r_s i_m^2}{2} \tag{2.7}$$

Now if we substitute the electrical power of (2.4) into the thermal equation (2.2) and neglect A_2 we obtain the solution for the laser diode temperature as:

$$T = T_0 + R_T \left\{ P_{DC} + \frac{A_1 \cos\left(\omega_m t - \psi_\lambda^{th}\right)}{\sqrt{1 + \omega_m^2 \tau_{th}^2}} \right\} \tag{2.8}$$

where τ_{th} is the thermal time constant given by $\tau_{th} = R_T C_T$ and ψ_λ^{th} is the thermal phase lag, $\psi_\lambda^{th} = \tan^{-1}(\omega_m \tau_{th})$. Note that ψ_λ^{th} is defined as being positive with the negative sign in (2.8) indicating the phase lag of the wavelength modulation.

The change in the laser's output wavelength may then be related to the change in temperature through the laser's temperature tuning coefficient, $\Delta\lambda/\Delta T = (\alpha_l + \alpha_n)\lambda$, where α_l and α_n are thermal linear expansion and refractive index coefficients, respectively. Hence the thermal DC tuning and wavelength modulation are given by:

$$\lambda = \lambda_0 + \left(\frac{\Delta\lambda}{\Delta T}\right) R_T \left\{ P_{DC} + \frac{A_1 \cos\left(\omega_m t - \psi_\lambda^{th}\right)}{\sqrt{1 + \omega_m^2 \tau_{th}^2}} \right\} \tag{2.9}$$

For one-dimensional heat flow from the active layer to the heat sink, the thermal resistance, R_T, is given by $R_T = d_s/kA$, where d_s is the thickness of the substrate (distance of the active layer from the heat sink), A is the area of the active layer and k is the thermal conductivity (~0.68 W cm K^{-1} for InP at 300 K). More realistically, the quasi 2-D heat flow from an active layer of width w and length l to a heat sink at a distance, d_s gives a thermal resistance of [21]:

$$R_T = \frac{\ln(4d_s/w)}{\pi k l} \tag{2.10}$$

2.4.2 DC Thermal Tuning

The DC thermal tuning characteristic of the laser diode follows from (2.5) and (2.9) as:

$$(\lambda - \lambda_0) = \left(\frac{\Delta\lambda}{\Delta T}\right) R_T \left\{ V_j I_b - \eta(I_b - I_{th}) + r_s \left[I_b^2 + \frac{i_m^2}{2} \right] \right\} \tag{2.11}$$

Note the slight non-linear dependence on the DC current (or current scan) and the small DC offset from the sinusoidal modulation current as a result of joule heating from the ohmic resistance, r_s.

The various parameters in (2.11) may be obtained from measurement of the DC characteristics of the particular laser used in a TDLS system. Figure 2.5 shows the DC tuning characteristic (wavelength shift as a function of the DC bias current) for typical values of laser parameters, as given in Table 2.1 with $i_m = 0$. A typical pressure-broadened absorption line at atmospheric pressure has a full linewidth of ~5 GHz or ~0.04 nm, so from Figure 2.5 it is clear that a current scan of ~40 mA is sufficient to sweep across the entire line. The actual operation wavelength of the laser may be set by adjustment of the steady-state laser temperature using the TEC unit (typical laser temperature tuning coefficient of ~0.1 nm $°C^{-1}$).

2.4.3 Thermal Modulation of the Optical Frequency

The magnitude of the wavelength modulation $\delta\lambda$, given by (2.9), is often described in terms of the magnitude of the equivalent optical frequency modulation,

Table 2.1 Typical values for parameters of semiconductor lasers

Parameter	Typical value
Junction voltage, V_j	0.9 V
Series resistance, r_s	4 Ω
Thermal resistance, R_T	30 °C W^{-1}
Temperature tuning coefficient, $\Delta\lambda/\Delta T$	0.1 nm °C^{-1}
Threshold current, i_{th}	10 mA
Slope efficiency, η	0.1 W A^{-1}
Injection efficiency, η_i	0.85
Spontaneous emission fraction, β	10^{-5}
Photon lifetime, τ_p	~1 ps
Carrier lifetime, τ_n	~2 ns
Differential gain, g_0	~10^{-19} to 10^{-20} m^2
Gain compression factor, ε	~10^{-22} to 10^{-23} m^3
Optical confinement factor, Γ	
BDH laser	~1
MQW laser	~0.01
Mode volume, V_p	~10^{-15} m^3
Linewidth enhancement factor, α	~2 to 6
DC thermal tuning coefficient, H_0	0.5 GHz mA^{-1}
Thermal cut-off frequency, f_{cs}	10 kHz
Carrier tuning coefficient, ξ_c	0.05 GHz mA^{-1}

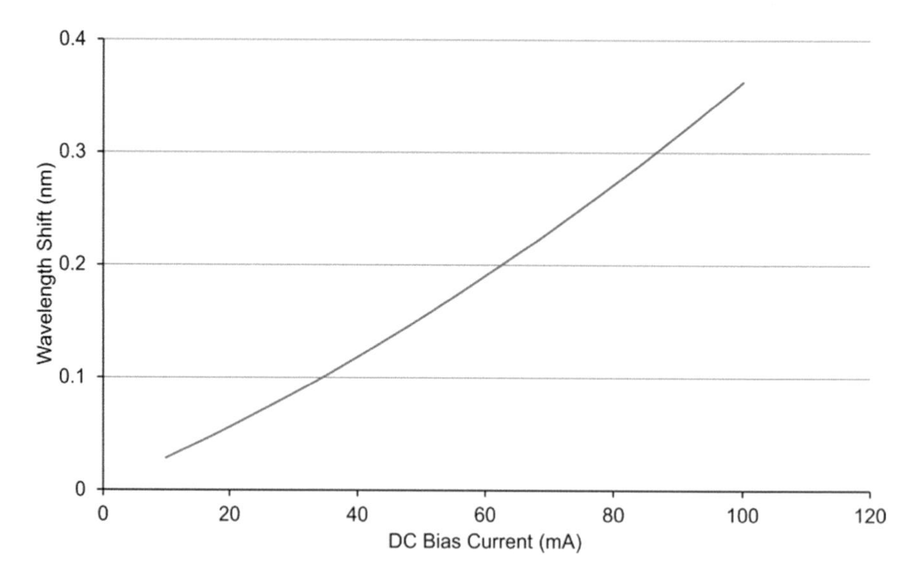

Figure 2.5 Typical DC current tuning characteristic of a DFB laser.

$\delta\nu = -\left(c/\lambda_0^2\right)\delta\lambda$, the negative sign indicating that the wavelength and equivalent optical frequency modulation differ in phase by π. From (2.6) and (2.9), a thermal tuning coefficient, ξ_{th} may therefore be defined as:

$$|\xi_{th}| = \frac{\delta\nu}{i_m} = \left(\frac{c}{\lambda_0^2}\right)\frac{\delta\lambda}{i_m} = \frac{H_0}{\sqrt{1 + (f_m/f_c)^2}} \tag{2.12}$$

where H_0 is the DC thermal tuning coefficient given by:

$$H_0 = \left(\frac{c}{\lambda_0^2}\right)\left(\frac{\Delta\lambda}{\Delta T}\right)R_T\left\{\left(V_j - \eta\right) + 2r_s I_b\right\} \tag{2.13}$$

and:

$$\psi_\lambda^{th} = \tan^{-1}(f_m/f_c) \tag{2.14}$$

where f_m is the modulation frequency and f_c is a thermal cut-off frequency, $f_c = 1/(2\pi R_T C_T)$.

As a reminder, note that ψ_λ^{th} is defined as being positive and the wavelength modulation has a phase angle of $-\psi_\lambda^{th}$ with respect to the current modulation, i.e. it lags the current modulation. Hence the phase of the equivalent thermal frequency modulation, ψ_f^{th}, relative to the current modulation is $\psi_f^{th} = \left(\pi - \psi_\lambda^{th}\right)$.

Figure 2.6 shows: (a) the magnitude of thermal tuning coefficient and (b) the thermal phase lag ψ_λ^{th} as a function of the modulation frequency at a bias current of 50 mA for the same laser parameters used in Figure 2.5 from Table 2.1 and with a thermal cut-off frequency of 10 kHz. As shown, the magnitude and phase follow a first-order filter response of the form $(1 + jf_m/f_c)^{-1}$, with magnitude diminishing with modulation frequency and phase lag increasing from 0° to a maximum of 90° with respect to the modulation current phase. The importance of selecting the modulation frequency and knowledge of the phase lag will become clear in Chapter 3, where we discuss in detail the various harmonics that arise in the output signal when using WMS.

Note from (2.13) that the tuning coefficient increases with I_b and hence will increase when scanning the wavelength through an absorption line; the strength of this effect is dependent on the slope resistance of the laser. (The physical reason for this is that the same magnitude of the sinusoidal modulation current produces a greater heat power variation at a higher bias current since the heat generated is dependent on the square of the *total* current). Also, as explained later, since the thermal contribution is combined with a contribution from the carrier density modulation by phasor addition, there will be a small change in the resultant phase of the overall wavelength modulation during a wavelength scan by a current ramp which may need to be taken into account when processing the harmonic signals obtained with WMS [11] – this will be discussed further in Chapter 3.

Finally, note that we have ignored the small wavelength modulation at the second harmonic of the modulation frequency, as indicated by (2.4) and (2.7).

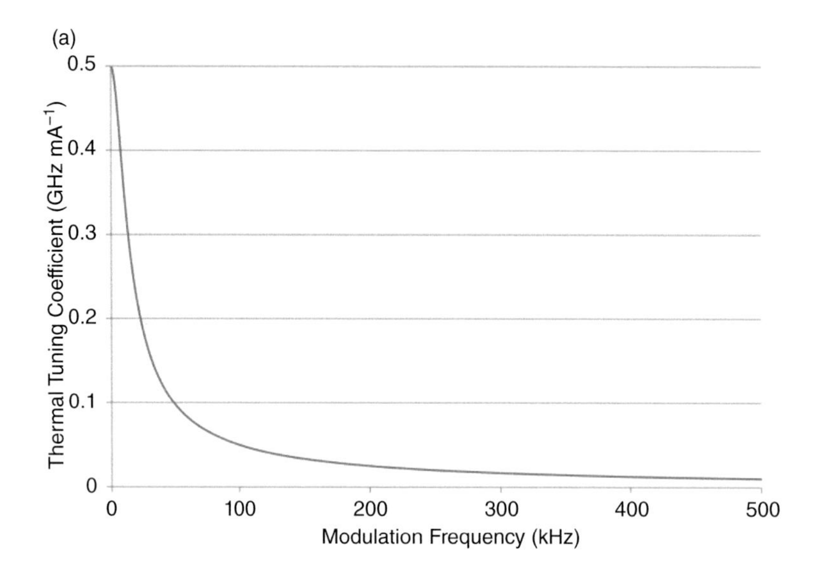

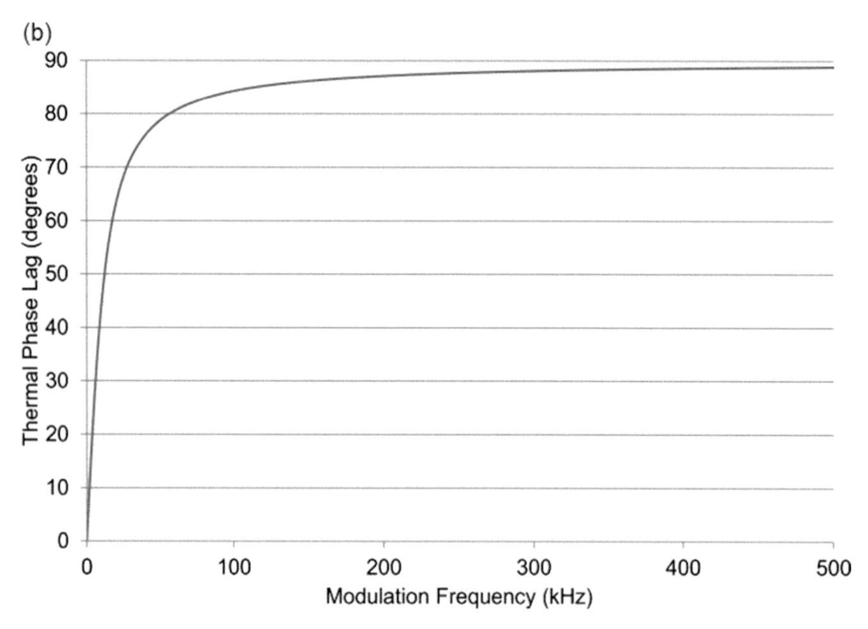

Figure 2.6 Typical thermal response of a DFB laser based on the RC model as a function of modulation frequency: (a) thermal tuning coefficient and (b) thermal phase lag of wavelength modulation.

2.4.4 1-D Thermal Model from Heat Conduction Equation

Though widely used, the RC model outlined above has a number of limitations. It is based on a 1-D approach, with the assumption of a linear temperature distribution from the active layer to the heat sink. Also, it has been observed that a number of lasers exhibit a significant departure from a first-order-type phase response, with the thermal phase lag increasing more slowly at increasing modulation frequency than would be expected from the RC model [11, 15–17, 24, 35]. Some lasers exhibit a dip in the magnitude of the tuning coefficient at certain frequencies. In order to explain these effects, various 1-D analytical or semi-analytical thermal models have been derived from the heat conduction equation. Dilwali [17] gives a model based on heat generation in an active layer positioned centrally in a three-layer model of the laser chip, but using an averaged temperature over the whole chip. Hangauer [24] gives a semi-analytical model specifically for VCSELs in the form of a Fourier transform equation for numerical integration.

In Appendix 2.1 we derive a 1-D analytical solution of the heat conduction equation based on the assumption of uniform temperature in a thin active layer, which gives a better description of the thermal properties than the simple RC model. From this model the thermal tuning coefficient at sub-megahertz modulation frequencies is approximated by:

$$\xi_{th} \cong \left\{ \frac{-H_0}{(jf_m/f_{cs})^{\frac{1}{2}} \coth (jf_m/f_{cs})^{\frac{1}{2}}} \right\} \tag{2.15}$$

This expression may be compared with the RC thermal model written in phasor notation as:

$$\xi_{th} = \left\{ \frac{-H_0}{1 + jf_m/f_{cs}} \right\} \tag{2.16}$$

Figure 2.7 shows: (a) the magnitude of the thermal tuning coefficient and (b) the thermal phase lag of the wavelength modulation from (2.15) compared with the RC model of (2.16) for the same parameters noted earlier, $H_0 = 0.5$ GHz mA^{-1} and cut-off frequency $f_{cs} = 10$ kHz. Note that for $f_m \gg f_{cs}$, the response described by (2.15) has an inverse square root behaviour of the form: $1/(jf_m/f_{cs})^{1/2}$ so that, compared with the RC model, the thermal phase lag approaches 45° rather than 90°, and the high-frequency roll-off is slower for the same cut-off frequency. Of course the model has limitations since it is based on simple 1-D heat flow in a three-layer structure, but further refinements require numerical solution of the heat conduction equation [12–14] with detailed knowledge of the specific laser structure involved.

2.5 Intensity and Frequency Modulation from Carrier Effects

In semiconductor lasers, the light output arises from stimulated transitions in the active region of injected carriers (electrons) in the valence band to holes in the conduction band. The operation of the laser may be simply described through two rate equations [9, 21, 22], one for the carrier density and one for the photon density:

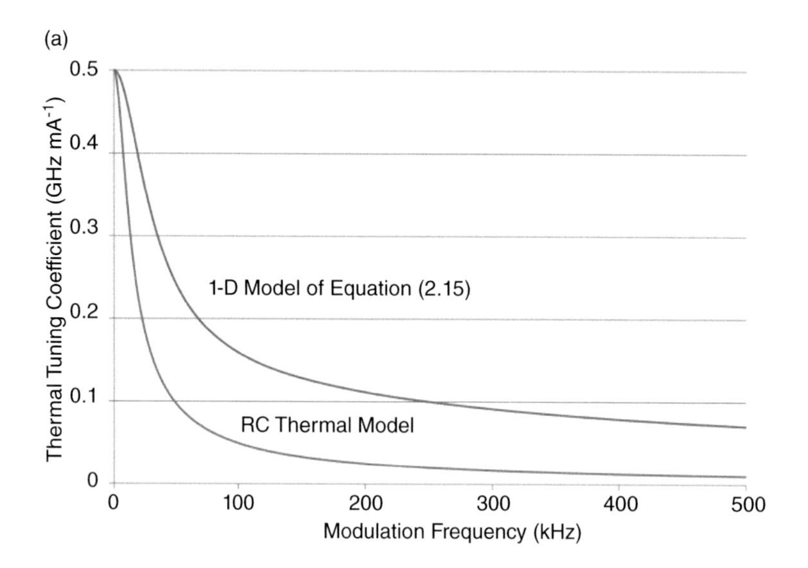

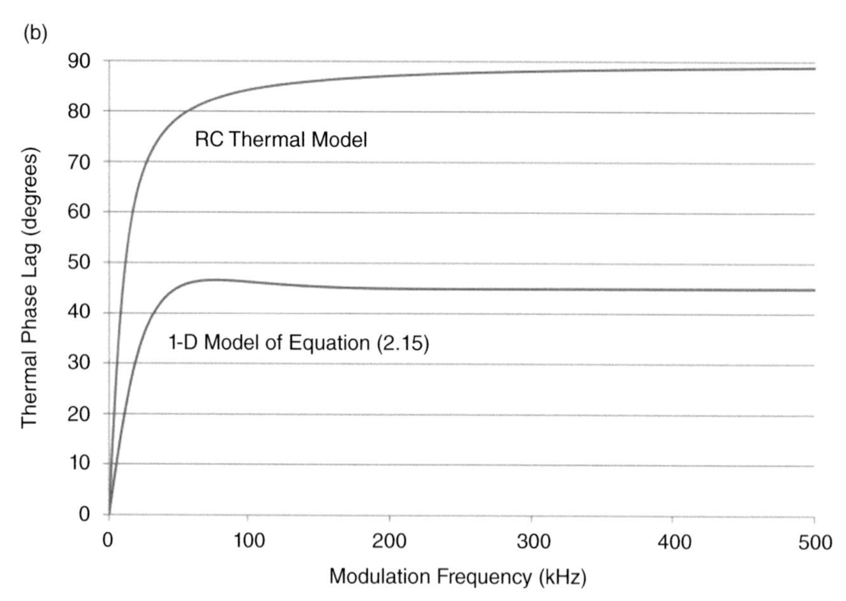

Figure 2.7 Comparison of thermal modulation response (2.15) with (2.16): (a) magnitude of the thermal tuning coefficient and (b) thermal phase lag of wavelength modulation.

$$\frac{\mathrm{d}N}{\mathrm{d}t} = \frac{\eta_i I}{q V_a} - \frac{N}{\tau_n} - v_g G(N,S) S \tag{2.17}$$

Rate of change of carrier density in the active region | Carrier injection rate | Spontaneous emission rate | Stimulated emission rate

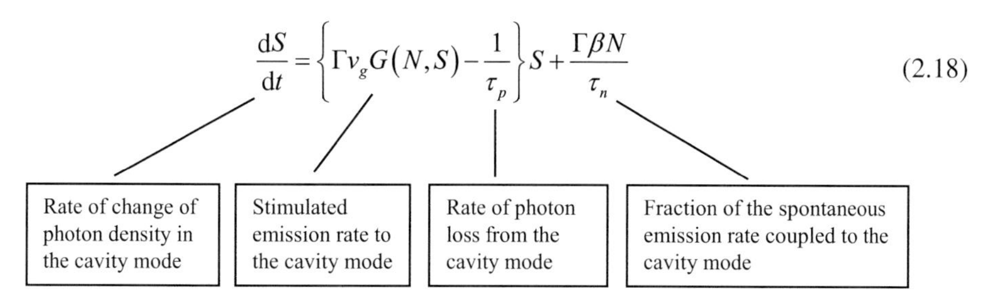

where N is the carrier density, S is the photon density, I is the injection current, η_i is the injection efficiency, τ_n is the carrier lifetime, τ_p is the photon lifetime, V_a is the volume of the active region, Γ is the optical confinement factor (ratio of active region volume, V_a, to optical mode volume, V_p), β is the fraction of spontaneous emission coupled to the mode, q is the electronic charge, G is the gain (which is dependent on the carrier density N and may also be dependent on the photon density S) and v_g is the group velocity. The group velocity converts the gain per unit length to gain per unit time. For simplicity, we have assumed that the carrier and photon densities are uniform in the lateral direction and non-radiative recombination is ignored.

2.5.1 Steady-State DC Analysis

We first consider the steady-state solutions of the rate equations, where carrier and photon density are constant in time so that $d/dt = 0$ on the left side of (2.17) and (2.18). Eliminating the gain term, $G(N,S)$, from both equations gives the steady-state photon density in the cavity, S_0, as:

$$S_0 = \frac{\Gamma \eta_i}{q V_a} \tau_p [I - I_{th}] \qquad (2.19)$$

and the threshold current I_{th} as:

$$I_{th} = (1 - \beta) \left(\frac{q V_a}{\eta_i} \right) \frac{N_0}{\tau_n} \cong \left(\frac{q V_a}{\eta_i} \right) \frac{N_0}{\tau_n} \qquad (2.20)$$

where N_0 is the steady-state carrier density.

The steady-state output power, P_0, is equal to the rate at which photons escape from the cavity to form the output beam multiplied by the energy per photon, $h\nu$. Since the total number of photons in the cavity mode is $V_p S_0$ and if we assume that power escapes from one end only (with the other end 100% reflecting) then we have:

$$P_0 = h\nu (v_g \alpha_r)(V_p S_0) \qquad (2.21)$$

where $v_g \alpha_r = -\{v_g/2l\} \ln(1 - T) \approx -\{v_g/2l\} T$ represents the escape rate of photons through transmission, T, at the output end of the cavity with round trip length, l.

The total photon loss rate of the cavity is given by the inverse photon lifetime, $\tau_p^{-1} = v_g(\alpha_r + \alpha_c)$, where α_c represents the internal cavity losses. An optical efficiency,

η_o, may be defined as the ratio of the photon loss rate to the output beam to the total photon loss rate:

$$\eta_o = \frac{v_g \alpha_r}{v_g(\alpha_r + \alpha_c)} = v_g \alpha_r \tau_p \tag{2.22}$$

Combining (2.19), (2.21) and (2.22) we obtain:

$$P_0 = \eta_o \eta_i \frac{h\nu}{q}[I - I_{th}] = \eta[I - I_{th}] \tag{2.23}$$

where η is the slope efficiency (slope of the laser power versus current characteristic).

Above threshold, (2.23) predicts a linear increase in the output optical power with current, but clearly this trend cannot continue indefinitely and in practice a graph of output light power versus drive current (*LI* curve) will begin to flatten at higher currents, as indicated in Figure 2.8.

Sinusoidal modulation of the diode current produces intensity modulation of the laser output and this may be derived for low modulation frequencies from the *LI* curve, as shown in Figure 2.8. Depending on the degree of non-linearity, the magnitude of the intensity modulation will slightly decrease with increase in the DC current, as occurs when scanning through an absorption line by a current ramp. There will also be a component of intensity modulation at the second harmonic of the modulation frequency

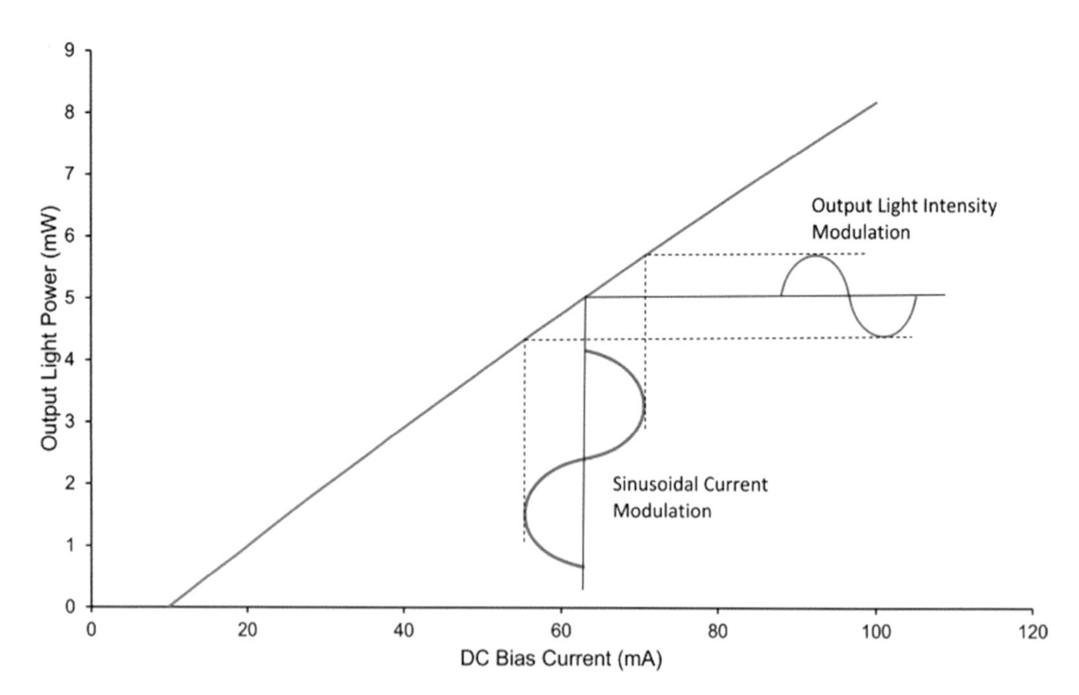

Figure 2.8 Light output power versus forward diode current for a typical DFB laser. Shown also is the intensity modulation generated from sinusoidal modulation of the diode current. Note the slight non-linearity in the *LI* curve.

(and at higher harmonics if there is a large degree of non-linearity in the *LI* curve) and this may need to be taken into account when analysing the harmonics arising in WMS, as discussed in Chapter 3. The non-linearity of the *LI* curve may be simply modelled by assuming that the slope efficiency (or carrier injection efficiency) is no longer constant, but dependent on the diode current. For example, considering second harmonic contribution only, we write:

$$\eta = \eta' - \eta''(I - I_{th}) \tag{2.24}$$

The optical power output is then given by:

$$P = \eta'(I - I_{th}) - \eta''(I - I_{th})^2 \tag{2.25}$$

The linear and non-linear coefficients may be obtained by fitting (2.25) to the experimentally measured *LI* characteristic for a particular laser. For example, the *LI* curve shown in Figure 2.8 corresponds to $\eta' = 0.1$ W A^{-1} and $\eta'' = 0.1$ W A^{-2}.

If the applied diode current has the form noted earlier, $I = I_b + i_m \cos(\omega_m t)$, then substitution in (2.25) gives the output power with first and second harmonic components as:

$$P = \left\{P_0 - \frac{1}{2}\eta'' i_m^2\right\} + \Delta P_1 \cos(\omega_m t) - \Delta P_2 \cos(2\omega_m t) \tag{2.26}$$

where:

$$\Delta P_1 = \{\eta' - 2\eta''(I_b - I_{th})\} i_m \tag{2.27}$$

$$\Delta P_2 = \frac{1}{2}\eta'' i_m^2 \tag{2.28}$$

Equation (2.27) shows clearly the reduction in the magnitude of the first harmonic component with increase in the DC diode current as a result of the non-linearity of the *LI* curve. The second harmonic component of (2.28) also indicates that there is a non-zero background when considering second harmonic signals arising from WMS, as will be discussed further in Chapter 3.

2.5.2 Perturbation Analysis for Effects of Current Modulation

The above DC analysis is based on the steady-state DC characteristics of the laser, where the carrier density is clamped at the steady-state value, N_{ss}, due to the lasing action. For an analysis of the effect of current modulation on the laser output in terms of both intensity and wavelength modulation at any modulation frequency and their relative phases, we may use small signal (linear perturbation) analysis of the dynamic rate equations (2.17) and (2.18), as outlined in Appendix 2.2, to take into account carrier density modulation. If we consider the situation above threshold, where the coupling of spontaneous emission to cavity modes can be neglected

($\beta \approx 0$) then the output power perturbation δP as a result of the current perturbation δI is, from (A2.2.8):

$$\delta P \cong \frac{\eta \delta I}{\left\{1-(\omega_m/\omega_0)^2+\varepsilon/\left(\tau_n v_g g_0\right)\right\}+j\omega_m\left\{\tau_p+1/\left(\omega_0^2\tau_n\right)+\varepsilon/\left(v_g g_0\right)\right\}} \tag{2.29}$$

where ε is a gain compression factor, g_0 is the differential gain, N_{tr} is the transparency carrier density, as defined in Appendix 2.2 and the relaxation oscillation frequency, ω_0, is given by:

$$\omega_0^2 \approx \frac{1}{\tau_n \tau_p}\left(\frac{N_0}{N_0-N_{tr}}\right)\left[\frac{I_0}{I_{th}}-1\right] \tag{2.30}$$

Using typical values for the parameters of semiconductor lasers, as listed in Table 2.1, we find that ω_0 is typically a few GHz and for modulation frequencies less than 100 MHz, all the terms (apart from the first) in the denominator of (2.29) are much less than unity, so that the intensity modulation is essentially in phase with the current modulation and given by: $\delta P \cong \eta \delta I$, as predicted by the above quasi-steady-state analysis.

The carrier density perturbation also perturbs the optical frequency of the output light and from the analysis in Appendix 2.2 the carrier tuning coefficient, ξ_c, is, from (A2.2.13):

$$\xi_c = \frac{\delta \nu}{\delta i} \cong \frac{\eta \alpha}{4\pi P_0}\left\{j\omega_m + \frac{\beta \eta I_{th}}{\tau_p P_0} + \frac{\varepsilon S_0}{\tau_p}\right\} \tag{2.31}$$

where α is the linewidth enhancement factor, as defined in Appendix 2.2.

As discussed in Appendix 2.2, an analysis of the magnitude of the various terms in (2.31) indicates that at less than ~100 MHz modulation frequencies, only the last term, representing gain compression or spectral hole-burning, is significant and hence we may write the tuning coefficient as:

$$\xi_{\text{spectral}} \cong \frac{\eta_i \alpha \varepsilon}{4\pi q V_p} \tag{2.32}$$

This spectral hole-burning tuning coefficient is independent of the modulation frequency and for the laser parameters given in Table 2.1 has typical values of 10–100 MHz mA^{-1} and is in phase with the current modulation.

There is, however, another effect, namely, spatial hole-burning, which can also contribute to the carrier tuning coefficient observed in DFB lasers. Spatial hole-burning arises from the distribution of the carrier density along the laser cavity affecting the round-trip phase and the feedback losses of the cavity [18, 22, 23, 26–29]. The total carrier tuning coefficient, combining both spectral and spatial hole-burning effects, can be approximated from (A2.2.18) as:

$$\xi_c = \xi_{\text{spectral}} + \xi_{\text{spatial}} \cong \frac{\eta_i \alpha \varepsilon}{4\pi q V_p} - \eta\left\{\frac{\alpha}{4\pi\tau_p^2}\frac{\partial \tau_p}{\partial P} + \frac{v_g}{\lambda}\frac{\partial n_r}{\partial P}\right\} \tag{2.33}$$

where the differential terms in (2.33) represent the effects of spatial hole-burning through both a power-dependent cavity loss and a power-dependent refractive index, as derived in Appendix 2.2.

Although the last term in (2.33) is a relatively simple expression for the effect of spatial-hole-burning on the carrier tuning coefficient, it is difficult to predict typical values from laser parameters. Detailed modelling and experimental tests of DFB lasers by Vankwikelberge et al. [23, 26] have indicated that its magnitude may vary significantly from device to device. An approximately flat response for the magnitude of the spatial-hole burning effect as a function of modulation frequency was predicted at modulation frequencies below 1 GHz, but with a strong bias current dependence, reducing from several GHz mA^{-1} at low bias currents to tens of MHz mA^{-1} at high bias currents. For some lasers, the spatial hole-burning contribution can be in anti-phase to that of spectral hole burning, resulting in cancellation at a certain bias level. Numerical modelling of spatial hole-burning effects by Schatz [27] for a typical DFB laser also indicates an approximately flat response at low modulation frequencies (<1 GHz) with values of around 20–40 MHz mA^{-1} for a bias current of ~100 mA.

2.6 Combined Carrier and Thermal Effects

The total frequency tuning coefficient, ξ, for DFB lasers is the complex sum of the carrier and thermal tuning coefficients:

$$\xi = \xi_c + \xi_{th} \tag{2.34}$$

Assuming that ξ_c is independent of frequency, as given by (2.33), then using (2.15) we can write:

$$\xi \cong \xi_c - \left\{ \frac{H_0}{(jf_m/f_{cs})^{\frac{1}{2}} \coth (jf_m/f_{cs})^{\frac{1}{2}}} \right\} \tag{2.35}$$

Here H_0 is the magnitude of the thermal tuning as $f_m \to 0$, as defined earlier by (2.13).

The simple RC thermal model of (2.16) is also useful to identify certain trends in the combined tuning coefficient. With the RC model an explicit expression for the magnitude and phase of the total tuning coefficient may be written as follows:

$$|\xi|^2 = \frac{H_0^2 - 2\xi_c H_0}{1 + (f_m/f_{cs})^2} + \xi_c^2 \tag{2.36}$$

with the phase, ψ_f, of the frequency modulation relative to the current modulation as:

$$\tan \psi_f = \frac{(f_m/f_{cs})H_0}{\xi_c\left\{1 + (f_m/f_{cs})^2\right\} - H_0} \tag{2.37}$$

Figure 2.9 shows the phase relative to the current modulation of the various components of the tuning coefficient in terms of frequency and wavelength modulation.

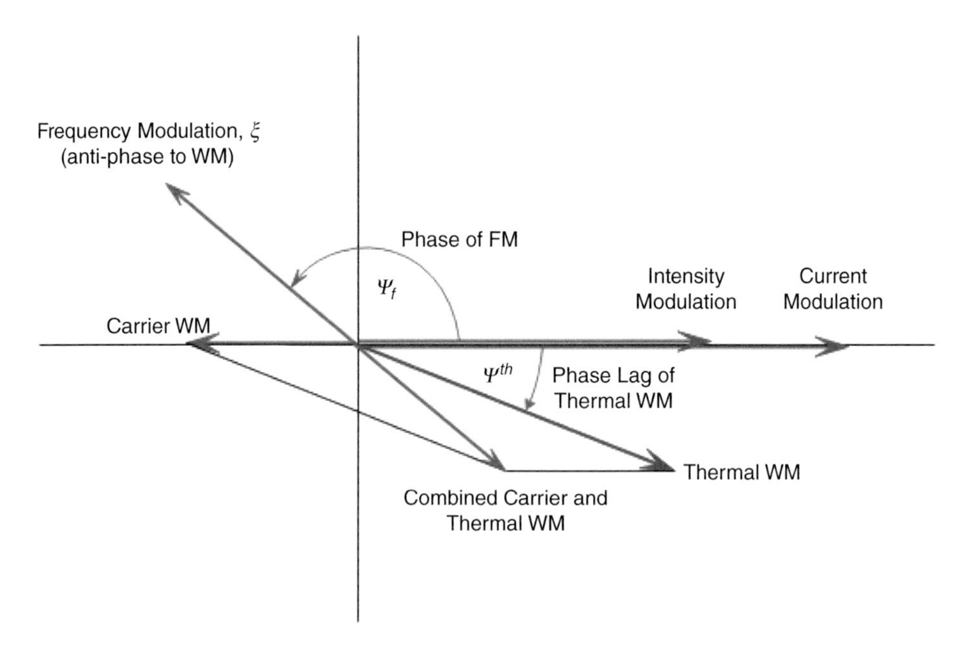

Figure 2.9 Contributions to the tuning coefficient and phase relative to the modulation current in terms of wavelength modulation (WM) and frequency modulation (FM).

Note from (2.36) that when $H_0 > 2\xi_c$ the magnitude of the tuning coefficient decreases with modulation frequency, which is normally the case, whereas when $H_0 < 2\xi_c$ the tuning coefficient increases with frequency. For $H_0 = 2\xi_c$ the magnitude of the tuning coefficient is independent of frequency, $\xi = \xi_c$, but the phase is still frequency dependent. Similar trends occur with (2.35), depending on the relative magnitudes of the thermal and carrier effects, but (2.35) also predicts a dip in the magnitude response in some cases, as observed in practice for certain lasers. In Figure 2.10 we compare: (a) the magnitude and (b) the phase lag of the wavelength modulation for the combined tuning coefficient from (2.35) and from (2.36) and (2.37) as a function of the modulation frequency, using the typical laser parameters given in Table 2.1. In particular, (2.35) gives a better description of the phase variation as a function of modulation frequency when compared with experimental measurements [35].

2.7 Measurement of DFB Laser Characteristics

For a specific DFB laser, a number of fundamental measurements may be carried out to predict and optimise the performance for near-IR spectroscopic systems. These include measurement of: (i) the drive current versus voltage (IV curve) to estimate the series resistance r_s and the junction voltage V_j, (ii) the output optical power versus drive current (*LI* curve) to determine the threshold current, slope efficiency and non-linearity in the slope, as described by (2.25) and (iii) measurement of the output

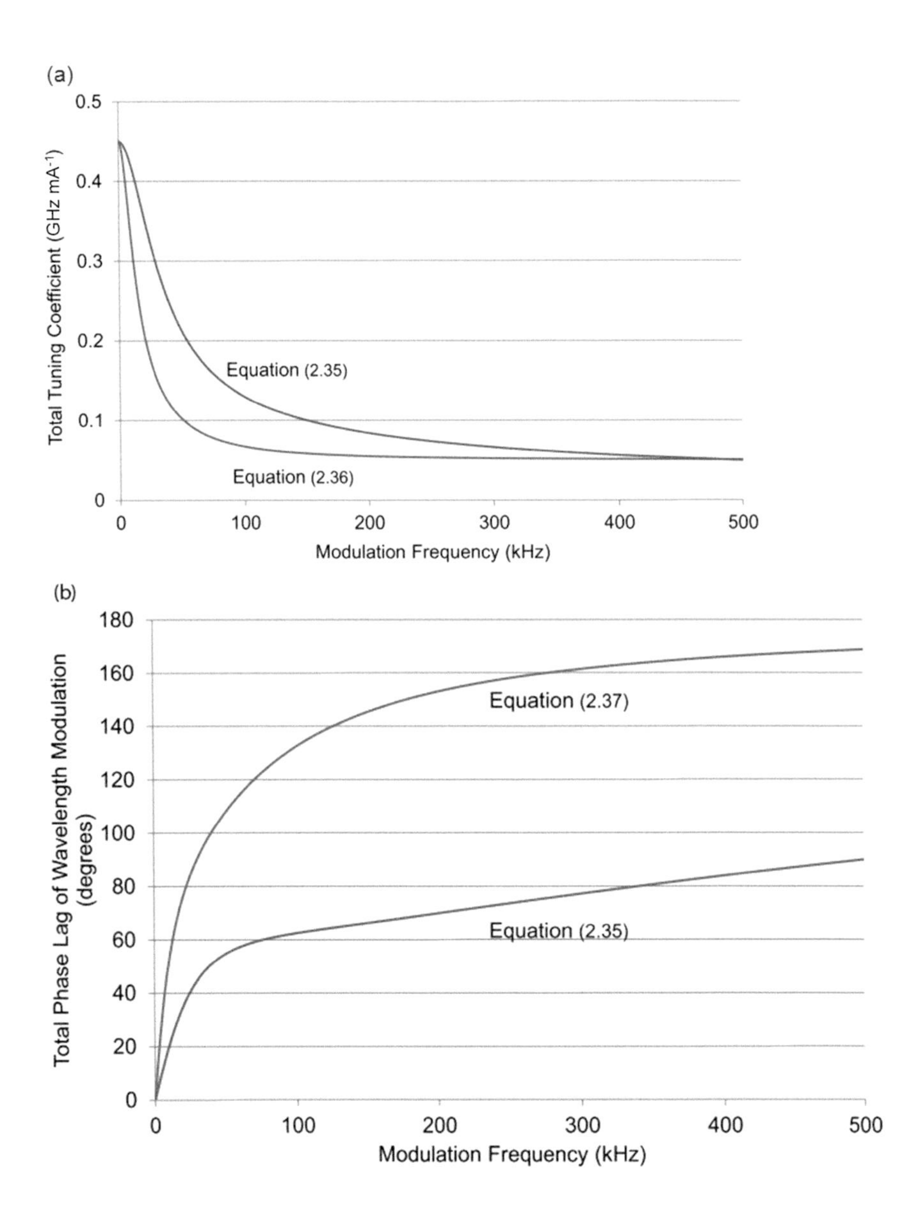

Figure 2.10 Total tuning coefficient for typical laser parameters: (a) magnitude (b) phase lag of the wavelength modulation.

wavelength as a function of drive current and temperature to determine the DC thermal tuning coefficient, H_0. These measured parameters may then be used to predict, from the theoretical relationships given in this chapter, the wavelength scanning and modulation characteristics of the laser when used for wavelength modulation spectroscopy. However, it may also be desirable to directly measure the frequency (or wavelength) modulation characteristics in terms of its amplitude and phase as a function of the modulation frequency. Some examples of direct measurement of the frequency response for spectroscopic applications are reported in the literature for various DFB lasers [11, 30–35].

For fibre optic systems, a convenient method for the experimental determination of the modulation characteristics and calibration of the wavelength scan is through use of a fibre optic ring resonator [11, 32–35]. A typical experimental arrangement for this measurement is illustrated in Figure 2.11a. The ring resonator consists of a fibre loop with length, l_f, of 0.5–2 m, with two fibre couplers of ~40% coupling ratio. The separation of the resonance peaks (free spectral range, FSR) is given by: $FSR = c/(n_e l_f)$, where the effective index $n_e \approx 1.45$ so the FSR is typically 100–400 MHz, depending on the chosen loop length. If a linear (sawtooth) current ramp is applied to the drive current of the DFB laser, then an output as shown in Figure 2.11b is obtained and the wavelength sweep can be obtained by counting the number of resonance peaks that are displayed over the course of the scan and multiplying by the FSR expressed in wavelength units as $FSR = \lambda_0^2/(n_e l_f)$. If a relatively large amplitude sinusoidal modulation is applied to the drive current then an output as shown in Figure 2.11c is obtained, which also shows the corresponding laser intensity modulation. The position of the peaks in the resonator trace, translated to a frequency scale by the FSR, may be plotted against their corresponding time points on the x-axis and the resulting curve fitted to a sinusoidal function of the form:

$$\nu = \nu_a + \Delta\nu \cos(\omega_m t + \phi_\nu) \tag{2.38}$$

where ν_a is an arbitrary constant, and $\Delta\nu$ and ϕ_ν are determined through the fitting procedure.

Similarly, the intensity modulation trace (displayed in units of voltage) may be fitted to a sinusoidal function of the form:

$$V = V_a + \Delta V_1 \cos(\omega_m t + \phi_1) + \Delta V_2 \cos(2\omega_m t + \phi_2) \tag{2.39}$$

which takes into account, if necessary, any non-linearity in the intensity modulation, as discussed previously and described by (2.26).

Using the above results, the magnitude of the frequency tuning coefficient may be determined from $|\xi| = \Delta\nu/i_m$ and its phase as $\psi_f = (\phi_\nu - \phi_1)$ where i_m is the amplitude of the current modulation used to generate the traces. (The intensity modulation is assumed to be in phase with the current modulation).

In the choice of the loop length for the fibre ring resonator, it should be noted that a longer length will give a better resolution in determining the sinusoidal fit to the frequency modulation response described by (2.38), since the resonances are more closely spaced with increase in loop length. However, a longer loop length restricts the maximum modulation frequency that can be used in the measurement since sufficient time is required for the resonance to build up from the circulating light in the loop and this is particularly acute when the sinusoidal modulation function is changing rapidly around its zero crossing points. The limitation on the loop length may be estimated by comparing the transit time of the loop (~5 ns per metre length of loop)

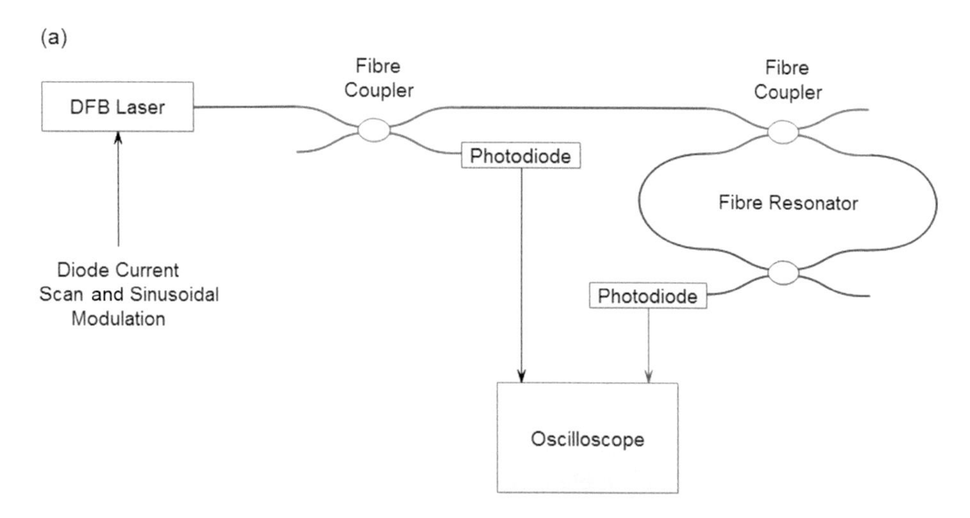

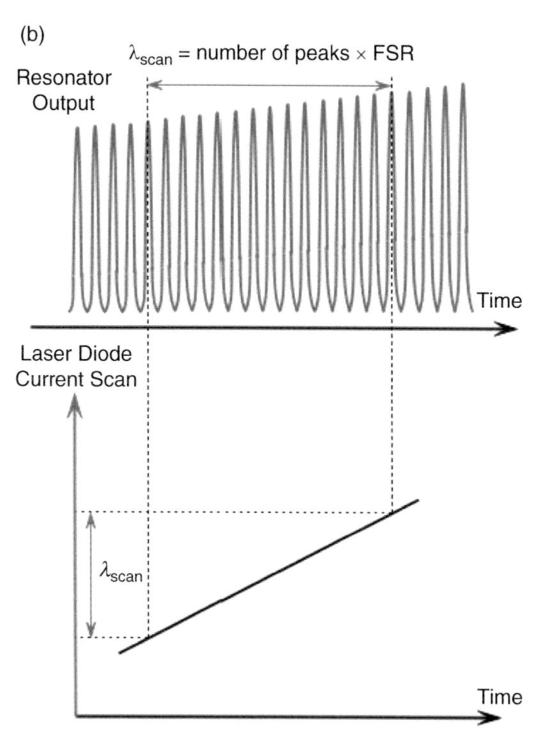

Figure 2.11 (a) Typical experimental arrangement using a fibre ring resonator to measure the modulation characteristics of a DFB laser, (b) calibration of a linear current scan from the resonator output, (c) calibration of the magnitude and phase of the frequency modulation (relative to the intensity modulation) from the resonator output with sinusoidal modulation of the drive current.

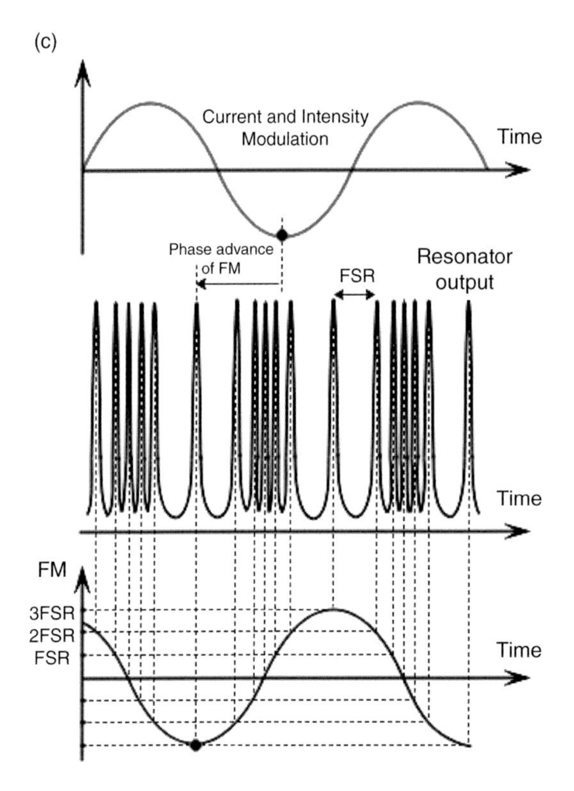

Figure 2.11 (cont.)

with the period of the sinusoidal modulation (for example, 10 ns for 100 MHz frequency). It is also important for stability to have appropriate mechanical and thermal isolation of the resonator. The resonator itself may be calibrated and the frequency positions of resonator peaks may be made absolute by reference to known gas absorption line(s) with precisely known centre frequencies [33].

2.8 Conclusion

This chapter has reviewed the basic characteristics of DFB lasers from the point of view of applications in near-IR spectroscopy. Of particular importance are the wavelength scanning and modulation properties, which have a direct bearing on the recovered signals in wavelength modulation spectroscopy. Detailed knowledge of the magnitude and relative phase of the wavelength and intensity modulation, as well as higher-order harmonics in the intensity modulation, are essential to understanding the various harmonic components that arise when interrogating a gas absorption line and we shall make extensive use of the relationships derived here in Chapter 3. Additionally, the detailed theoretical review of thermal and carrier effects provided here may be useful for the future design of DFB lasers tailor-made for spectroscopic applications, as opposed to standard DFB lasers for the fibre optic communications sector, which have somewhat different criteria for their design. Finally, although our focus has been on DFB lasers in the near-IR, the principles discussed apply equally well to a range of other types of lasers used in spectroscopy and several references on the general principles of semiconductor laser devices are included in the reference list [20–22, 36–39].

Appendix 2.1 Analytical 1-D Thermal Model of a Diode Laser

Consider the 1-D planar structure shown in Figure A2.1.1, where we assume uniform heat generation from electrical power dissipation $P_{el}(t)$ in the thin active layer of thickness d_a, positioned at a distance, d_s from the heat sink, with a top buffer or confinement layer of thickness d_b.

The heat generated in the active layer from (2.4), neglecting A_2 and written in exponential notation is:

$$P_{el}(t) \approx P_{DC} + A_1 e^{j\omega_m t} \tag{A2.1.1}$$

We assume that the temperature throughout the thin active layer (and over the optical profile) is approximately uniform if $d_a \ll d_s$ and is given by $T_{ss} + T_a(t)$ where:

$$T_a(t) = T_a e^{j(\omega_m t + \varphi)} \tag{A2.1.2}$$

$$T_{ss} = R_T P_{DC} + T_0 \tag{A2.1.3}$$

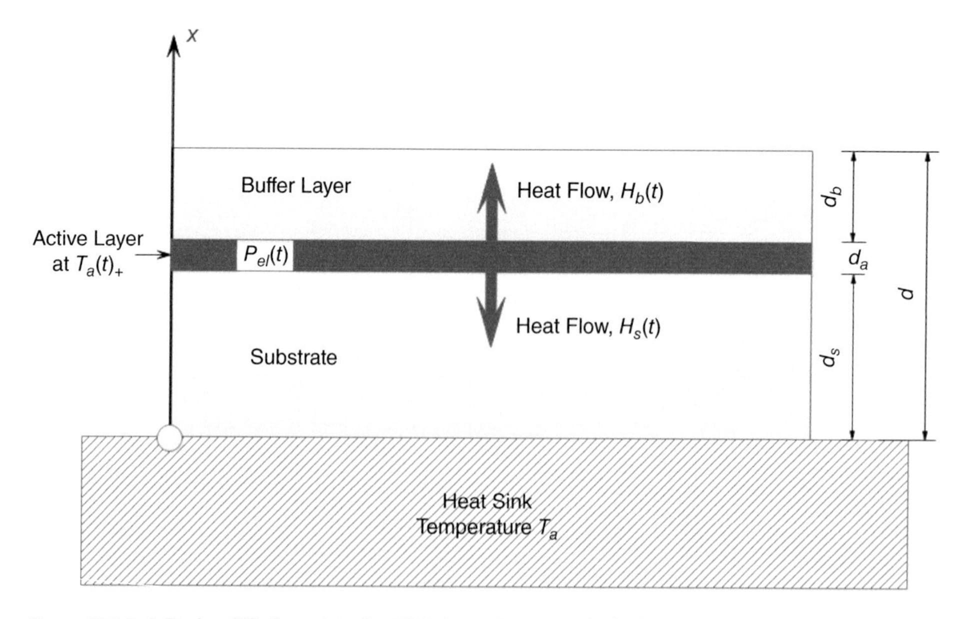

Figure A2.1.1 1-D simplified model of a DFB laser for an analytical solution of the heat conduction equation.

T_{ss} is the steady-state temperature of the active region, T_0 is the heat sink temperature and R_T is the thermal resistance to the heat sink.

For boundary conditions, we assume a fixed temperature of T_0 at $x = 0$ (heat sink) and we further assume that there is no heat loss from the upper surface so that $\partial T/\partial x = 0$ at $x = d$.

In the substrate and buffer regions, where we assume that no heat generation occurs, the heat conduction equation [12] has the form:

$$\frac{dT}{dt} = \kappa \frac{d^2T}{dx^2} \tag{A2.1.4}$$

which has the following time-varying solutions:

$$T(x,t) = T_a(t)\left[\frac{e^{qx} - e^{-qx}}{e^{qd_s} - e^{-qd_s}}\right], \qquad 0 \le x \le d_s \tag{A2.1.5}$$

$$T(x,t) = T_a(t)\left[\frac{e^{q(x-d)} + e^{-q(x-d)}}{e^{qd_b} + e^{-qd_b}}\right], \quad (d_s + d_a) \le x \le d \tag{A2.1.6}$$

where $q = \sqrt{j\omega_m/\kappa}$, $\kappa = k/(\rho C_p)$ is the thermal diffusivity, k is the thermal conductivity, ρ is the density and C_p is the specific heat capacity (typically, ρ ~ 4.81 g cm^{-3}, C_p ~ 0.31 J g^{-1}K^{-1} and κ ~ 0.46 cm^2 s^{-1} for InP at 300 K).

Since heat flow is given by: $H = -kA(\partial T/\partial x)$, it follows from (A2.1.5) and (A2.1.6) that the heat flow out of the active layer (see Figure A2.1.1) is given by:

$$H_s(t) = kAqT_a(t) \coth(qd_s) \tag{A2.1.7}$$

$$H_b(t) = kAqT_a(t) \tanh(qd_b) \tag{A2.1.8}$$

where A is the area of the active layer.

Energy balance in the active layer requires that the rate of (electrical) energy supplied to the active layer must equal the rate of energy loss from the active layer (by outward heat flow) plus the rate of energy storage in the active layer. Hence:

$$C_a \frac{dT_a(t)}{dt} + H_s(t) + H_b(t) = P_{el}(t) \tag{A2.1.9}$$

where $C_a = \rho C_p V_a$ is the thermal capacity of the active layer and V_a is its volume.

Combining (A2.1.1), (A2.1.2), (A2.1.7), (A2.1.8) with (A2.1.9) we obtain:

$$T_a(t) = \frac{A_1 e^{j\omega_m t}}{j\omega_m C_a + kAq[\coth(qd_s) + \tanh(qd_b)]} \tag{A2.1.10}$$

From (A2.1.2), (A2.1.3) and (A2.1.10), the active layer temperature is:

$$T = T_0 + R_T\left\{P_{DC} + \frac{A_1 e^{j\omega_m t}}{j\omega_m R_T C_a + qd_s[\coth(qd_s) + \tanh(qd_b)]}\right\} \tag{A2.1.11}$$

If we define cut-off frequencies associated with the substrate, buffer and active layers by: $\omega_{cs} = \kappa/d_s^2$, $\omega_{cb} = \kappa/d_b^2$ and $\omega_{ca} = (d_s/d_a)\omega_{cs}$, respectively, then the thermal tuning coefficient (in magnitude and phase) from (2.6), (2.13) and (A2.1.11) is:

$$\xi_{th} = \left\{ \frac{-H_0}{j\omega_m/\omega_{ca} + (j\omega_m/\omega_{cs})^{1/2}\left[\coth (j\omega_m/\omega_{cs})^{1/2} + \tanh (j\omega_m/\omega_{cb})^{1/2}\right]} \right\} \tag{A2.1.12}$$

where again the negative sign indicates that the modulation, in terms of frequency, is in anti-phase with the current modulation (in-phase in terms of wavelength) as $\omega_m \to 0$.

Typically the substrate layer thickness d_s is around three orders of magnitude greater than the active layer thickness, d_a so that $\omega_{ca} \gg \omega_{cs}$ and the first term in the denominator may be neglected for sub-megahertz modulation frequencies. Similarly, if the buffer (top) layer thickness is of the order of a few microns, then $\omega_{cb} \sim$ 10 MHz and the last term may be neglected at low frequencies. This gives the simplified expression for the thermal tuning coefficient at sub-megahertz modulation frequencies:

$$\xi_{th} \cong \left\{ \frac{-H_0}{(jf_m/f_{cs})^{\frac{1}{2}} \coth (jf_m/f_{cs})^{\frac{1}{2}}} \right\} \tag{A2.1.13}$$

Appendix 2.2 Perturbation Analysis of the Laser Rate Equations

The effect of current modulation on the intensity and wavelength of the laser output can be derived through perturbation analysis of the dynamic rate equations (2.17) and (2.18). For this, we assume that the current modulation gives rise to modulation of the carrier and photon densities and the gain according to:

$$I = I_0 + \delta I e^{j\omega_m t} \tag{A2.2.1a}$$

$$N = N_0 + \delta N e^{j\omega_m t} \tag{A2.2.1b}$$

$$S = S_0 + \delta S e^{j\omega_m t} \tag{A2.2.1c}$$

$$G = G_0 + \delta G e^{j\omega_m t} \tag{A2.2.1d}$$

where I_0, N_0, S_0 and G_0 are the steady-state values and δI, δN, δS, δG are the complex perturbations (including both magnitude and phase).

Substitution of (A2.2.1) into the rate equation (2.18) and using the steady-state condition:

$$\left\{\Gamma v_g G_0 - \frac{1}{\tau_p}\right\} S_0 + \frac{\Gamma \beta N_0}{\tau_n} = 0$$

gives:

$$\delta G = \left\{\frac{j\omega_m}{\Gamma v_g S_0} + \frac{\beta N_0}{\tau_n v_g S_0^2}\right\} \delta S \tag{A2.2.2}$$

where terms containing the products $\delta G \delta S$ and $\beta \delta N$ are small and have been neglected.

The gain is dependent on the carrier density, but may also be weakly dependent on the photon density, the latter being described as spectral hole-burning or gain compression [6, 7]. Hence we may write in general:

$$\delta G = \frac{\partial G}{\partial N} \delta N + \frac{\partial G}{\partial S} \delta S \tag{A2.2.3}$$

Substituting in (A2.2.2) and rearranging we obtain the carrier density perturbation as:

$$\delta N = \frac{1}{\partial G / \partial N} \left\{\frac{j\omega_m}{\Gamma v_g S_0} + \frac{\beta N_0}{\tau_n v_g S_0^2} - \frac{\partial G}{\partial S}\right\} \delta S \tag{A2.2.4}$$

To obtain the partial derivatives in (A2.2.3) and (A2.2.4), an explicit expression for the gain in terms of carrier and photon density may be used. Typically, the gain in a semiconductor laser may be represented by an expression of the form:

$$G(N,S) \cong \frac{g_0}{1+\varepsilon S}(N_{tr}+N_s)\ln\left\{\frac{N+N_s}{N_{tr}+N_s}\right\} \tag{A2.2.5}$$

where ε is a gain compression factor, g_0 is the differential gain, N_{tr} is the transparency carrier density ($G = 0$ when $N = N_{tr}$) and N_s is a gain curve-fitting parameter [21].

For $\varepsilon S \ll 1$ and $(N - N_{tr}) \ll (N_{tr} + N_s)$, a simple linear approximation to (A2.2.5) is frequently used [9, 21, 22]:

$$G(N,S) \approx g_0(1 - \varepsilon S)(N - N_{tr}) \tag{A2.2.6}$$

An expression for the photon density perturbation may also be obtained by substitution of (A2.2.1) into the rate equation (2.17) to give:

$$\delta S = \frac{(\eta_i/qV_a)\delta I - v_g S_0 \delta G - (j\omega_m + 1/\tau_n)\delta N}{v_g G_0} \tag{A2.2.7}$$

Expressions for δG and δN from (A2.2.2) and (A2.2.4) may then be substituted into (A2.2.7) to give the photon density perturbation in terms of the injection current perturbation. If we consider the situation above threshold, where coupling of spontaneous emission to cavity modes can be neglected ($\beta \approx 0$) then after some manipulation we obtain:

$$\delta P \cong \frac{\eta \delta I}{\left\{1 - (\omega_m/\omega_0)^2 + \varepsilon/(\tau_n v_g g_0)\right\} + j\omega_m\left\{\tau_p + 1/(\omega_0^2\tau_n) + \varepsilon/(v_g g_0)\right\}} \tag{A2.2.8}$$

where we have used $\Gamma v_g G_0 \approx 1/\tau_p$ from the steady-state condition when $\beta \approx 0$. Equations (2.21), (2.22) and (2.23) have been used to convert δS to output power δP and (A2.2.6) to obtain the partial derivatives of the gain.

The relaxation oscillation frequency, ω_0, is given by:

$$\omega_0^2 = \left(\frac{S_0}{\tau_p}\right) v_g \frac{\partial G}{\partial N} \approx \frac{1}{\tau_n \tau_p}\left(\frac{N_0}{N_0 - N_{tr}}\right)\left[\frac{I_0}{I_{th}} - 1\right] \tag{A2.2.9}$$

where again we have used (A2.2.6) for the partial derivative and (2.19) and (2.20) for S_0 and I_{th}.

We now consider the effect of the carrier density perturbation on the optical frequency of the output light. The perturbation in the carrier density, δN, as given by (A2.2.4), results in a perturbation, δn_r, in the real part of the refractive index of the active region, which in turn causes a perturbation, $\delta \nu$, in the frequency of the lasing mode. Maintaining the Bragg condition for the lasing mode, written in terms of optical frequency, $\nu n_e = kc/2\Lambda$, where k is the diffraction order, requires that $\Delta(\nu n_e) = 0$ and hence the fractional shift in the optical frequency can be written as:

$$\frac{\delta \nu}{\nu} = -\frac{\delta n_e}{n_e} \cong -\Gamma \frac{v_g}{c}\frac{\partial n_r}{\partial N}\delta N \tag{A2.2.10}$$

Note that the refractive index decreases with increasing carrier concentration [25] so that $\partial n_r/\partial N$ is negative in the above equation and hence the change in frequency is in phase with the change in carrier concentration and drive current.

Combining (A2.2.4) and (A2.2.10) gives:

$$\delta \nu = \frac{\alpha}{4\pi}\left\{\frac{j\omega_m}{S_0} + \frac{\Gamma\beta N_0}{\tau_n S_0^2} - \Gamma v_g \frac{\partial G}{\partial S}\right\}\delta S \qquad \text{(A2.2.11)}$$

where α is the linewidth enhancement factor [9, 21] defined as the ratio of change in the real part to the imaginary part of the index with carrier density:

$$\alpha = -\frac{4\pi\nu}{c}\frac{\partial n_r}{\partial N} \Big/ \frac{\partial G}{\partial N} = \left|\frac{\partial n_r}{\partial N} \Big/ \frac{\partial n_i}{\partial N}\right| \qquad \text{(A2.2.12)}$$

where the change in gain is related to the change in the imaginary index by $\delta G = 2k_0 \delta n_i = (4\pi\nu/c)\delta n_i$.

Using (2.21), (2.22), (2.23) and (A2.2.6), we can express (A2.2.11) in terms of a carrier tuning coefficient, ξ_c, as:

$$\xi_c = \frac{\delta\nu}{\delta i} \cong \frac{\eta\alpha}{4\pi P_0}\left\{j\omega_m + \frac{\beta\eta I_{th}}{\tau_p P_0} + \frac{\varepsilon S_0}{\tau_p}\right\} \qquad \text{(A2.2.13)}$$

The relative magnitude of the various terms in (A2.2.13) may be estimated using the typical values for semiconductor laser parameters given in Table 2.1. From this we find that the first term (associated with relaxation oscillation) is only significant at high modulation frequencies $>$100 MHz, the second term (arising from the coupling of spontaneous emission to the lasing mode resulting in the gain being slightly less than loss) is only significant near threshold when the output power, $P_0 \sim 0$. Hence at modulation frequencies $<$100 MHz only the last term, representing gain compression or spectral hole-burning, is significant and may be written as:

$$\xi_{\text{spectral}} \cong \frac{\eta\alpha}{4\pi P_0}\left\{\frac{\varepsilon S_0}{\tau_p}\right\} = \frac{\eta_i \alpha \varepsilon}{4\pi q V_p} \qquad \text{(A2.2.14)}$$

Spatial hole-burning may also contribute to the carrier tuning coefficient observed in DFB lasers. Spatial hole-burning effects are complex but may be approximately modelled by adding a perturbation of the form $\tau = \tau_p + \delta\tau_p e^{j\omega_m t}$ to the cavity loss which is dependent on the photon density. Repeating the above analysis with the extra perturbation term, then (A2.2.11) becomes:

$$\delta \nu = \frac{\alpha}{4\pi}\left\{\frac{j\omega_m}{S_0} + \frac{\Gamma\beta N_0}{\tau_n S_0^2} - \Gamma v_g \frac{\partial G}{\partial S} - \frac{1}{\tau_p^2}\frac{\partial \tau_p}{\partial S}\right\}\delta S \qquad \text{(A2.2.15)}$$

Spatial hole-burning also affects the phase of the distributed reflections [24], which can be represented as a power dependence of n_r, giving an additional frequency shift as:

$$\frac{\delta\nu}{\nu} = -\frac{v_g}{c}\frac{\partial n_r}{\partial S}\delta S \qquad \text{(A2.2.16)}$$

Adding this contribution to (A2.2.15) gives the total optical frequency deviation as a function of the photon density perturbation as:

$$\delta\nu = \frac{\alpha}{4\pi}\left\{\frac{j\omega_m}{S_0} + \frac{\Gamma\beta N_0}{\tau_n S_0^2} - \Gamma v_g \frac{\partial G}{\partial S}\right\}\delta S - \left\{\frac{\alpha}{4\pi\tau_p^2}\frac{\partial \tau_p}{\partial S} + \frac{v_g}{\lambda}\frac{\partial n_r}{\partial S}\right\}\delta S \qquad \text{(A2.2.17)}$$

where the last bracketed term is the contribution from spatial hole-burning.

Hence the carrier tuning coefficient including spectral and spatial hole burning effects is:

$$\xi_c = \xi_{\text{spectral}} + \xi_{\text{spatial}} \cong \frac{\eta_i \alpha \varepsilon}{4\pi q V_p} - \eta\left\{\frac{\alpha}{4\pi\tau_p^2}\frac{\partial \tau_p}{\partial P} + \frac{v_g}{\lambda}\frac{\partial n_r}{\partial P}\right\} \qquad \text{(A2.2.18)}$$

References

1. B. Kögel, H. Halbritter, S. Jatta, et al., Simultaneous spectroscopy of NH3 and CO using a > 50 nm continuously tunable MEMS-VCSEL, *IEEE Sens. J*, 7, (11), 1483–1489, 2007.
2. C. Massie, G. Stewart, G. McGregor and J. R. Gilchrist, Design of a portable optical sensor for methane gas detection, *Sens. Actuators B Chem.*, 113, 830–836, 2006.
3. Nanosystems & Technologies GmbH. Distributed feedback lasers. 2019. [Online]. Available: www.nanoplus.com/en/products/distributed-feedback-lasers/ (accessed March 2020)
4. Eblana Photonics Ltd. Optical sensing. 2019. [Online]. Available: www.eblanaphotonics .com/ (accessed March 2020)
5. Sacher Lasertechnik GmbH Laser diodes. 2019. [Online]. Available: www.sacher-laser.com/ home/laser-diodes/distributed_feedback_laser/dfb/single_mode.html?gclid=CMXh-Jvm0s8CFUKVGwodbdMGKw#specifications/?gad=DFBLaser (accessed March 2020)
6. S. O. Kasap, Stimulated emission devices: lasers, in *Optoelectronics and Photonics: Principles and Practice, Upper Saddle River*, Upper Saddle River, NJ, Prentice-Hall, ch. 4. 159–216, 2001.
7. R. Syms and J. Cozens, Optoelectronic devices, in *Optical Guided Waves and Devices*, London, McGraw-Hill, ch. 12, 326–389, 1992.
8. J. Singh, *Semiconductor Optoelectronics: Physics and Technology*, Singapore, McGraw-Hill, 1995.
9. R. S. Tucker, High-speed modulation of semiconductor lasers, *J. Lightw. Techn.*, LT-3, (6), 1180–1192, 1985.
10. O. Doyle, P. B. Gallion and G. Debarge, Influence of carrier non-uniformity on the phase relationship between frequency and intensity modulation in semiconductor lasers, *IEEE J. Quant. Electron.*, 24, (1), 516–522, 1998.
11. T. Benoy, M. Lengden, G. Stewart and W. Johnstone, Recovery of absorption line-shapes with correction for the wavelength modulation characteristics of DFB lasers, *IEEE Photon.*, 8, (3), 2016.
12. H. S. Carslaw and J. C. Jaeger, *Conduction of Heat in Solids*, Oxford University Press, 1959.
13. W. B. Joyce and R. W. Dixon, Thermal resistance of heterostructure lasers, *J Appl. Phys.*, 46, (2), 855–862, 1975.
14. M. Ito and T. Kimura, Stationary and transient thermal properties of semiconductor laser diodes, *IEEE J. Quant. Electron.*, QE–17, (5), 787–795, 1981.

15. G. S. Pandian and S. Dilwali, On the thermal response of a semiconductor laser diode, *IEEE Photon. Techn. Lett.*, 4, (2), 130–133, 1992.
16. P. Correc, O. Girard and I. F. Defaria, On the thermal contribution to the FM response of DFB lasers: theory and experiment, *IEEE J. Quant. Electron.*, 30, (11), 2485–2490, 1994.
17. S. Dilwali, Transfer function of thermal FM, FSK step response and the dip in the FM response of laser diodes, *Opt. Quant. Electron.*, 24, 661–676, 1992.
18. M. Funabashi, H. Nasu, T. Mukaihara, et al., Recent advances in DFB lasers for ultra-dense WDM applications, *IEEE J. Quant. Electron.*, 10., (2), 312–320, 2004.
19. X. Li and W. Huang, Simulation of DFB semiconductor lasers incorporating thermal effects, *IEEE J. Quant. Electron.*, 31, (10), 1848–1855, 1995.
20. J. Buus, M. C. Amman and D. J. Blumenthal, *Tunable Laser Diodes and Related Optical Sources*, New Jersey, John Wiley & Sons, 102–104, 2005.
21. L. A. Coldren and S. W. Corzine, *Diode Lasers and Photonic Integrated Circuits*, New York, John Wiley & Sons, 1995, 185–213.
22. Y. Yamamoto, *Coherence, Amplification and Quantum Effects in Semiconductor Lasers*, New York, John Wiley & Sons, 1991, 122–126.
23. G. Morthier and P. Vankwikelberge, *Handbook of Distributed Feedback Laser Diodes*, 2nd edn., Norwood, MA, USA, Artech House, 2013.
24. A. Hangauer, J. Chen, R. Strzoda and M. C. Amann, The frequency modulation response of vertical-cavity surface-emitting lasers: experiment and theory, *IEEE J. Sel. Top. Quant. Electron.*, 17, (6), 1584–1993, 2011.
25. S. Kobayashi, Y. Yamamoto, M. Ito and T. Kimura, Direct frequency modulation in AlGaAs semiconductor lasers, *IEEE J. Quant. Electron.*, 18, (4), 582–595, 1982.
26. P. Vankwikelberge, F. Buytaert, A. Franchois, et al., Analysis of the carrier-induced FM response of DFB lasers: theoretical and experimental case studies, *IEEE J. Quant. Electron.*, 25, (11), 2239–2254, 1989.
27. R. Schatz, Dynamics of spatial hole burning effects in DFB lasers, *IEEE J. Quant. Electron.*, 31, (11), 1981–1993, 1995.
28. J. Kinoshita, Modelling of high-speed DFB lasers considering the spatial hole-burning effect using three rate equations, *IEEE J. Quant. Electron.*, 30, (4), 929–938, 1994.
29. J. Kinoshita, Transient chirping in distributed feedback lasers: effect of spatial hole-burning along the laser axis, *IEEE J. Quant. Electron.*, 24, (11) 2160–2169, 1988.
30. R. Phelan, J. O'Carroll, D. Byrne, et al., $In_{0.75}Ga_{0.25}As$/InP multiple quantum-well discrete mode laser diode emitting at 2μm, *IEEE Photon. Techn. Lett.*, 24, (8), 652–654, 2012.
31. S. Schilt and L. Thevenaz, Experimental method based on wavelength-modulation spectroscopy for the characterization of semiconductor lasers under direct modulation, *Appl. Opt.*, 43, (22), 4446–4453, 2004.
32. H. Li, G. B. Rieker, X. Liu, J. B. Jeffries and R. K. Hanson, Extension of wavelength modulation spectroscopy to large modulation depth for diode laser absorption measurements in high pressure gases, *Appl. Opt.*, 45, (5), 1052–1061, 2006.
33. W. Johnstone, A. J. McGettrick, K. Duffin, A. Cheung and G. Stewart, Tuneable diode laser spectroscopy for industrial process applications: system characterisation in conventional and new approaches, *IEEE Sens. J.*, 8, (7), 1079–1088, 2008.
34. G. Stewart, W. Johnstone, J. R. P. Bain, K. Ruxton and K. Duffin, Recovery of absolute gas absorption line shapes using tuneable diode laser spectroscopy with wavelength modulation – Part I: theoretical analysis, *IEEE J. Lightw. Techn.*, 29, (6), 811–821, 2011.

35. J. R. P. Bain, W. Johnstone, K. Ruxton, et al. Recovery of absolute gas absorption line shapes using tuneable diode laser spectroscopy with wavelength modulation – part 2: experimental investigation *IEEE J. Lightw. Techn.*, 29, (7), 987–996, 2011.
36. P. Bhattacharya, *Semiconductor Optoelectronic Devices*, New Delhi, Prentice Hall of India, 2006.
37. J. Carroll, J. Whiteaway, D. Plumb, *Distributed Feedback Semiconductor Lasers, SPIE Optical Engineering Press*, 1998.
38. P. S. Zory, *Quantum Well Lasers*, New York, Academic Press, 1993.
39. K Petermann, *Laser Diode Modulation and Noise*, Netherlands, Kluwer Academic Publishers, 1988.

3 Wavelength Modulation Spectroscopy with DFB Lasers

3.1 Introduction

In this chapter we derive the theoretical harmonic signals that arise when the modulated output of a DFB laser is scanned across a gas absorption line. These harmonic signals may be measured experimentally in magnitude and phase, and used to obtain the gas parameters of concentration, pressure or temperature. However, the extraction of this information from the harmonic signals is not entirely straightforward, especially as a result of the simultaneous intensity and wavelength modulation of the DFB laser, as discussed in Chapter 2. Furthermore, in practical applications it is desirable to have calibration-free operation and/or robust methods for referencing to ensure reliable and consistent operation in the face of factors such as drift of laser parameters with time or with aging. Here we examine in detail two methods for recovery of lineshape information, the first based on the intensity modulation of the DFB laser and the second based on the harmonic signals generated from the wavelength modulation. In each case, a number of factors need to be taken into account in order to ensure accurate recovery of the true lineshape, especially under conditions of varying concentration, pressure or temperature. We also compare the methods in terms of sensitivity and signal-to-noise ratio, with a view to identifying when a particular technique might be used.

3.2 Techniques for Gas Absorption Spectroscopy

In Chapter 1, we presented the Beer–Lambert law in various forms, described by (1.2) to (1.8). In particular, let us repeat here (1.8), written in the form:

$$(P_{\mathrm{in}} - P_{\mathrm{out}})/P_{\mathrm{in}} = \Delta P/P_{\mathrm{in}} = \{1 - \mathrm{e}^{-\alpha Cl}\} \approx \alpha Cl \tag{3.1}$$

where the approximation applies for small absorbance and the absorption coefficient, α, is related to the absorption lineshape function by (1.7):

$$\alpha(\nu, p, T) = N_0 S(T)\phi(\nu, p, T) \tag{3.2}$$

It is the fundamental objective of absorption spectroscopy to measure $\Delta P/P_{\mathrm{in}}$ as accurately as possible (in the presence of various noise sources) and hence determine, through (3.1) and (3.2), the lineshape function and gas parameters such as concentration, pressure and temperature, as discussed in detail in Section 1.3 of Chapter 1. There are several well-established techniques for this measurement as follows [1–4]:

(i) The simplest method is direct absorption spectroscopy (DAS), where the CW output from a suitable laser is passed through an absorption cell of length, l, and the output intensity from the cell is measured as the laser wavelength is scanned through an absorption line, as illustrated in Figure 3.1a. By dividing the output intensity by the background (off-line) level, a normalised transmission and absorption lineshape is obtained. The problem with these DC measurements is that the absorbance signal, ΔI, from the gas is very small, particularly for near-IR spectroscopy, compared with the large background, and so the signal is easily overwhelmed by the background noise. The situation may be improved by intensity modulation (IM) of the input beam (for example, by using an external modulator or a beam chopper) in which case phase-sensitive detection by a lock-in amplifier may be employed to greatly improve the signal-to-noise ratio. Even so, the absorbance signal is still appearing on a large background level and there is the additional requirement of an external modulator.

(ii) In wavelength modulation spectroscopy (WMS), the wavelength of the laser input to the gas cell is sinusoidally modulated, while the central laser wavelength is slowly scanned through the absorption line, as illustrated in Figure 3.1b. For DFB lasers the wavelength modulation and scanning is conveniently performed by modulation of the laser drive current, although this also produces intensity modulation of the output, as discussed in detail in Chapter 2. WMS is characterised by typical modulation frequencies of 10 kHz–1 MHz, very much less than the linewidth of the absorption line, with the amplitude of the wavelength modulation similar to the linewidth, typically ~0.04 nm, or several GHz in terms of optical frequency modulation.

(iii) Frequency modulation spectroscopy (FMS) is similar to WMS except that the modulation frequency is much greater than the linewidth (typically in the GHz range) so that only a single sideband interacts with the absorption line and the amplitude of the modulation is relatively small. A variant of FMS is two-tone frequency modulation spectroscopy (TTFMS), where two closely spaced frequencies are used for modulation with detection at the difference (beat) frequency so that only a modest detector bandwidth is required.

Compared with DAS with intensity modulation, WMS has the advantage that second and higher harmonics of the modulation frequency obtained by phase-sensitive detection have a zero or low background level, thus permitting higher sensitivity in gas detection. Greatest sensitivity may be obtained by FMS due to lower noise, but with the disadvantage of greater expense and complexity. For our purposes, our primary focus will be the use of WMS with DFB lasers, where a reasonable compromise can be obtained in terms of simplicity and cost against

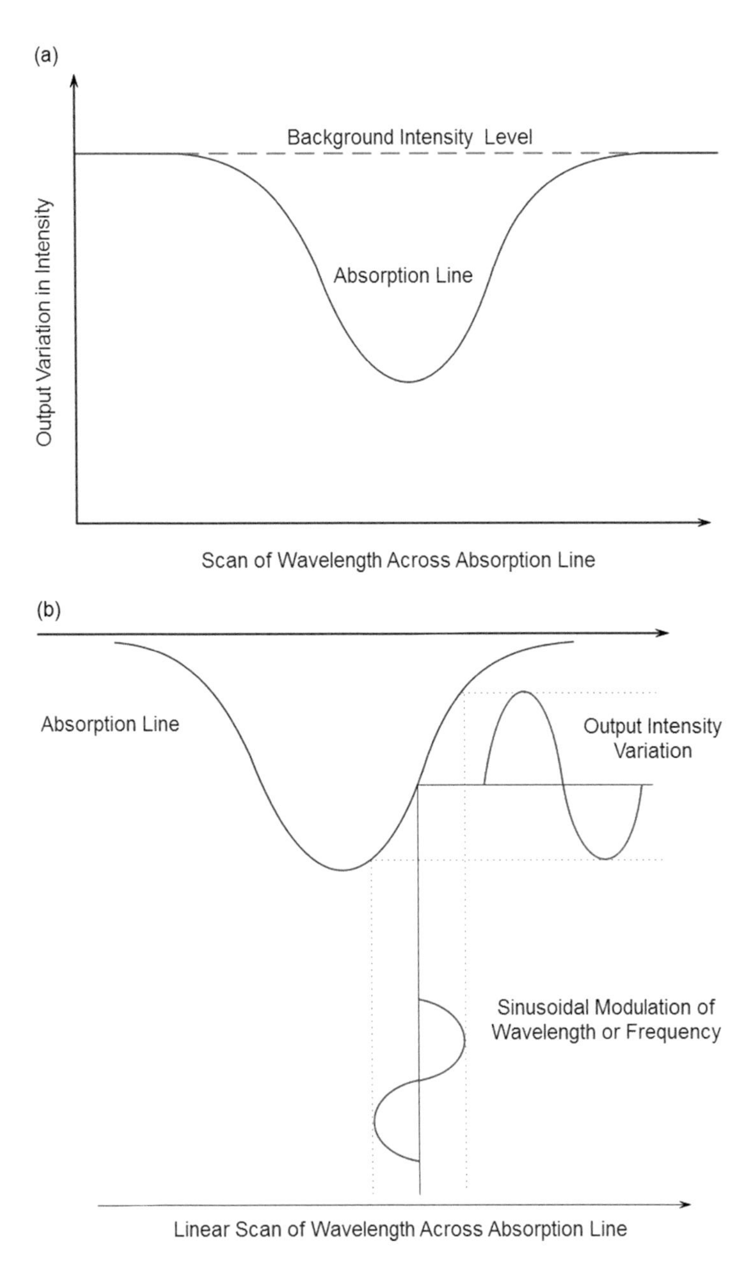

Figure 3.1 Techniques for absorption spectroscopy: (a) direct absorption spectroscopy, (b) wavelength or frequency modulation spectroscopy.

sensitivity. Clearly, WMS and FMS are essentially the same technique in different frequency regimes, so that some of the sensitivity advantage of FMS can be gained in the traditional WMS approach by appropriate choice of the modulation parameters.

3.3 Theoretical Description of Wavelength Modulation Spectroscopy

The basic procedure for WMS with DFB lasers is as follows (see Figure 3.1b):

(i) The operation wavelength of the laser is thermally tuned to the vicinity of the target absorption line using the in-built thermo-electric cooler (TEC) and thermistor, along with an external temperature controller unit.

(ii) The laser diode current is adjusted to a suitable DC bias level above threshold and a linear current scan (typically a sawtooth or triangular waveform at 1–10 Hz frequency) is applied to the diode current to scan the laser wavelength through the absorption line.

(iii) The laser diode current is simultaneously modulated by a sinusoidal function, typically in the frequency range 10 kHz–1 MHz, to give a wavelength (or frequency) modulation with amplitude similar to the linewidth.

(iv) After detection of the output from the gas cell by a photodiode, a lock-in amplifier, referenced to the modulation frequency, is used to extract one or more harmonics of the modulation frequency from which the gas parameters are extracted. Rather than use of hardware instrumentation, extraction and processing of the harmonic signals from the detector output may be carried out entirely by digital signal-processing techniques.

3.3.1 Harmonic Signals Arising from WMS

A theoretical description of the harmonics arising from FMS and WMS has been given by a number of authors [5–8]. For FMS, the analysis is based on an electric field approach, but for WMS a simpler approach based on the intensity analysis is sufficient.

As discussed in detail in Chapter 2, current modulation of DFB lasers gives rise to both intensity and wavelength modulation of the output, with a phase difference between the two. The forward injection current, i_f, supplied to the laser for WMS may be written as:

$$i_f = I_b + i_m \cos(\omega_m t) \tag{3.3}$$

where $I_b = I_{\text{bias}} + I_{\text{scan}}$ is the DC bias current and scan current, i_m is the modulation current amplitude and ω_m is the modulation (angular) frequency.

In order to simplify the analysis at this stage, we shall assume that the *LI* characteristic of the laser is approximately linear so that the output power and intensity modulation from the laser as a result of the applied current can be simply written from (2.26) in Chapter 2 with $\Delta P_2 \sim 0$ as:

$$P_l(\nu, t) = P(\nu) + \Delta P(\nu) \cos(\omega_m t) \tag{3.4}$$

Note that the DC current scan increases the laser power as well as sweeping the laser wavelength (or optical frequency) through the absorption line, so we have written the CW laser power, $P(\nu)$, as a function of the optical frequency, ν. Similarly any slight non-linearity in the *LI* characteristic will mean that the intensity modulation, $\Delta P(\nu)$, will also change across the current scan, as described by (2.27), so that it also is a function of the optical frequency.

In Chapter 2, Section 2.6 we defined a frequency tuning coefficient, ξ, for DFB lasers as the complex sum of the carrier and thermal tuning coefficients, as shown in Figure 2.9. Hence for the applied current of (3.3) we can write the optical frequency, ν_l, of the laser output as:

$$\nu_l = \nu + \delta\nu \cos(\omega_m t + \psi_f) \tag{3.5}$$

where $\delta\nu = |\xi| i_m$ and ψ_f is the argument of ξ, that is, the phase advance of the frequency modulation relative to the current modulation. Note, as explained in Chapter 2, that the *phase advance* of the frequency modulation is related to the *phase lag* of the equivalent wavelength modulation by $\psi_f = (\pi - \psi_\lambda)$, where ψ_λ is defined as being positive, i.e. the wavelength modulation has a phase angle of $-\psi_\lambda$ with respect to the current modulation.

If we assume that the absorbance is small, then using (3.4) and (3.5) in the Beer–Lambert law gives the optical power from a gas cell of length l as:

$$P_{\text{out}}(\nu, t) \cong \{P(\nu) + \Delta P(\nu) \cos(\omega_m t)\}\{1 - \alpha(\nu_l)Cl\} \tag{3.6}$$

It is convenient at this point to define, from (3.2), a dimensionless lineshape function $f(\nu, \gamma)$ through the expression:

$$\alpha(\nu, p, T) = N_0 S(T)\phi(\nu, p, T) = \alpha_0(\gamma, T) f(\nu, \gamma) \tag{3.7}$$

where $\alpha_0(\gamma, T)$ is defined as the absorption coefficient at the line centre. Note that the half-linewidth, γ, depends on pressure and temperature, as discussed in Chapter 1, Section 1.2.2. In the following, for convenience, we shall write $\alpha_0(\gamma, T)$ simply as α_0 and $f(\nu, \gamma)$ simply as $f(\nu)$, but their dependence on pressure and temperature should be remembered.

From (1.11) for a Gaussian lineshape this gives:

$$f(\nu) = \exp\left\{-(\nu - \nu_0)^2/\gamma_D^2\right\} = \exp\{-\Delta^2\} \tag{3.8}$$

where $\alpha_0 = N_0 S(T)/(\gamma_D\sqrt{\pi})$ and $\Delta = (\nu - \nu_0)/\gamma_D$ is the normalised deviation from the line centre.

From (1.14) for a Lorentzian lineshape:

$$f(\nu) = \frac{1}{\left\{1 + (\nu - \nu_0)^2/\gamma_c^2\right\}} = \frac{1}{\{1 + \Delta^2\}} \tag{3.9}$$

where $\alpha_0 = N_0 S(T)/(\pi\gamma_c)$ and $\Delta = (\nu - \nu_0)/\gamma_c$.

Considering (3.5), the function $f(\nu_l)$ can be expanded as a Fourier series in harmonics of the modulation frequency as:

$$f(\nu_l) = a_0 + \sum\nolimits_{n=1}^{\infty} a_n \cos\left(n\omega_m t + n\psi_f\right) = a_0 + \sum\nolimits_{n=1}^{\infty} a_n \cos n\theta \tag{3.10}$$

The Fourier coefficients are given by:

$$a_0 = \frac{1}{\pi}\int_0^{\pi} f(\nu_l)\mathrm{d}\theta \tag{3.11}$$

$$a_n = \frac{2}{\pi}\int_0^{\pi} f(\nu_l)\cos n\theta \mathrm{d}\theta \tag{3.12}$$

where for a Lorentzian lineshape:

$$f(\nu_l) = \frac{1}{\left\{1 + (\Delta + m\cos\theta)^2\right\}} \tag{3.13}$$

and for a Gaussian lineshape:

$$f(\nu_l) = \exp\left\{-(\Delta + m\cos\theta)^2\right\} \tag{3.14}$$

where $m = \delta\nu/\gamma$ is the index of the optical frequency modulation.

Note that $a_0 \approx f(\nu)$ for $m \ll 1$, i.e. it follows the lineshape function, as illustrated for a Lorentzian lineshape in Figure 3.2. However, as m increases, a_0 flattens out due to the 'smearing' effect of the optical frequency modulation. In general, when m is very small, the Fourier coefficient a_n is proportional to the corresponding derivative $f^n(\nu)$, as may be shown from a Taylor series expansion (see Appendix 3.1), $a_0(\nu) \approx f(\nu)$, $a_1(\nu) \approx f'(\nu)\delta\nu$ and $a_2(\nu) \approx f''(\nu)\delta\nu^2/4$, etc. For this reason WMS is sometimes referred to as 'derivative spectroscopy'. For larger values of m, the shape deviates from the derivative function and Figures 3.3 and 3.4 show a_1 and a_2 for various values of m

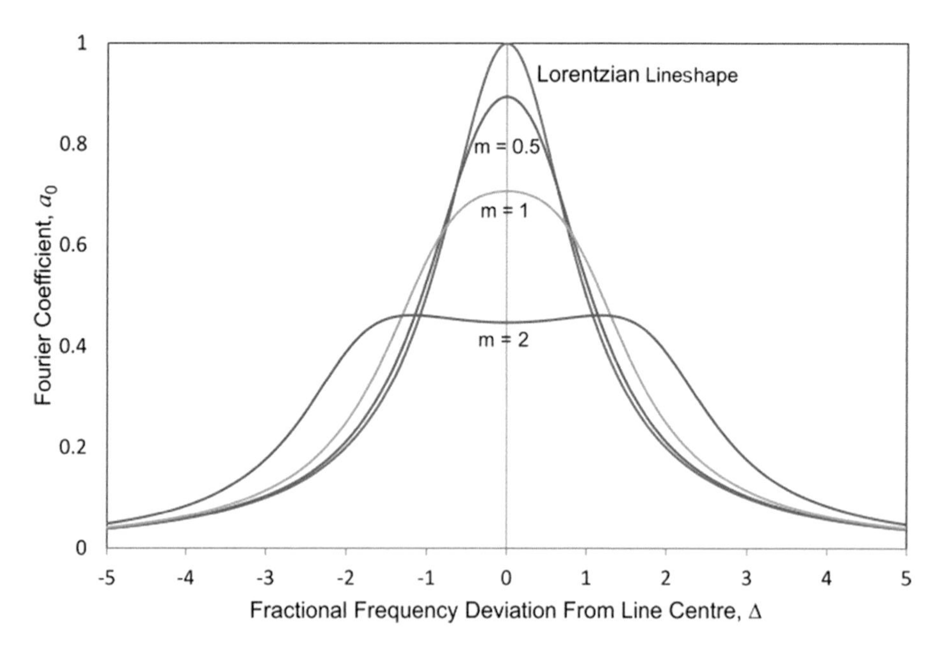

Figure 3.2 Fourier coefficient a_0 for increasing frequency modulation index, m, showing deviation from the Lorentzian lineshape function.

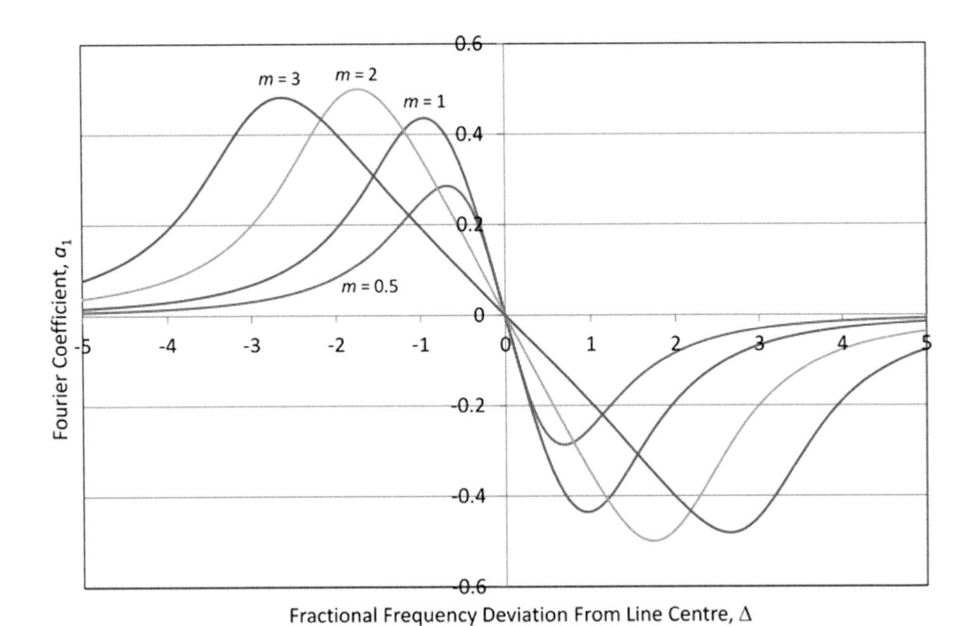

Figure 3.3 Fourier coefficient a_1 for a Lorentzian lineshape function for various values of the frequency modulation index, m.

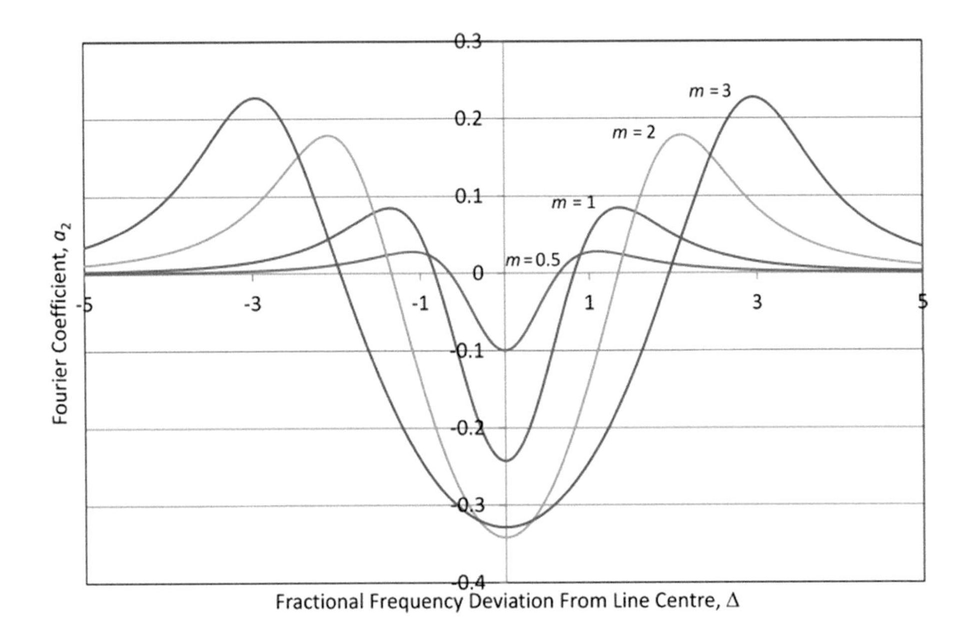

Figure 3.4 Fourier coefficient a_2 for a Lorentzian lineshape function for various values of the frequency modulation index, m.

with a Lorentzian lineshape. Note that the peaks on a_1 occur at $\Delta = \pm\{\sqrt{3m^2+4}-1\}/\sqrt{3} = \pm\sqrt{3}$ for $m = 2$, where they attain the maximum value of $a_1 = \pm 0.5$. For a_2 the central peak at $\Delta = 0$ attains a maximum value of $|a_2| = 0.343$ when $m = 2.2$.

Combining (3.6), (3.7) and (3.10) we obtain the harmonics that arise in WMS as follows:

$$P_{\text{out}}(\nu, t) \cong \{P(\nu) + \Delta P(\nu)\cos(\omega_m t)\}\left[1 - A_0\left\{a_0 + \sum_{n=1}^{\infty} a_n \cos\left(n\omega_m t + n\psi_f\right)\right\}\right] \tag{3.15}$$

where A_0 is the observed absorbance at the line centre for a given cell length and gas concentration, $A_0 = \alpha_0 Cl$ and, for small absorbance, is the same as the depth, d_c, as defined in Chapter 1, Section 1.3.

As noted earlier, (3.15) and the Fourier coefficients defined by (3.11) and (3.12) are derived on the basis of small absorbance using the linear approximation to the exponential term in the Beer–Lambert law. For the non-linear case, the Fourier coefficients are dependent on the absorbance and are defined in Appendix 3.2.

Splitting (3.15) into the two parts arising from $P(\nu)$ and $\Delta P(\nu)$ gives:

$$\begin{aligned} P_{\text{out}}(\nu,t) \cong {} & P(\nu)(1-A_0 a_0) - A_0 P(\nu)\sum_{n=1}^{\infty} a_n \cos\left(n\omega_m t + n\psi_f\right) \\ & + \Delta P(\nu)(1-A_0 a_0)\cos(\omega_m t) - A_0 \Delta P(\nu)\cos(\omega_m t)\sum_{n=1}^{\infty} a_n \cos\left(n\omega_m t + n\psi_f\right) \end{aligned} \tag{3.16}$$

The first term of the first part is direct absorption on the laser intensity (DAS, as discussed in Section 3.2) except that a_0 progressively deviates from the lineshape function with increase in the frequency modulation index, m, as was illustrated in Figure 3.2. The second term describes the harmonics arising from the optical frequency (or wavelength) modulation of the laser output, that is, the conversion of wavelength modulation to intensity modulation (IM), as shown in Figure 3.1b. These are the desired signals of WMS and would be the only harmonic terms if pure wavelength modulation of the laser could be achieved. However, the simultaneous IM of the laser gives rise to additional harmonic terms, as shown in the second part of (3.16).

The first term of the second part is DAS on the laser's IM, again with a_0 progressively deviating from the lineshape function with increasing m. The second term represents an IM of the WMS harmonics. This term can be expanded into sum and difference frequencies using the trigonometric relation for the product of two cosine functions:

$$\begin{aligned} \cos(\omega_m t)\sum_{n=1}^{\infty} a_n \cos\left(n\omega_m t + n\psi_f\right) = {} & \frac{1}{2}\sum_{n=1}^{\infty} a_n \{\cos\left([n+1]\omega_m t + n\psi_f\right) \\ & + \cos\left([n-1]\omega_m t + n\psi_f\right)\} \end{aligned} \tag{3.17}$$

Hence for any particular selected harmonic, in addition to the primary WMS component from the first part of (3.16), we have two additional components from the second part of (3.16) arising from the IM of the laser.

Using a direct output from the signal generator at the modulation frequency as the reference input to a lock-in amplifier, phase-sensitive detection allows low noise extraction of one or more of the harmonics of the modulation frequency from the total output signal. The lock-in output may be in the form of the Cartesian components (X, Y) or the corresponding magnitude (R) of the signal and its phase (θ) with respect to a reference. Since the intensity modulation is always present on the laser output, even in the absence of gas, it is convenient in practice to establish the reference phase from the intensity modulation for measurements of the harmonics. We have in fact followed this procedure in the above analysis where we have assumed in (3.4) that the intensity modulation is in phase with the current modulation (3.3) and that the frequency modulation described by (3.5) is phase-shifted relative to the modulation current. In the following sections we examine in detail the signal components at particular harmonics of the modulation frequency and present phasor diagrams with the intensity modulation as the reference phase. Some authors [5, 22] make a different choice and use the primary harmonics as the reference phase, which corresponds to making the time transformation $t = t' - \psi_f/\omega_m$ in the above analysis. Of course, the final results are the same whatever choice is made for the reference phase and, as we shall see, the lock-in phase may be adjusted as required to null particular components.

3.3.2 The First Harmonic Signal

We first consider the output signal as measured by a lock-in amplifier at the modulation frequency. From (3.16) and (3.17) the three components at ω_m are:

(i) direct absorption on the IM: $\Delta P(\nu)(1 - A_0 a_0)\cos(\omega_m t)$
(ii) the primary WMS first harmonic: $-A_0 P(\nu) a_1 \cos(\omega_m t + \psi_f)$
(iii) IM on second harmonic, $n = 2$ in (3.17): $-\frac{1}{2} A_0 \Delta P(\nu) a_2 \cos(\omega_m t + 2\psi_f)$

From the first term we see that the total first harmonic output is dominated by a large background signal, $\Delta P(\nu)\cos(\omega_m t)$, from the intensity modulation of the laser, sometimes referred to as the residual amplitude modulation (RAM). Note also that the three components each differ in phase by ψ_f, which means that they can be separated, at least to some extent, by choice of detection phase of the lock-in amplifier.

The total first harmonic signal may be written in phasor notation as:

$$h_1(\nu) = \Delta P(\nu) - A_0[\Delta P(\nu) a_0 - P(\nu) a_1 e^{-j\psi_\lambda} + 0.5\Delta P(\nu) a_2 e^{-2j\psi_\lambda}] \tag{3.18}$$

where, for convenience, it is written in terms of ψ_λ, using the definition given earlier, $\psi_f = (\pi - \psi_\lambda)$, and is plotted on the phasor diagram in Figure 3.5.

Figure 3.6 shows the magnitude, $|h_1(\nu)|$, of the total first harmonic signal for typical laser and modulation parameters, as given in Chapter 2. The various parameters used for Figure 3.6 are determined as follows. The total diode current, including the current scan through the absorption line, is related to the fractional frequency deviation, Δ, from line centre through: $I_d = I_0 - (\gamma\Delta)/H_0$ where I_0 is the diode current at line centre and H_0 is the DC current tuning coefficient of the laser diode. Hence, using (2.25) and (2.27) in

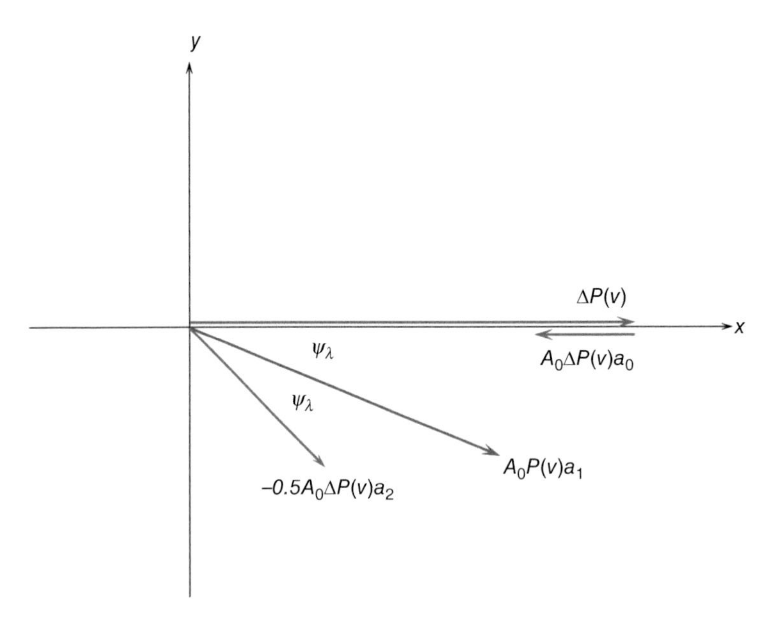

Figure 3.5 Phasor diagram showing the components of the first harmonic signal.

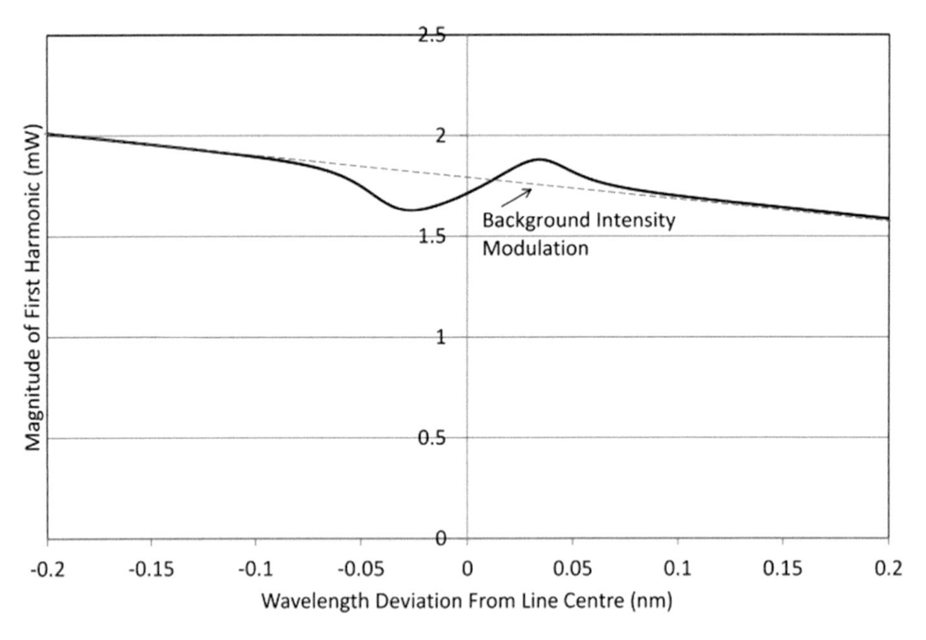

Figure 3.6 Magnitude of the total first harmonic signal (see text for parameters used).

Chapter 2, the variation in laser power, $P(\nu)$, and the intensity modulation, $\Delta P(\nu)$, across the scan can be determined as a function of Δ from $P(\nu) = \eta'(I_d - I_{th}) - \eta''(I_d - I_{th})^2$ and $\Delta P(\nu) = \{\eta' - 2\eta''(I_d - I_{th})\}i_m$, where i_m is the modulation current amplitude. For Figure 3.6, we use $H_0 = 0.5$ GHz mA^{-1}, γ =2.5 GHz, $(I_0 - I_{th}) = 50$ mA, $i_m = 20$ mA,

$\eta' = 0.1$ W A^{-1} and $\eta'' = 0.1$ W A^{-2}. From Figure 2.10 in Chapter 2 we assume a tuning coefficient of ~0.25 GHz mA^{-1} with phase lag of 50° at a typical modulation frequency of 40 kHz giving a modulation index of $m = 2$ for 20 mA current modulation. To calculate the Fourier coefficients, a Lorentzian lineshape is assumed with a line centre absorbance of $A_0 = 0.1$ at a wavelength of ~1500 nm.

Note that the direction of the x-axis in Figure 3.6 is reversed compared to Figures 3.2–3.4 and is plotted in terms of wavelength deviation, $(\lambda - \lambda_0)$, from the line centre. The reason for the change is that in experimental plots, the x-axis normally points in the direction of increasing laser diode current, which corresponds to increasing wavelength, as explained in Chapter 2. (Note that the fractional wavelength deviation from the line centre is defined by $\Delta_\lambda = (\lambda - \lambda_0)/\gamma_\lambda = -\Delta$, where $\gamma_\lambda = (\lambda^2/c)\gamma$ is the half-linewidth in wavelength units.)

Figure 3.6 shows clearly that the total first harmonic is dominated by the background level, which is sloping due to the slight non-linearity of the *LI* curve of the laser diode. Also, the shape of the signal arising from the gas absorption line is significantly distorted from the perfectly anti-symmetric form of the a_1 component, as shown in Figure 3.3, due to the presence of the other components and the increase in laser power across the wavelength scan.

3.3.3 The Second Harmonic Signal

For lock-in detection at twice the modulation frequency, the three components from (3.16) and (3.17) at $2\omega_m$ are as follows:

(i) IM on first harmonic, $n = 1$ in (3.17): $-½A_0\Delta P(\nu)a_1 \cos(2\omega_m t + \psi_f)$
(ii) the primary WMS second harmonic: $-A_0P(\nu)a_2 \cos(2\omega_m t + 2\psi_f)$
(iii) IM on third harmonic, $n = 3$ in (3.17): $-½A_0\Delta P(\nu)a_3 \cos(2\omega_m t + 3\psi_f)$

Compared with the first harmonic, the second harmonic output has the advantage of a zero or low background level, assuming that the *LI* characteristic of the laser is only slightly non-linear, and, again, each component differs in phase by ψ_f.

3.3.4 Effect of a Non-Linear *LI* Curve

The above analysis was made under the simplifying assumption that the *LI* characteristic of the laser was approximately linear, so that any intensity modulation from the laser at twice the modulation frequency was ignored. To include the effect of $\Delta P_2 \neq 0$ for a laser with a highly non-linear *LI* curve we need to use the full version of (2.26) from Chapter 2 in (3.4), (3.6) and (3.15), hence adding another part to (3.16), namely:

$$+\Delta P_2(\nu)\cos(2\omega_m t + \psi_2)(1 - A_0 a_0)$$
$$-A_0\Delta P_2(\nu)\cos(2\omega_m t + \psi_2)\sum_{n=1}^{\infty} a_n \cos(n\omega_m t + n\psi_f) \tag{3.19}$$

where $\psi_2 = \pi$ and $\Delta P_2(\nu) = \frac{1}{2}\eta'' i_m^2$ from (2.28).

As before, by expanding into sum and difference frequencies, a further set of contributions to the harmonic signals are obtained, which in many cases may be ignored if $\Delta P_2 \ll \Delta P_1$. However, an important consequence is that the second harmonic signal now has a small background level of: $\Delta P_2(\nu)\cos(2\omega_m t + \psi_2)$.

Including this background level with the three components described in Section 3.3, the total second harmonic signal may then be written in phasor notation (again in terms of ψ_λ) as:

$$h_2(\nu) = \Delta P_2(\nu)e^{j\psi_2} + A_0[0.5\Delta P(\nu)a_1 e^{-j\psi_\lambda} - P(\nu)a_2 e^{-2j\psi_\lambda} + 0.5\Delta P(\nu)a_3 e^{-3j\psi_\lambda}] \quad (3.20)$$

These components are plotted on the phasor diagram in Figure 3.7.

Figure 3.8 shows the magnitude, $|h_2(\nu)|$, of the total second harmonic signal for the same parameters as given for Figure 3.6, with the background level calculated from the model described by (2.26), as noted above. Note again the change from the pure symmetrical shape of the a_2 component (as shown in Figure 3.4) due to the presence of the other components and the variation in the laser power and intensity modulation across the scan, For example, for this set of modulation parameters, the first peak is larger than the third peak (even though the laser power increases across the scan), while for other parameters the situation may be reversed.

Note from (3.16) and (3.17) that the same pattern emerges for all further harmonics, i.e. the output signal at $n\omega_m$ from a lock-in amplifier consists of a primary WMS signal with phase $n\psi_f$ and components on either side of the primary component with phases of $(n \pm 1)\psi_f$. Hence, ignoring any small background level, the nth harmonic ($n \geq 2$) may be written as:

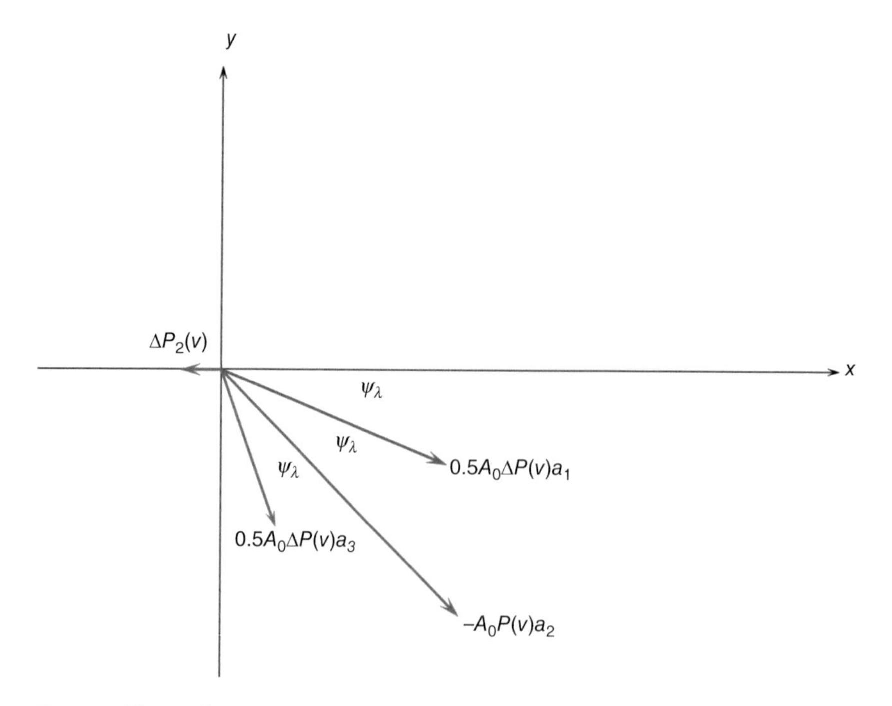

Figure 3.7 Phasor diagram showing the components of the second harmonic signal.

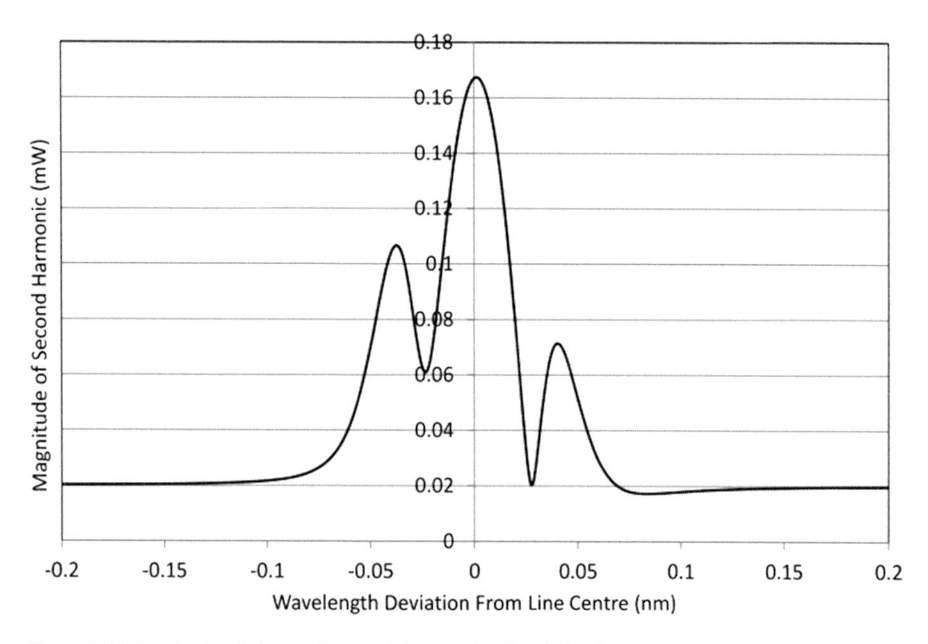

Figure 3.8 Magnitude of the total second harmonic signal for the same parameters as used in Figure 3.6.

$$h_n(\nu) = \pm A_0[0.5\Delta P(\nu)a_{n-1}\mathrm{e}^{-(n-1)j\psi_\lambda} - P(\nu)a_n\mathrm{e}^{-nj\psi_\lambda} + 0.5\Delta P(\nu)a_{n+1}\mathrm{e}^{-(n+1)j\psi_\lambda}] \tag{3.21}$$

where the plus sign applies for n even and the minus sign for n odd.

Although not explicitly shown in the above equations, all the Fourier coefficients, $a_n(\nu)$ are functions of ν and also depend on A_0 when the approximation for small absorbance is no longer valid – see (A3.2.4) and (A3.2.5) in Appendix 3.2 for a more general definition of $a_n(\nu, A_0)$. As explained in Appendix 3.2, some authors [21–24] define alternative $H_n(\nu, A_0)$ coefficients from the Fourier analysis. From (A3.2.8) and (A3.2.9) in Appendix 3.2, the harmonic signals may be written in terms of these coefficients as:

$$h_1(\nu) = [\Delta P(\nu)H_0 - P(\nu)H_1\mathrm{e}^{-j\psi_\lambda} + 0.5\Delta P(\nu)H_2\mathrm{e}^{-2j\psi_\lambda}] \tag{3.22}$$

$$h_n(\nu) = \pm[-0.5\Delta P(\nu)H_{n-1}\mathrm{e}^{-(n-1)j\psi_\lambda} + P(\nu)H_n\mathrm{e}^{-nj\psi_\lambda} - 0.5\Delta P(\nu)H_{n+1}\mathrm{e}^{-(n+1)j\psi_\lambda}] \tag{3.23}$$

3.4 Use of the Intensity Modulation of the Laser Output for Gas Measurements

As explained in detail in Chapter 1, accurate recovery of absorption lineshapes allows the measurement of gas concentration, pressure or temperature. Furthermore, it is desirable in practical and field applications to have calibration-free methods. We first examine the possibility of using the laser's intensity modulation at the first harmonic (based on the a_0 Fourier coefficient) for calibration-free measurement of gas parameters [9–13]. This approach is not really WMS, but is direct absorption spectroscopy with intensity modulation, as discussed in Section 3.2, but without the need for an external modulator. However, as we shall see, the method is not entirely straightforward since the intensity modulation is 'contaminated' with the wavelength modulation.

Consider again the phasor diagram in Figure 3.5. We may rotate the lock-in X-axis anti-clockwise to be at a phase angle of $\theta_L = 90° - \psi_\lambda$, as shown in Figure 3.9, so that the primary WMS first harmonic contribution is now *nulled* along the X-axis of the lock-in. Hence the measured signal from the X-axis will be:

$$X_{\text{lock-in}} = \Delta P(\nu)\left[1 - A_0\left(a_0 - \frac{1}{2}a_2\right)\right]\cos\theta_L \tag{3.24}$$

where we have used the identity, $\cos\theta_L = \sin\psi_\lambda$

Now if the frequency modulation index is small, $m < 0.1$, then $a_2 \sim 0$ and a_0 follows the lineshape, as shown in Figure 3.2. As in the case of DAS in Figure 3.1a, we may use the background, $\Delta P(\nu)$, to normalise the signal and obtain an absolute absorption lineshape. Essentially, we are isolating the intensity modulation on the output from the wavelength modulation and using direct absorption on the laser intensity modulation to recover the lineshape, but unfortunately the isolation is not complete because of the a_2 contribution. For this simple method to be effective we need to minimise the wavelength modulation to ensure that m is small and maximise the intensity modulation $\Delta P(\nu)$ for sufficient signal-to-noise ratio in the output signal. Since both the intensity and frequency modulation of a DFB laser are proportional to the amplitude of the modulation current, this is best achieved at higher modulation frequencies where the tuning coefficient drops off (see Figure 2.10a, Chapter 2), whereas the intensity

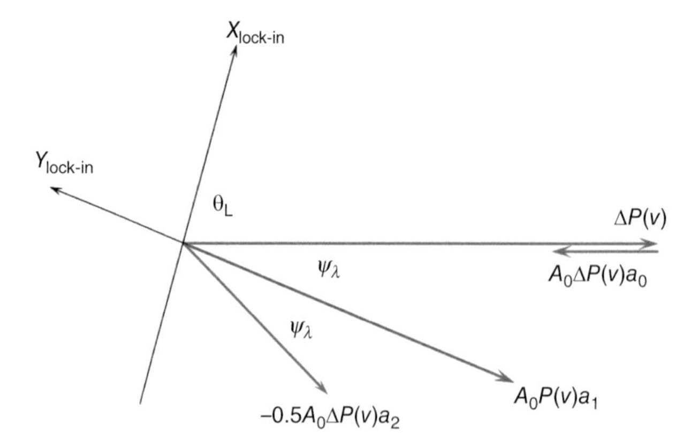

Figure 3.9 Nulling the contribution of the primary WMS first harmonic component along the X-axis of the lock-in by setting the lock-in phase, $\theta_L = 90° - \psi_\lambda$.

modulation remains independent of the modulation frequency for the range of frequencies of interest here.

However, there are several issues with this simple approach:

(i) As can be seen from Figure 3.9, the recovered intensity modulation signal is also reduced by its projection on the lock-in X-axis, that is, by the $\cos\theta_L$ term in Equation (3.24). One solution [14] would be to operate at a modulation frequency where $\psi_\lambda = 90^\circ$ (see Figure 2.10b, Chapter 2), but this may not be possible in practice.

(ii) To attain an adequate signal-to-noise ratio for low absorbance, it may be necessary to work at higher m values but now a_2 and the deviation of a_0 from the lineshape can no longer be ignored.

(iii) The method suffers from the usual problem of DAS, namely, a high background level.

Let us consider how each of these issues may be mitigated, at least, to some extent.

3.4.1 Lock-In Measurements from Both Axes

Rather than aligning the lock-in phase to null the primary WMS contribution, as in Figure 3.9, the x-axis of the lock-in may be aligned with the background intensity (RAM) signal, as in Figure 3.5, and both the x- and y-axis outputs of the lock-in recorded synchronously.

From Figure 3.5, the signal magnitudes measured along the x- and y-axis will therefore be:

$$X_{\text{lock-in}} = \Delta P(\nu)(1 - A_0 a_0) + A_0 P(\nu) a_1 \cos\psi_\lambda - \frac{1}{2} A_0 \Delta P(\nu) a_2 \cos 2\psi_\lambda \tag{3.25}$$

$$Y_{\text{lock-in}} = -A_0 P(\nu) a_1 \sin\psi_\lambda + \frac{1}{2} A_0 \Delta P(\nu) a_2 \sin 2\psi_\lambda \tag{3.26}$$

If for each measurement pair, as the laser wavelength is scanned across an absorption line, we compute the expression: $X_{\text{lock-in}} + Y_{\text{lock-in}}/\tan\psi_\lambda$, then from (3.25) and (3.26) we have, after some straightforward trigonometry:

$$\left\{X_{\text{lock-in}} + \frac{Y_{\text{lock-in}}}{\tan\psi_\lambda}\right\} = \Delta P(\nu)\left[1 - A_0\left(a_0 - \frac{1}{2}a_2\right)\right] \tag{3.27}$$

This is the same as (3.24) but without the $\cos\theta_L$ term so the reduction in signal strength is avoided. This approach has been described as the phasor decomposition method (PDM) [11, 13]. There is, however, the need to know ψ_λ in advance for the particular laser and modulation frequency used – see Chapter 2, Section 2.7 for measurement of ψ_λ. Indeed for some lasers, ψ_λ may vary across the scan, as explained in Chapter 2, Section 2.4.3, so this varying phase needs to be applied when computing the above ratio [15].

It is also possible to use PDM without the need to align the x-axis of the lock-in by using an iteration process [15]. Suppose the lock-in x-axis has an unknown and arbitrary orientation, θ_L, with respect to the phase of the intensity modulation. In this case we now compute the expression: $X_{\text{lock-in}} + Y_{\text{lock-in}}/\tan(\theta + \psi_\lambda)$,where θ is iterated from 0 to 180°, typically in intervals of 0.2°. Only at $\theta = \theta_L$ is the a_1 component completely eliminated from this expression and this is recognised in the iteration process by the point at which there is the least difference between the non-absorbing baseline and the maximum absorption point.

3.4.2 Approximations for Higher Modulation Indices

If the frequency modulation index, m, is chosen to be greater than ~0.1, then the effects of the wavelength modulation can no longer be ignored and the term $(a_0 - 0.5a_2)$ in (3.27) needs to be evaluated. This may be done directly using the expressions for the Fourier coefficients, (3.11) and (3.12), provided m is known in advance. Figure 3.10 shows the change of the term $(a_0 - 0.5a_2)$ with increasing m for Lorentzian and Gaussian lineshapes.

Alternatively, under the assumption of small absorbance and modulation index less than unity, the following approximation may be used (see Appendix 3.1):

$$\left(a_0 - \frac{1}{2}a_2\right) \approx f(\nu) + \frac{1}{8}f''(\nu)\delta\nu^2 + \frac{1}{192}f''''(\nu)\delta\nu^4 + \cdots \tag{3.28}$$

For the particular cases of Lorentzian and Gaussian lineshapes, this can be evaluated using the derivatives of the lineshape function to give an explicit expression in terms of the modulation index:

$$\left(a_0 - \frac{1}{2}a_2\right) \approx f(\nu)\left\{1 + \frac{1}{4}Bm^2 + \frac{1}{8}Dm^4 + \cdots\right\} \tag{3.29}$$

where $f(\nu) = 1/(1 + \Delta^2)$ for a Lorentzian lineshape, $f(\nu) = \exp(-\Delta^2)$ for a Gaussian lineshape and B, D are defined in Appendix 3.1 for each case. This approximation is also plotted in Figure 3.10, which shows that it is indistinguishable from the exact solution at $m = 0.5$, but deviates as m approaches unity (but less so for a Gaussian profile).

Provided m is known, by fitting the exact solution (or the approximation of (3.29) for $m < 1$) to the measured lineshape, point by point as a function of Δ, the true Lorentzian or Gaussian profile may be recovered. However, the big disadvantage is the requirement to know m in advance. In practice the linewidth changes under conditions of varying pressure or temperature, so m will not be constant and may also change as a result of aging of the laser. In principle, an iteration procedure could be used in the fitting algorithm to find the values of m and A_0 that give the optimum fit. An alternative method is to take a measurement at small m initially and hence determine the line centre absorbance, A_0, from the peak height of the measured profile since (3.29) at line centre is approximately unity when m is small. A simple

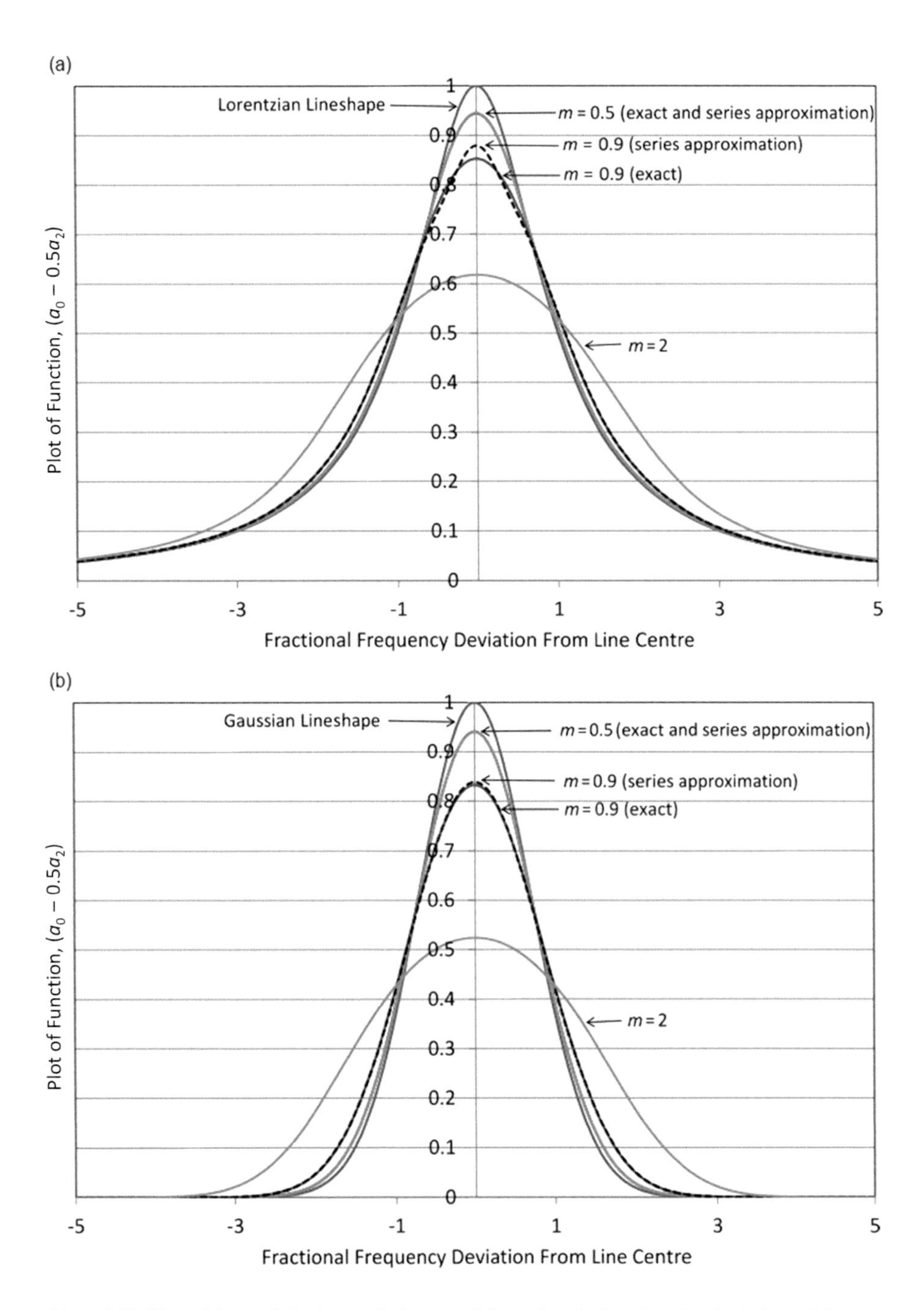

Figure 3.10 Plot of $(a_0 - 0.5a_2)$ as m is increased from 0 to 2 showing the deviation of the measured lineshape from the true lineshape due to wavelength modulation for: (a) a Lorentzian lineshape and (b) a Gaussian lineshape. Also shown is the approximate series solution of (3.29) for $m = 0.5$ and 0.9.

least-mean-squares fitting procedure may then be applied to obtain the value of m from the measured profile at a higher value of m or it may be determined from the peak height at the higher modulation index using the exact or approximate solution of (3.29) at $\Delta = 0$. In essence, the principle of the method is to use a small m to obtain the line centre absorbance and then a large m to accurately re-construct the full lineshape. The method depends, however, on the line centre absorbance being sufficiently high so that it can be measured at a small m value.

3.4.3 Elimination of the Background Intensity Modulation

Using the intensity modulation for gas measurements has the disadvantage that the relatively small absorption signal is superimposed on the high background level. On one hand, the no-gas background level is essential for normalizing the signal, but on the other hand it is a disadvantage, since the dynamic range of the detection system is dictated by the background rather than the signal. In phase-sensitive detection, elimination of the background to leave the small absorption signal on a near-zero background level means that the gain of the lock-in amplifier can be significantly increased without overloading the output. However, as we shall see, the background can be eliminated by either optical or electronic cancellation techniques, while preserving the ability to normalise the signal, but at the expense of greater system complexity

Optical cancellation [16–18] may be performed by an optical fibre delay line, as illustrated in Figure 3.11.

Here the modulated output from the laser is split by typically a 50/50 fibre coupler into two paths, one through the gas cell and the other forming a delay line. A second 50/50 coupler combines the output from the two paths. If the length of the fibre delay line is chosen so that the delay time, $t_D = n_e L_D/c$, of its output, relative to that of the direct path, is equal to half the period, $T_m/2 = \pi/\omega_m$, of the intensity modulation, then the intensity modulation from the two paths will be in anti-phase at the second coupler. This gives a delay line length of $L_D = \pi c/(n_e\omega_m) = c/(2n_e f_m)$, where $n_e \sim 1.45$ is the effective index of the guided mode in the fibre. Typically, at a modulation frequency of 100 kHz, this

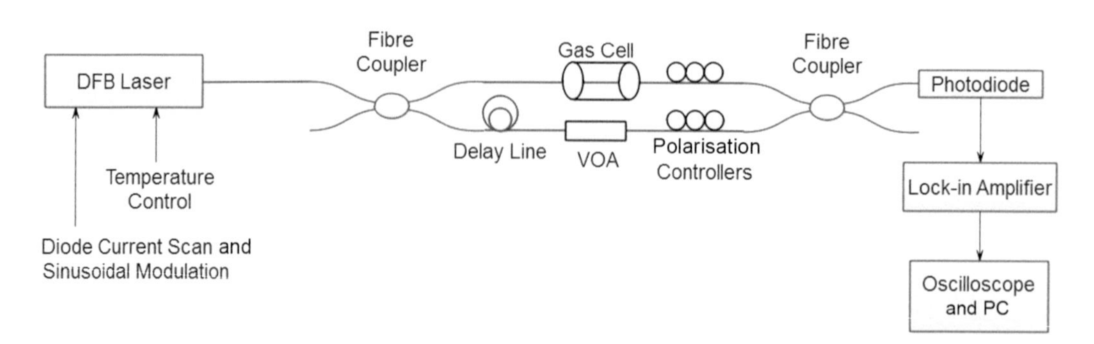

Figure 3.11 Optical nulling of the background intensity modulation by use of an optical delay line.

corresponds to a fibre length of ~1 km. In practice the modulation frequency may be finely tuned to match the actual fibre length and hence attain the exact π shift in the modulation phase. The variable optical attenuator (VOA) in the delay line is used to equalise the amplitude of the intensity modulation from the two paths for total cancellation of the background at the output coupler. Note that the background may not remain perfectly zero across a wavelength scan as a result of wavelength-dependent components in the system [17]. To avoid introducing optical interference noise, the delay line should be substantially longer than the coherence length of the (modulated) laser and orthogonal polarisation states established at the output from the two paths by adjustment of the polarisation controllers. For normalisation and compensation for fluctuation in system losses or laser power, a fibre optic switch may be used to temporarily interrupt the delay line and hence obtain the background level, or a fibre tap with known and stable coupling ratio may be inserted into the direct path before the output coupler.

Electronic cancellation [19, 20] may be performed by the Hobb's balanced ratio detector (BRD) circuit shown in Figure 3.12.

This circuit configuration is also known as the auto-balanced dual-beam noise canceller and cancels common-mode noise to near-shot noise levels. In the standard mode of operation, the output from the laser is split into two paths, one through the gas cell to the signal photodiode and the other as a reference to the reference photodiode. The reference beam power, P_r, is chosen to be greater than the measurement beam power, P_m, so that the DC photodiode current from the reference photodiode is greater than that from the signal photodiode. With switch S1 closed, an imbalance in currents at the summing junction, S, creates a DC voltage at the linear output of amplifier A1. This voltage appears at the input of the integrator

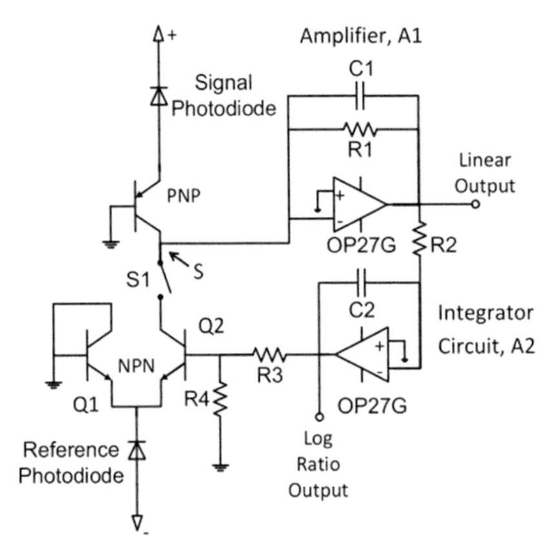

Figure 3.12 Hobbs dual-beam, auto-balancing receiver circuit. © [2016] IEEE. Reprinted, with permission, from [20].

circuit A2 and in turn its output is applied to the base of transistor Q2. The voltage at the base of Q2 controls the split ratio of the DC photodiode current from the reference photodiode between the two transistors Q1 and Q2. Consequently this integrator servo-feedback loop automatically reduces the DC voltage on the linear output to zero by adjusting the Q2 base voltage until the current flowing downwards from the summing junction S is equal to the photodiode current flowing into S from the signal photodiode, with zero current to the input of amplifier A1. If the DC intensity of either the measurement or the reference beam changes or fluctuates (at a rate less than the bandwidth of the integrator) the circuit automatically finds the new balance position, maintaining the linear output at zero, while the Q2 base voltage gives a measure of the beam ratio, $V_b \cong -(kT/q) \ln \{P_r/P_m - 1\}$. Most importantly, as a result of the current balance, noise and modulation common to both beams are automatically cancelled at the linear output, up to very high frequencies limited by the transistors' bandwidth. However, the harmonic signals arising from WMS are only generated in the measurement beam and are faster than the integrator bandwidth so they do appear on the linear output and may be detected by a lock-in amplifier. At first sight it might seem that, with the nulling of the intensity modulation background on the linear output, the direct absorption signal on the intensity modulation, $a_0 A_0 \Delta P(\nu)$, which again is only generated in the measurement beam, would also appear on the linear output with zero background. However the CW measurement beam also carries the direct absorption line signature and if the wavelength scan through the absorption line is slower than the bandwidth of the integrator, then auto-balancing occurs continuously across the scan and the absorption line signature appears on the DC voltage, V_b, at the base of Q2. This auto-balancing nulls the $a_0 A_0 \Delta P(\nu)$ contribution on the linear output. This may be prevented if either the bandwidth of the integrator is reduced or the scan rate increased so that the integrator loop is too slow to respond to the direct absorption signature on the CW beam intensity. The $a_0 A_0 \Delta P(\nu)$ contribution now appears on the linear output along with the other WMS components. It should be noted that reduction of the integrator bandwidth does not affect the common-mode noise cancellation properties of the circuit, but the frequency range for cancellation of non-common mode noise (such as fluctuation of power in a single beam from drift) is restricted to within the response time of the integrator. For normalisation purposes, the background intensity modulation may be obtained on the linear output by opening the switch S1 to temporarily disable the auto-balancing operation of the circuit.

3.5 Use of WMS Harmonics for Gas Measurements

We now turn our attention to the use of the primary WMS harmonics for lineshape recovery, based on the a_1, a_2 or a_3 Fourier coefficients. This is the traditional, well-established approach, but again is not entirely straightforward because the wavelength modulation is 'contaminated' by the intensity modulation, and lineshapes are not directly recovered from the harmonic, but careful data processing is required to extract the key parameters of line depth and width. In general the magnitude of the harmonic diminishes with the order of the harmonic so the first harmonic is the largest, but as we have seen there is a large background intensity modulation at the first harmonic frequency. Hence the second harmonic at $2\omega_m$, which normally has a near-zero background level has been the favourite choice for WMS, with the use of the first harmonic as a reference for normalisation and calibration. Of course, as is evident from (3.20), if the laser has a strongly non-linear *LI* characteristic, the second harmonic will have a significant background level, which needs to be taken into account, or alternatively a higher-order harmonic may be considered.

As discussed in Section 3.3.3 and shown in Figure 3.7, there are three components of the output signal at $2\omega_m$, one being the primary WMS signal and the other two arising from the effects of the intensity modulation on the first and third harmonics. As was done in Section 3.4.1 with the first harmonic, synchronous recording of measurements from both lock-in axes may be used to eliminate one of these components. For example if the x-axis of the lock-in can be aligned with the primary WMS signal at a phase lag of $2\psi_\lambda$, then, analogous to (3.27), the combination may be formed:

$$\left\{X_{\text{lock-in}} \pm \frac{Y_{\text{lock-in}}}{\tan\psi_\lambda}\right\} = -A_0\left[a_2 P(\nu) - \Delta P(\nu) a_n \cos\psi_\lambda\right] \tag{3.30}$$

where $n = 1$ for the plus sign and $n = 3$ for the minus sign.

This eliminates one component, but the primary WMS signal, $a_2 A_0 P(\nu)$, is still not fully isolated. Again we emphasise that the Fourier coefficients depend on the optical modulation index and the linewidth, so extraction of the line depth and width of a gas absorption line is not straightforward, especially under conditions of varying temperature and pressure. Hence we now consider a calibration-free approach, which allows retrieval of the absorption line parameters, provided the laser characteristics are known. We focus on the ratio of the second harmonic to the first harmonic, *2f/1f*, but the approach is equally valid with higher harmonics, *3f/1f*, *4f/1f*, etc. It can also be applied, as we shall see, to the first harmonic itself.

3.5.1 The *2f/1f* Technique

The basis of the method [21–26] is to simulate the harmonic signals that are generated from a particular WMS system, knowing all the key laser source parameters. The theoretical magnitudes of the total first and second harmonic signals, including all their components, as given in Sections 3.3.2 and 3.3.3, are computed along with the ratio of the second to the first harmonic. This theoretical ratio is least-squares fitted to the measured *2f/1f* ratio with free parameters of absorbance, A_0, optical modulation index, $m = \delta\nu/\gamma$ (or linewidth using the known $\delta\nu$ from the laser tuning coefficient and the modulation current) and centre wavelength of the absorption line. The *2f/1f* ratio ensures that power fluctuation from transmission changes or beam drift is factored out and the desired free parameters are determined from the best fit. The *1f* signal is normally dominated by the background level, except for high absorbance, in which

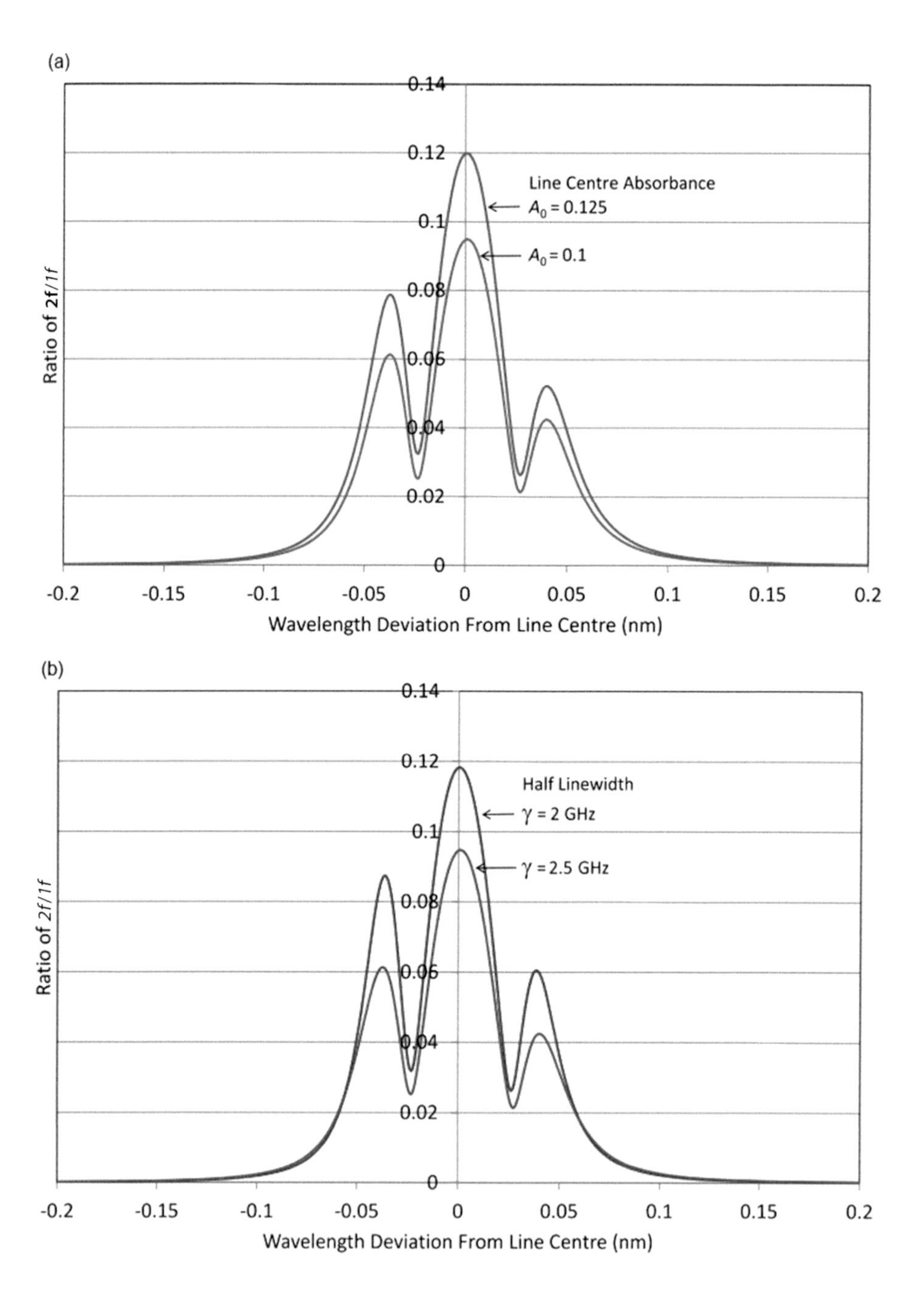

Figure 3.13 The theoretical *2f/1f* ratio showing: (a) the effect of changing the target gas concentration at fixed pressure and temperature, (b) the effect of changing the pressure (and hence linewidth) at constant concentration and temperature (same parameters as for Figures 3.6 and 3.8 except where otherwise indicated).

case an off-line extrapolated value may be required if the total *1f* signal becomes zero or close to zero at certain points across the absorption line scan. In principle the *2f/1f* method can also be used with non-isolated absorption lines.

Using (3.18) and (3.20) in Sections 3.3.2 and 3.3.3, Figure 3.13a illustrates, for a Lorentzian lineshape, how the theoretical *2f/1f* magnitude ratio, $|h_2(\nu)|/|h_1(\nu)|$, changes with a 25% increase in target gas concentration at constant pressure and temperature, with the same parameters as used for Figures 3.6 and 3.8, except as indicated. In the simulations of Figure 3.13, the *X*- and *Y*-components of the *2f* background have been subtracted before forming the *2f/1f* magnitude ratio, as would be done in practical measurements. In practical applications, the modulation current amplitude is normally fixed, but temperature or pressure fluctuations will cause the absorption linewidth to change, as described in Chapter 1 by (1.12), (1.15) and (1.16). As a consequence, with a fixed modulation current, the modulation index will also change. As an example, consider the case where a ~20% reduction in pressure causes the half-linewidth to change from 2.5 GHz to 2 GHz, with the gas concentration remaining constant. This reduction in linewidth causes an increase in the line centre absorbance by a factor of 2.5/2 = 1.25 since, from (3.9), A_0 is related to the inverse of the half-linewidth by: $A_0 = \alpha_0 Cl = \{N_0 S(T)/(\pi\gamma_c)\}Cl$. This situation is illustrated in Figure 3.13b. As evident from Figure 3.13b, changes in the linewidth do not significantly change the overall width of the *2f/1f* signal, which may seem surprising at first sight. However, this arises from two competing effects – a reduction in the linewidth reduces the extent of the absorption line, but the consequent increase in the modulation index increases the extent of the a_2 component, as shown in Figure 3.4.

Figures 3.13a and 3.13b have illustrated the effects on the *2f/1f* ratio of changing the concentration (at constant pressure and temperature) and changing the pressure at fixed concentration. Comparing the two figures we see that changes in concentration have the simple effect of scaling the signal, as indicated by (3.20), if the *2f* background level is low, whereas changes in pressure or temperature at constant concentration affect the magnitudes of the side peaks relative to the main peak. Of course, in practice all three parameters may be changing simultaneously and the challenge is to accurately extract these parameters using appropriate fitting algorithms with precise modelling and knowledge of the laser and modulation parameters.

For a more comprehensive model of the theoretical *2f/1f* ratio than the semi-analytical description above, the simulated harmonics and *2f/1f* ratio may be obtained entirely by digital signal processing of the theoretical time evolution function of the system. For example, extending (3.5) to (3.7) to include the scan function through the absorption line, $f_s(t)$, and the effects of non-linear absorption and a non-linear *LI* curve for the laser, the theoretical time evolution of the output intensity may be written as:

$$P_{out}(t) \cong P_{in}(t) \exp\{-A_0 f(\nu_l)\} \tag{3.31}$$

where $f(\nu_l)$ is the usual lineshape function as defined earlier, $P_{in}(t)$ is the input power:

$$P_{in}(t) = \{P(\nu_l) + \Delta P_s f_s(\omega_s t - \varphi_s) + \Delta P_1(\nu_l)\cos(\omega_m t) + \Delta P_2(\nu_l)\cos(2\omega_m t + \psi_2)\} \tag{3.32}$$

and ν_l is the laser frequency, which is a function of time:

$$\nu_l(t) = \nu_0 + \Delta\nu_s f_s(\omega_s t - \psi_s) + \delta\nu \cos(\omega_m t + \psi_f) \tag{3.33}$$

Here ΔP_s and $\Delta\nu_s$ are the amplitudes of the optical power and frequency sweep of the scan, respectively, and ω_s, φ_s, ψ_s are the scan frequency and phases, respectively.

All the laser and modulation parameters in (3.32) and (3.33) are assumed to be known or measured, as is the absorption lineshape function, $f(\nu)$. This leaves the free parameters of absorbance A_0, line position ν_0, and linewidth γ to be determined by fitting to the experimental *2f/1f* ratio. The magnitudes of the experimental harmonics and hence the experimental *2f/1f* ratio may be measured in the usual way via a lock-in amplifier or by digital processing of the experimental signal obtained direct from the photodiode detector. Quadrature reference signals may be used to extract the *X*- and *Y*-components and hence the magnitude of a particular harmonic.

As evident from the above discussion, the main disadvantage of the *2f/1f* method is the requirement to have prior knowledge of the laser parameters, which may change over time. However it is possible in certain circumstances to monitor these parameters in situ and in real-time where the gas absorption line under investigation is well isolated so that non-absorbing wings are available within the scan range of the laser wavelength [27, 28]. Hence $\Delta P_1(\nu_l)$ and $\Delta P_2(\nu_l)$ may be monitored continuously for each sweep from the magnitudes of the first and second harmonic signals off-line in the non-absorbing wings and by interpolation across the absorption line. Similarly, ΔP_s may be monitored from the optical power variation across the scan using low-pass filtering of the detector output. A fibre ring resonator, as described in Chapter 2, Section 2.7, connected into the system through a fibre tap from the laser output, may be used to monitor $\delta\nu$ across the scan.

3.5.2 The $1f/1f_x$ Technique

As noted, the *2f/1f* approach could be equally applied with higher harmonics, *3f/1f*, *4f/1f*, etc., but since the magnitude of a harmonic diminishes with its order, it is not surprising that *2f/1f* is the most common method used. However, it is possible to apply a similar technique with only the *1f* harmonic [29] in order to take advantage of its increased magnitude compared with the *2f* harmonic. In this case the magnitude of the first harmonic as described by (3.18) is measured from the *X*- and *Y*-components of the lock-in output with its axes aligned, as shown in Figure 3.9, and is then normalised by its *X*-component, $1f_x$, as described by (3.24). As with the *2f/1f* method, knowledge, or real-time characterisation, of all the laser parameters allows this $1f/1f_x$ ratio to be simulated and compared with the experimental ratio for extraction of the gas parameters. A further advantage of the technique is the reduced signal processing and data acquisition required since only one harmonic is used. Figure 3.14 illustrates the theoretical $1f/1f_x$ ratio for the same laser parameters as used in Figure 3.13a.

Note the constant no-gas background level in Figure 3.14 as compared with the changing level of Figure 3.6. This arises because the $1f/1f_x$ ratio is simply given by

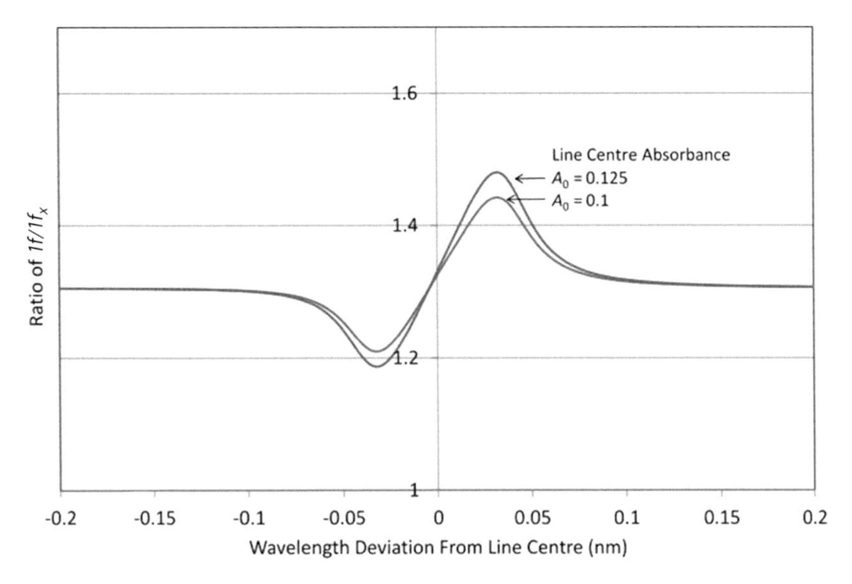

Figure 3.14 The theoretical $1f/1f_x$ ratio showing the effect of changing the target gas concentration at fixed pressure and temperature (same parameters as for Figure 3.13a).

$(\sin\psi_\lambda)^{-1}$ from (3.18) and (3.24) for $A_0 = 0$ (no gas) and hence is constant if ψ_λ is constant across a scan. This background level may be subtracted from the measured signal ratios to give the output on a zero background level.

3.6 Comparison of Methods

It is useful to compare the relative merits of the intensity modulation method discussed in Section 3.4 with the *2f/1f* method (or more generally the *nf/1f* methods) of Section 3.5 in terms of raw signal amplitudes and signal-to-noise considerations.

Consider first the intensity modulation method with a Lorentzian absorption line-shape. From (3.11) to (3.13) and (3.27), the change in the magnitude of the intensity modulation (i.e. the deviation from the background level) due to the absorption at the line centre is:

$$S_{IM} = A_0 \Delta P(\nu_0) \left\{ \frac{1}{\pi} \int_0^{\pi} \frac{(1 - \cos 2\theta)}{\{1 + m^2 \cos^2\theta\}} \mathrm{d}\theta \right\} \leq A_0 \Delta P(\nu_0) \tag{3.34}$$

where $\Delta P(\nu_0)$ is the background level at the line centre.

The integral term is approximately unity for $m < 0.4$ and decreases to ~0.6 at $m = 2$. As noted earlier, we can minimise the value of m for a given modulation current by working at higher modulation frequencies where the tuning coefficient drops off. Assuming a tuning coefficient of ~0.05 GHz mA^{-1} at high frequencies (see Figure 2.10a,

Chapter 2), then $m \sim 0.4$ for a half-linewidth of 2.5 GHz and a modulation current amplitude of 20 mA. With a typical laser slope efficiency of ~0.1 W A^{-1}, then $\Delta P(\nu_0)$ ~2 mW and hence the change in magnitude of the intensity modulation at line centre is ~$2A_0$ (in units of mW).

Compare this with the magnitude of the raw *2f* signal. From Section 3.3.3, the primary WMS second harmonic has a magnitude of:

$$S_{2f} = A_0 P(\nu) a_2 \tag{3.35}$$

We noted in Section 3.3.1 that the central peak in this signal is maximised at $m = 2.2$, with the maximum value of $a_2 = 0.343$ for a Lorentzian profile. Hence with a typical laser power at line centre of 5 mW, the raw *2f* signal has a magnitude of ~$1.7A_0$ (in units of mW). In this case, the required modulation index of 2.2 may be obtained at the same modulation current amplitude of 20 mA by working at a lower modulation frequency of say ~20 kHz where the tuning coefficient is ~0.3 GHz mA^{-1}.

It is clear from the above discussion that the magnitudes of the raw signals from both methods can be similar, depending on the laser parameters and the chosen modulation parameters. Of course the optimum operating conditions for each case are different and the intensity modulation signal may be increased by increasing the modulation current (within the limits of the laser diode current range) whereas the *2f* signal may be increased by increasing the laser power.

Now let us compare both methods in regard to the signal-to-noise performance. Specifically consider the case of intensity noise caused by, for example, beam drift from vibrations or transmission variation through optical fibre systems and components. The effect of this (multiplicative) noise on the detected power may be simply represented by: $P(\nu)\{1 + \delta_n\}$ where δ_n represents the fractional variation from the noise around the mean level.

For the intensity modulation method, the total output using (3.27) with this noise included is:

$$S_T = \Delta P(\nu)\{1 + \delta_n\}\left[1 - A_0\left(a_0 - \frac{1}{2}a_2\right)\right] \tag{3.36}$$

From this we see that there is a large noise contribution of $\delta_n \Delta P(\nu)$ on the background intensity modulation. Comparing this with (3.34), where the maximum attainable value at line centre is $A_0 \Delta P(\nu_0)$, the contribution from the absorption will be lost in the noise if $\delta_n \geq A_0$. The situation may be greatly improved by use of the dual-beam noise-canceller circuit described in Section 3.4.3 which cancels common-mode noise affecting both reference and measurement beams. Drift noise in a single beam at frequencies below the bandwidth of the integrator loop will also be cancelled through the auto-rebalancing function of the circuit.

Compare the situation with the *2f* signal where the signal plus noise has the form:

$$S_{2f} = A_0 P(\nu)\{1 + \delta_n\} a_2 \tag{3.37}$$

Because of the zero (or near zero) background, the multiplicative noise on the *2f* signal, $A_0 P(\nu)\delta_n a_2$, is very much smaller than that of the intensity modulation case and the signal-to-noise ratio is fixed at $1/\delta_n$, so that small absorbance can be seen, even in the presence of relatively large beam fluctuations. Furthermore, the situation is improved in the *2f/1f* technique because both the total *2f* and *1f* signals detected are affected by this noise in the same way and both scaled by the same $\{1 + \delta_n\}$ factor. Hence this noise is essentially eliminated when the *2f/1f* ratio is computed point-by-point across the scan.

In view of the above, the question naturally arises as to why one would use the intensity modulation method when the *nf/1f* methods are much superior in terms of signal-to-noise performance and sensitivity. The answer depends very much on the particular application. The major drawback of the *nf/1f* methods are that they require considerable digital processing with careful fitting algorithms to accurately extract the lineshape properties in circumstances of changing pressure or temperature [24, 25]. Furthermore precise knowledge of all the laser and modulation parameters is required and these may drift over time or change with laser aging. Hence for applications where sensitivity is not an issue and high absorption levels are present, direct lineshape recovery by the intensity modulation method may be the favoured option.

3.7 Conclusion

This chapter has provided a detailed theoretical description of the harmonic signals arising in wavelength modulation spectroscopy with DFB laser sources. In general, the total harmonic signal at a particular frequency consists of a primary component from the wavelength modulation of the DFB laser source plus two other components as a result of the simultaneous intensity modulation, with the possibility of additional components becoming significant if the light versus current characteristic of the laser is highly non-linear. Hence careful analysis is required to extract information on the lineshape and the gas parameters of interest. We have examined in detail two methods to accomplish this, one based on the laser intensity modulation, with correction factors at higher modulation indices, and the other based on the laser wavelength modulation using the second or other harmonic signals normalised through the first harmonic. While both methods may give similar magnitudes of output signal for a given gas concentration, the *2f/1f* method is superior in terms of noise performance, but at the expense of considerable complexity in digital signal processing and uncertainty if the laser parameters are prone to drift. We shall illustrate the use of these methods in Chapter 5, where we discuss a number of practical systems employing wavelength modulation spectroscopy for near-IR gas sensing. The same techniques and principles also apply to mid-IR diode lasers, as will be discussed in Chapter 8 for mid-IR gas sensing.

Appendix 3.1 Approximations for Fourier Coefficients from a Taylor Series Expansion

The dimensionless lineshape function, $f(\nu)$, defined by (3.7) can be expanded as a Taylor series around any particular frequency in terms of its derivatives at that point. Using (3.5) for the laser frequency, $\nu_l = \nu + \delta\nu \cos\theta$, where $\theta(t) = (\omega_m t + \psi_f)$, we can write:

$$f(\nu_l) = f(\nu) + f'(\nu)\{\nu_l - \nu\} + \frac{1}{2}f''(\nu)\{\nu_l - \nu\}^2 + \frac{1}{6}f'''(\nu)\{\nu_l - \nu\}^3$$
$$+ \frac{1}{24}f''''(\nu)\{\nu_l - \nu\}^4 + \ldots$$
$$= f(v) + \delta v f'(v)\cos\theta + \frac{\delta v^2}{2}f''(v)\cos^2\theta + \frac{\delta v^3}{6}f'''(v)\cos^3\theta + \frac{\delta v^4}{24}f''''(v)\cos^4\theta + \cdots \quad \text{(A3.1.1)}$$

A series approximation for the Fourier coefficients (with the assumption of small absorbance) may be obtained by substitution of (A3.1.1) into (3.11) and (3.12) for the Fourier coefficients and integration term by term. The following identities are required for the evaluation of the integrals:

$$\cos n\theta = 2\cos\theta\cos(n-1)\theta - \cos(n-2)\theta$$

$$\int_0^{\pi} \cos^n\theta \mathrm{d}\theta = 0$$

if n is odd

$$\int_0^{\pi} \cos^n\theta \mathrm{d}\theta = \frac{n-1}{n}\int_0^{\pi} \cos^{n-2}\theta \mathrm{d}\theta$$

if n is even. (A3.1.2)

Hence the following series approximations for the Fourier coefficients are obtained, up to the fourth-order term:

$$
\begin{aligned}
a_0(\nu) &= f(\nu) + \frac{1}{4} f''(\nu)\delta\nu^2 + \frac{1}{64} f''''(\nu)\delta\nu^4 + \cdots \\
a_1(\nu) &= f'(\nu)\delta\nu + \frac{1}{8} f'''(\nu)\delta\nu^3 + \cdots \\
a_2(\nu) &= \frac{1}{4} f''(\nu)\delta\nu^2 + \frac{1}{48} f''''(\nu)\delta\nu^4 + \cdots \\
a_3(\nu) &= \frac{1}{24} f'''(\nu)\delta\nu^3 + \cdots
\end{aligned} \tag{A3.1.3}
$$

Note that a_n is proportional to the nth derivative for small modulation depth, $\delta\nu$.

An approximation for the term ($a_0 - ½a_2$) appearing in (3.27) is therefore:

$$
\left(a_0 - \frac{1}{2}a_2\right) \approx f(\nu) + \frac{1}{8} f''(\nu)\delta\nu^2 + \frac{1}{192} f''''(\nu)\delta\nu^4 + \cdots \tag{A3.1.4}
$$

Using derivatives of Lorentzian and Gaussian lineshape functions, (A3.1.4) may be evaluated to give an explicit expression in terms of the modulation index, $m = \delta\nu/\gamma$:

$$
\left(a_0 - \frac{1}{2}a_2\right) \approx f(\nu)\left\{1 + \frac{1}{4}Bm^2 + \frac{1}{8}Dm^4 + \cdots\right\} \tag{A3.1.5}
$$

where for a Lorentzian lineshape:

$$
\begin{aligned}
f(\nu) &= \frac{1}{(1+\Delta^2)} \\
B &= \frac{(3\Delta^2 - 1)}{(1+\Delta^2)^2} \\
D &= \frac{(5\Delta^4 - 10\Delta^2 + 1)}{(1+\Delta^2)^4}
\end{aligned} \tag{A3.1.6}
$$

and for a Gaussian lineshape:

$$
\begin{aligned}
f(\nu) &= \exp(-\Delta^2) \\
B &= 2\Delta^2 - 1 \\
D &= \frac{1}{6}(4\Delta^4 - 12\Delta^2 + 3)
\end{aligned} \tag{A3.1.7}
$$

and $\Delta = (\nu - \nu_0)/\gamma$.

Appendix 3.2 Fourier Coefficients for Non-Linear Absorption

In the linear approximation to the Beer–Lambert law for small absorbance, the exponential term is replaced by:

$$\exp\{-\alpha(\nu_l)Cl\} = \exp\{-A_0 f(\nu_l)\} \approx \{1 - A_0 f(\nu_l)\} \tag{A3.2.1}$$

where $f(\nu)$ is the dimensionless lineshape function defined in (3.7) and A_0 is the line centre absorbance, $A_0 = \alpha_0 Cl$.

For non-linear absorption where this approximation is no longer valid we write:

$$\exp\{-A_0 f(\nu_l)\} = \{1 - A_0 f_{nl}(\nu_l)\} \tag{A3.2.2}$$

where the function $f_{nl}(\nu)$ is defined by:

$$f_{nl}(\nu) = \frac{1 - \exp\{-A_0 f(\nu)\}}{A_0} \tag{A3.2.3}$$

Of course when A_0 is small $f_{nl}(\nu) \to f(\nu)$

Hence the Fourier coefficients in the non-linear case are:

$$a_0 = \frac{1}{\pi A_0}\int_0^{\pi} [1 - \exp\{-A_0 f(\nu_l)\}]\mathrm{d}\theta \tag{A3.2.4}$$

$$a_n = \frac{2}{\pi A_0}\int_0^{\pi} [1 - \exp\{-A_0 f(\nu_l)\}]\cos n\theta \mathrm{d}\theta \tag{A3.2.5}$$

where the function $f(\nu_l)$ is defined as before by (3.13) and (3.14) for Lorentzian and Gaussian profiles.

Using the transmission function, $P_{out}/P_{in} = \exp\{-A_0 f(\nu_l)\}$, some authors define H coefficients as:

$$H_0 = \frac{1}{\pi}\int_0^{\pi} \exp\{-A_0 f(\nu_l)\}\mathrm{d}\theta \tag{A3.2.6}$$

$$H_n = \frac{2}{\pi}\int_0^{\pi} \exp\{-A_0 f(\nu_l)\}\cos n\theta \mathrm{d}\theta \tag{A3.2.7}$$

Hence the relationship between our definition of the Fourier coefficients as given above and the H coefficients are:

$$a_0 A_0 = 1 - H_0 \tag{A3.2.8}$$

$$a_n A_0 = -H_n \tag{A3.2.9}$$

References

1. J. A. Silver, Frequency-modulation spectroscopy for trace species detection: theory and comparison among experimental methods, *Appl. Opt.*, 31, (6), 707–717, 1992.
2. J. M. Supplee, E. A. Whittaker and W. Lenth, Theoretical description of frequency modulation and wavelength modulation spectroscopy, *Appl. Opt.*, 33, (27), 6294–6302, 1994.
3. V. G. Avetisov and P. Kauranen, High-resolution absorption measurements by use of two-tone frequency-modulation spectroscopy with diode lasers, *Appl. Opt.*, 36, (18), 4043–4054, 1997.
4. C. S. Goldenstein, R. M. Spearrin, J. B. Jeffries and R. K. Hanson, Infrared laser-absorption sensing for combustion gases, *Prog. Energy Combust. Sci.*, 60, 132–176, 2016.
5. P. Kluczynski and O. Axner, Theoretical description based on Fourier analysis of wavelength-modulation spectrometry in terms of analytical and background signals, *Appl. Opt.*, 38, (27), 5803–5815, 1999.
6. P. Kluczynski, A. M. Lindberg and O. Axner, Characterization of background signals in wavelength modulation spectrometry in terms of a Fourier based formalism, *Appl. Opt.*, 40, (6), 770–782, 2001.
7. S. Schilt, L. Thevenaz and P. Robert, Wavelength modulation spectroscopy: combined frequency and intensity laser modulation, *Appl. Opt.*, 42, (33), 6728–6738, 2003.
8. G. Stewart, W. Johnstone, J. R. P. Bain, K. Ruxton and K. Duffin, Recovery of absolute gas absorption line shapes using tuneable diode laser spectroscopy with wavelength modulation – part 1: theoretical analysis, *IEEE J. Lightwave Technol.*, 29, (6), 811–821, 2011.
9. J. R. P. Bain, W. Johnstone , K. Ruxton, et al., Recovery of absolute gas absorption line shapes using tunable diode laser spectroscopy with wavelength modulation – part 2: experimental investigation, *IEEE J. Lightwave Technol.*, 29, (7), 987–996, 2011.
10. K. Duffin, A. J. McGettrick, W. Johnstone, G. Stewart and D. G. Moodie, Tunable diode laser spectroscopy with wavelength modulation: a calibration-free approach to the recovery of absolute gas absorption line-shapes, *IEEE J. Lightwave Technol.*, 25, (10), 3114–3131, 2007.
11. A. J. McGettrick, K. Duffin, W. Johnstone, G. Stewart and D. G. Moodie, Tunable diode laser spectroscopy with wavelength modulation: a phasor decomposition method for calibration free measurements of gas concentration and pressure, *IEEE J. Lightwave Technol.*, 26, (4), 432–440, 2008.
12. W. Johnstone, A. J. McGettrick, K. Duffin, A. Cheung and G. Stewart, Tunable diode laser spectroscopy for industrial process applications: system characterisation in conventional and new approaches, *IEEE Sens. J.*, 8, (7), 1079–1088, 2008.
13. K. Ruxton, A. L. Chakraborty, W. Johnstone, et al., Tunable diode laser spectroscopy with wavelength modulation: elimination of residual amplitude modulation in a phasor decomposition approach, *Sens. Actuators, B: Chem.*, 150, (1), 367–375, 2010.
14. A. Upadhyay and A. L. Chakraborty, Residual amplitude modulation method implemented at the phase quadrature frequency of a 1650nm laser diode for line shape recovery of methane, *IEEE Sens. J.*, 15, (2), 1153–1160, 2015.
15. T. Benoy, M. Lengden, G. Stewart and W. Johnstone, Recovery of absorption line-shapes with correction for the wavelength modulation characteristics of DFB lasers, *IEEE Photon.*, 8, (3), 2016.
16. A. L. Chakraborty, K. Ruxton, W. Johnstone, M. Lengden and K. Duffin, Elimination of residual amplitude modulation in tuneable diode laser wavelength modulation spectroscopy using an optical fiber delay line, *Opt. Express*, 17, (12), 9602–9607, 2009.

17. A. L. Chakraborty, K. Ruxton and W. Johnstone, Influence of the wavelength-dependence of fiber couplers on the background signal in wavelength modulation spectroscopy with RAM nulling, *Opt. Express*, 18, (1), 267–280, 2010.
18. A. L. Chakraborty, K. Ruxton, W. Johnstone, Suppression of intensity modulation contributions to signals in second harmonic wavelength modulation spectroscopy, *Opt. Lett.*, 35, (14), 2400–2402, 2010.
19. P. C. Hobbs, Ultra-sensitive laser measurements without tears, *Appl. Opt.*, 36, (4), 903–920, 1997.
20. J. R. P. Bain, M. Lengden, G. Stewart and W. Johnstone, Recovery of absolute absorption line shapes in tunable diode laser spectroscopy using external amplitude modulation with balanced detection, *IEEE Sens. J.*, 16, (3), 675–680, 2016.
21. H. Li, G. B. Rieker, X. Liu, J. B. Jeffries and R. K. Hanson, Extension of wavelength modulation spectroscopy to large modulation depth for diode laser absorption measurements in high-pressure gases, *Appl. Opt.*, 45, (5), 1052–1061, 2006.
22. G. B. Rieker, J. B. Jeffries and R. K. Hanson, Calibration-free wavelength-modulation spectroscopy for measurements of gas temperature and concentration in harsh environments, *Appl. Opt.*, 48, (29), 5546–5560, 2009.
23. K. Sun, X. Chao, R. Sur, et al., Analysis of calibration-free wavelength-scanned wavelength modulation spectroscopy for practical gas sensing using tunable diode lasers, *Meas. Sci. Technol.*, 24, (12), 125203–125214, 2013.
24. C. S. Goldenstein, C. L. Strand, I. A. Schultz, et al., Fitting of calibration-free scanned-wavelength-modulation spectroscopy spectra for determination of gas properties and absorption lineshapes, *Appl. Opt.*, 53, (3), 356–367, 2014.
25. C. S. Goldenstein, Thesis: Wavelength-modulation spectroscopy for determination of gas properties in hostile environments, 2014. [Online]. Available: https://purl.stanford.edu/fg346yx4996 (accessed April 2020)
26. T. Benoy, D. Wilson, M. Lengden, et al., Measurement of CO_2 concentration and temperature in an aero-engine exhaust plume using wavelength modulation spectroscopy, *IEEE Sens. J.*, 17, (19), 6409–6417, 2017.
27. A. Upadhyay and A. L. Chakraborty, Calibration-free 2f WMS with in situ real-time laser characterization and 2f RAM nulling, *Opt. Lett.*, 40, (17), 4086–4089, 2015.
28. A. Upadhyay, D. Wilson, M. Lengden, et al., Calibration-free WMS using a cw-DFB-QCL, a VCSEL, and an edge-emitting DFB laser with in-situ real-time laser parameter characterization, *IEEE Photon.*, 9, (2), 6801217, 2017.
29. A. Upadhyay, M. Lengden, D. Wilson, et al., A new RAM normalized 1f -WMS technique for the measurement of gas parameters in harsh environments and a comparison with 2f/1f, *IEE Photon.*, 10, (6), 6804611, 2018.

4 Photoacoustic Spectroscopy with DFB Sources

4.1 Introduction

Instead of detecting gas absorption through the small decrease in the intensity of an optical beam after passage through the gas, as was considered in detail in Chapter 3, an alternative way is to detect the pressure change or acoustic wave that arises in the gas from the heat generated by gas absorption. Under the right conditions, photoacoustic spectroscopy (PAS) can prove to be a highly sensitive technique and has found application in a number of areas, including environmental and pollution monitoring, agricultural applications, process control and in medical diagnostics such as human breath analysis. In the near-IR, the acoustic signal can be generated using the same DFB laser sources and modulation techniques [1, 2] as was presented in detail in Chapters 2 and 3. Extension to the mid-IR is straightforward with replacement of the near-IR source by a suitable mid-IR laser, modulated to generate similar acoustic signals. However, the same issues as those discussed in Chapter 3 arise in the interpretation of the output acoustic signal harmonics as a result of the simultaneous intensity and wavelength modulation with DFB sources, but there are also significant differences compared with standard absorption spectroscopy. In this chapter, we will first review the fundamentals of PAS and the use of acoustic cells to act as resonators for signal enhancement. In particular, we derive expressions for the acoustic harmonic signals that are generated in resonant acoustic cells by gas absorption using DFB laser sources. The design of acoustic cells for operation in longitudinal or radial modes is considered in detail, identifying the key parameters affecting the performance, with examples of typical photoacoustic cell designs in both bulk and miniaturised form and the expected output signal levels. As an alternative to the use of photoacoustic cells, the technique of quartz-enhanced photoacoustic spectroscopy (QEPAS) is also briefly reviewed.

4.2 Fundamentals of Photoacoustic Spectroscopy

4.2.1 Theoretical Description

The absorption of light by a gas results in the production of heat due to the relaxation of the excited states to lower levels through molecular collisions. At fixed gas volume, a release of heat, dQ, results in an increase in gas temperature, dT = dQ/nC_v and, from the ideal gas law, the increase in gas pressure is given by:

$$\mathrm{d}p = \left(\frac{nR}{V}\right)\mathrm{d}T = \left(\gamma_{sp} - 1\right)\frac{\mathrm{d}Q}{V} \tag{4.1}$$

where $\gamma_{sp} = C_p/C_v$ is the ratio of specific heat of the gas at constant pressure to that at constant volume, n is the number of moles within the gas volume V and R is the universal gas constant, $R = (C_p - C_v)$.

If the amount of optical energy absorbed is modulated, either through modulation of the input light intensity or through wavelength modulation over a gas absorption line, then dQ and dp will experience modulation at the same frequency and hence an acoustic wave will be generated in the gas, the strength of which will be dependent on the gas absorption characteristics.

The propagation of the acoustic wave is described by a standard wave equation of the form:

$$\nabla^2 p - \frac{1}{v^2}\frac{\partial^2 p}{\partial t^2} = 0 \tag{4.2}$$

where $p(r, \varphi, z, t)$ represents the spatial and time variation of the acoustic wave, v is the acoustic velocity and the Laplacian operator in cylindrical coordinates (r, φ, z) is given by:

$$\nabla^2 = \frac{\partial^2}{\partial r^2} + \frac{1}{r}\frac{\partial}{\partial r} + \frac{1}{r^2}\frac{\partial^2}{\partial \varphi^2} + \frac{\partial^2}{\partial z^2} \tag{4.3}$$

If the rate of heat generation per unit volume in the gas through absorption is defined by: $H(r, \varphi, z, t) = (1/V)\partial Q/\partial t$, then the resultant acoustic signal is determined by adding a source term to the above wave equation as follows [3]:

$$\nabla^2 p - \frac{1}{v^2}\frac{\partial^2 p}{\partial t^2} = -\frac{\left(\gamma_{sp} - 1\right)}{v^2}\frac{\partial H}{\partial t} \tag{4.4}$$

$H(t)$ may be calculated from the Beer–Lambert law for absorption, $I_{\mathrm{out}} = I_{\mathrm{in}} \exp(-\alpha C z)$, as given in Chapter 1. From differentiation of the Beer–Lambert law, the optical power absorbed over a propagation length element dz and cross-sectional area element dA is given by, dP = $\alpha C(I\mathrm{d}A)\mathrm{d}z$ and, assuming than all the absorbed energy is converted to heat, then:

$$H(t) = \frac{\mathrm{d}P}{\mathrm{d}V} = \alpha(\nu)CI(t) \tag{4.5}$$

If the laser output is intensity modulated then the intensity $I(t)$ will be a function of time and similarly if the laser wavelength is modulated, the absorption $\alpha(\nu)$ will be a function of time. In the following analysis, we assume that the rate of transfer of the absorbed optical energy to heat is faster than the modulation rate so that $H(t)$ does not lag $I(t)$.

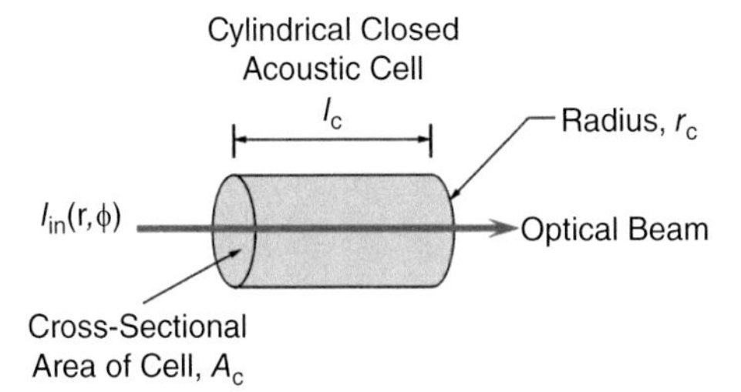

Figure 4.1 Closed photoacoustic cell with central passage of optical beam.

4.2.2 Non-Resonant Solution

We shall consider later a full solution to (4.4), including cell resonance, but first let us consider the simplest case of the low-frequency, non-resonant situation where the acoustic wavelength is much greater than the cell dimensions (for acoustic frequencies <1 kHz, the acoustic wavelength is >30 cm). Hence, with the additional assumption of small optical absorption, the spatial variation of p in (r, φ, z) is negligible. Figure 4.1 shows a closed cylindrical acoustic cell of cross-sectional area, A_c, with the laser beam passing centrally through the cell.

Setting the spatial term $\nabla^2 p = 0$ in (4.4) gives the simplified equation for the closed cell, where the rate of change of acoustic pressure is proportional to the rate of heat generation per unit volume:

$$\frac{dp}{dt} = \left(\gamma_{sp} - 1\right)H(t) \tag{4.6}$$

Integration of (4.6) and using (4.5) gives the photoacoustic signal:

$$p(t) = \{(\gamma_{sp} - 1)C\} \cdot \int \alpha(\nu)I(t)dt \tag{4.7}$$

Equation (4.7) may be used to calculate the low-frequency, non-resonant, acoustic signal for pure intensity modulation, where only $I(t)$ is sinusoidally modulated in time, for wavelength modulation, where $\alpha(\nu)$ is a function of time, for combined intensity and wavelength modulation, where both are functions of time or for other functions of time such as pulsed excitation or a single sweep through an absorption line.

We shall consider later the case of combined intensity and wavelength modulation, as occurs with a DFB laser source, but first let us consider pure intensity modulation of the laser output. We use standard phasor notation, where the intensity modulation and the subsequent output signals are the real part of the various complex functions involved. The intensity modulation of the laser output is described by:

$$I(t) = I + \Delta I e^{j\omega_m t} = I(1 + \kappa e^{j\omega_m t}) \tag{4.8}$$

where $\kappa = \Delta I/I$ is the intensity modulation index.

Substitution in (4.7) and neglecting the DC term gives:

$$p(t) = \frac{(\gamma_{sp} - 1)\alpha(\nu)C\Delta P}{j\omega_m A_c} \cdot e^{j\omega_m t} \tag{4.9}$$

where A_c is the cross-sectional area of the cell and ΔP is the laser power modulation, $\Delta P = A_c \Delta I = A_c \kappa I$. This in fact assumes that the intensity is uniform over the cross-sectional area of the cell, but the same result is derived later without this assumption.

Compared with the signal obtained by optical detection, as discussed in Chapter 3, note the following differences in the simple photoacoustic signal described by (4.9):

(i) There is, in principle, no background signal with intensity modulation.
(ii) The signal is inversely dependent on the cross-sectional area of cell and independent of length.
(iii) There is an inverse dependence on $j\omega_m$ (due to the integration) giving a 90° phase shift and the possibility of large signals at low modulation frequencies.

However, the above analysis has neglected losses in the cell. In practice, the low-frequency signal is limited by heat conduction from the gas to the walls of the cell, characterised by a thermal damping time, $\tau = \tau_{th}$. With damping, (4.6) is modified to read:

$$\frac{dp}{dt} + \frac{p}{\tau} = (\gamma_{sp} - 1)H(t) \tag{4.10}$$

which has the solution:

$$p(t) = \{(\gamma_{sp} - 1)C\} \cdot e^{-t/\tau} \int \alpha(\nu)I(t)e^{t/\tau}dt \tag{4.11}$$

Integration as above gives:

$$p(\nu, t) = \frac{(\gamma_{sp} - 1)\alpha_0 C\Delta P}{A_c} \left\{ \frac{1}{j\omega_m + \tau_{th}^{-1}} \right\} \cdot f(\nu)e^{j\omega_m t} \tag{4.12}$$

where the absorption coefficient has been written as $\alpha(\nu) = \alpha_0 f(\nu)$ from (3.7) in Chapter 3.

Hence the low-frequency response is limited by the thermal damping factor and reaches a maximum magnitude of $\{\tau_{th}(\gamma_{sp} - 1)/A_c\}\alpha_0 C\Delta P$ as $\omega_m \to 0$.

4.2.3 Acoustic Resonant Modes of Cells

The photoacoustic signal may be greatly enhanced by operating at an acoustic resonance of the gas detection cell. The resonant modes [3–7] of a lossless cylindrical cell of radius r_c and length l_c may be obtained by solution of the wave equation (4.2), giving the kth eigenmode as:

$$p_k(r, \varphi, z, t) = p_k(r, \varphi, z)e^{j\omega_k t} \tag{4.13}$$

where the spatial distribution of the mode, characterised by the integers (l, m, q), is given by:

$$p_k(r,\varphi,z) = B_k\left\{J_m\left(\frac{\pi\alpha_{lm}r}{r_c}\right)\cos(m\varphi)\cos\left(\frac{q\pi z}{l_c}\right)\right\} \tag{4.14}$$

and B_k is a dimensionless constant determined by normalisation of the eigenfunctions (see (A4.1.15) in Appendix 4.1) and ω_k is the resonant frequency of the mode, obtained by substitution of (4.13) and (4.14) into the wave equation (4.2) to give:

$$\omega_k = v\sqrt{\left(\frac{\pi\alpha_{lm}}{r_c}\right)^2 + \left(\frac{q\pi}{l_c}\right)^2} \tag{4.15}$$

Here $J_m(x)$ is the Bessel function of order m and α_{lm} is the lth value of (x/π) for which $\partial J_m(x)/\partial x = 0$. Approximate values for α_{lm} are given in Table 4.1.

The fundamental types of modes supported by the cell, namely, radial, azimuthal and longitudinal, are determined by the Bessel and cosine functions, respectively, in (4.14) and are illustrated in Figure 4.2.

Note that the solution of (4.14) is for a closed, rigid gas cell where the acoustic velocity must be zero at the cell boundaries. Since the acoustic velocity is proportional to the gradient of p, the boundary conditions, $\partial p/\partial z = 0$ at $z = 0$ and at l_c and $\partial p/\partial r = 0$ at $r = r_c$ have been applied, with pressure anti-nodes at the ends. For an open-ended cell with pressure nodes at the ends, the pressure variation along the cell length follows a

Table 4.1 Approximate values of α_{lm}

	$m = 0$	$m = 1$	$m = 2$
$l = 0$	0	0.59	0.97
$l = 1$	1.22	1.70	2.13
$l = 2$	2.23	2.72	3.17

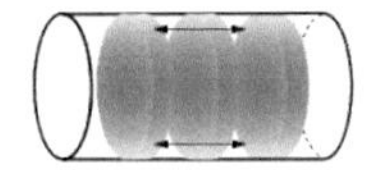

(a) Longitudinal Mode

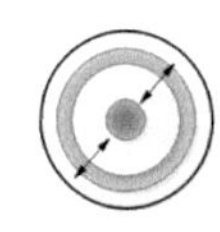

(b) Radial Mode

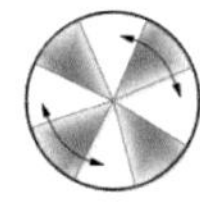

(c) Azimuthal Mode

Figure 4.2 Acoustic modes of closed cell showing direction of acoustic vibrations: (a) longitudinal mode, (b) radial mode, (c) azimuthal mode.

sine function so, for this case, the cosine function in (4.14) is replaced by a sine function.

In general, the total acoustic field in a cell is a summation of the eigenmodes:

$$p(r, \varphi, z, t) = \sum_k A_k(t) p_k(r, \varphi, z) \tag{4.16}$$

where $A_k(t)$ represents the time-varying amplitude of each mode and has units of pressure, N m^{-2} or equivalently J m^{-3}. Equation (4.16) may be used to describe any form of mode excitation, such as through modulation of the source or pulsed operation.

4.2.4 Excitation of Resonant Modes by Intensity Modulation

Consider first the simpler case of pure intensity modulation of the laser output producing modulation of the heat source at the same frequency. From (4.5) and (4.8) we can write for this situation:

$$H(r, \varphi, z, t) = H(r, \varphi, z) e^{j\omega_m t} = \alpha(\nu) C \cdot \{\kappa I(r, \varphi, z)\} e^{j\omega_m t} \tag{4.17}$$

where we have neglected the DC term as before, and the absorption coefficient may be written as $\alpha(\nu) = \alpha_0 f(\nu)$ from (3.7) in Chapter 3.

The modulation of the heat source excites an acoustic field in the cell and the total field can be represented as a sum of the eigenmodes, as represented by (4.16). These eigenmodes are forcibly driven at the modulation frequency, so we have:

$$p(r, \varphi, z, t) = e^{j\omega_m t} \sum_k A_k p_k(r, \varphi, z) \tag{4.18}$$

Substitution of (4.17) and (4.18) into the acoustic wave equation gives an expression for the amplitude A_k of each eigenmode excited by the intensity modulation. The mathematical derivation, including losses, is given in Appendix 4.1, giving the expression:

$$A_k = \frac{j\omega_m (\gamma_{sp} - 1) H_k}{\{(\omega_k^2 - \omega_m^2) + j\omega_k \omega_m / Q_k\}} \cdot \{e^{j\omega_m t}\} \tag{4.19}$$

where the Q factor for the kth mode is $Q_k = \omega_k \tau_k / 2 = \omega_k / \Delta\omega_k$, τ_k represents the decay time due to thermal and viscous losses [3] and $\Delta\omega_k$ is the $1/\sqrt{2}$ full-width of the resonance.

In (4.19), the quantity H_k represents the coupling or spatial overlap of the heat distribution with the acoustic mode distribution and is given by:

$$H_k = \frac{\iiint \{p_k(r, \varphi, z) H(r, \varphi, z)\} r \mathrm{d}r \mathrm{d}\varphi \mathrm{d}z}{V_c} \tag{4.20}$$

where the normalisation condition is applied to the eigenfunctions:

$$\iiint |p_k(r, \varphi, z)|^2 r \mathrm{d}r \mathrm{d}\varphi \mathrm{d}z = V_c \tag{4.21}$$

with the integrations performed over the cell volume, V_c.

Now the optical power input to the cell is the integral of the input intensity over the cross-sectional area of the cell:

$$P_{\text{in}} = \iint I_{\text{in}}(r,\varphi) r\mathrm{d}r\mathrm{d}\varphi \tag{4.22}$$

Using this expression along with the expression for $H(r,\varphi,z)$ from (4.17) we can write the amplitude A_k of each eigenmode from (4.19) as:

$$A_k(\nu,t) = \frac{(\gamma_{sp}-1)\alpha_0 C\kappa P_{in}}{A_c} \cdot \frac{Q_k}{\omega_k} \cdot \left\{\frac{1}{1-jQ_k\Delta_k}\right\} F_k \cdot f(\nu) \mathrm{e}^{j\omega_m t} \tag{4.23}$$

where $f(\nu)$ is the lineshape function as defined by (3.7) in Chapter 3, $\Delta_k = (\omega_k^2 - \omega_m^2)/(\omega_m\omega_k)$ is the deviation of the modulation frequency from the resonant frequency and F_k is the dimensionless spatial overlap of the optical intensity distribution with the acoustic mode distribution:

$$F_k = \frac{\iiint \{p_k(r,\varphi,z) I(r,\varphi,z)\} r\mathrm{d}r\mathrm{d}\varphi\mathrm{d}z}{l_c \iint I_{\text{in}}(r,\varphi) r\mathrm{d}r\mathrm{d}\varphi} \tag{4.24}$$

We see from the various terms in (4.23) that, among other things, the amplitude of the eigenmode depends on the input power, the deviation from resonance and the overlap factor. A particular mode may be preferentially excited by choosing the modulation frequency to be coincident with, or close to, the resonant frequency of the mode and by selection of the beam profile and/or position within the cell to maximise the overlap factor between the mode profile and the beam profile.

Note that for small absorption, where the intensity is approximately uniform over the cell length, then $I(r,\varphi,z) \cong I_{\text{in}}(r,\varphi)$. If, further, the beam profile has no azimuthal variation in φ and passes centrally through the cell, then the overlap factor is zero for all azimuthal modes and only modes with $m = 0$ can be excited.

Modes with no azimuthal and additionally no radial variation ($m = 0$ and $\alpha_{lm} = 0$) are pure longitudinal acoustic modes and the overlap factor of (4.24) reduces to:

$$F_q \cong \frac{1}{l_c} \int p_q(z)\mathrm{d}z \tag{4.25}$$

where $p_q(z)$ represents the longitudinal variation of the mode (sine or cosine function) and the integration is performed over the cell length.

For a closed cell with a cosine distribution of pressure in the longitudinal direction, as indicated in (4.14), the integral of (4.25) is zero for all values of q, so these modes cannot be excited by a uniform beam over the cell length. However odd-order pure longitudinal modes may be excited in an open-ended cell where the pressure has a sine distribution:

$$p_q(z) = \sqrt{2}\,\sin\left(\frac{q\pi z}{l_c}\right) \tag{4.26}$$

and the normalisation factor of $\sqrt{2}$ is obtained using (4.21).

Substitution of (4.26) into (4.25) gives the overlap factor as:

$$F_q = \frac{2\sqrt{2}}{q\pi} \text{ for } q = 1, 3, 5, \ldots$$
$$F_q = 0 \text{ for } q = 2, 4, 6, \ldots \tag{4.27}$$

Hence, only the odd-order longitudinal modes can be excited in the open-ended cell under the above-described conditions.

To illustrate these results we consider and compare two cases, namely the non-resonant situation discussed in Section 4.2.2 which corresponds to the (000) mode in a closed cell and the excitation of the lowest-order longitudinal mode (001) in an open-ended cell.

For the (000) mode, setting $\alpha_{lm} = m = q = 0$ in (4.14) gives $p_0(r, \varphi, z) = B_0$. From the normalisation condition of (4.21), we have $B_0 = 1$ and the overlap factor is unity.

Hence, by first substituting $Q_k = \omega_k \tau_{th}$ into (4.23) and then setting $\omega_k = 0$ from (4.15), the amplitude of the (000) mode is:

$$A_0 = \frac{(\gamma_{sp} - 1)\alpha_0 C\kappa P_{\text{in}}}{A_c} \left\{ \frac{1}{j\omega_m + \tau_{th}^{-1}} \right\} \cdot f(\nu) e^{j\omega_m t} \tag{4.28}$$

This is the same as the solution derived earlier, see (4.12), with $\kappa P_{in} = \Delta P$.

Consider now the (001) pure longitudinal mode for an open-ended cell, $q = 1$ in (4.26) and (4.27). If the modulation frequency is set at the resonance frequency, the amplitude of the mode from (4.23) is:

$$A_1 = \frac{(\gamma_{sp} - 1)\alpha_0 C\kappa P_{\text{in}}}{A_c} \left\{ \frac{Q_1}{\omega_m} \right\} \frac{2\sqrt{2}}{\pi} \cdot f(\nu) e^{j\omega_m t} \tag{4.29}$$

Comparing (4.28) and (4.29), we see that at frequencies when $\omega_m \gg \tau_{th}^{-1}$ the first longitudinal mode gives an enhanced PAS signal by a factor of ~ Q over the non-resonant situation.

4.2.5 Photoacoustic Signal with a DFB Laser Source

We now consider the PAS signals generated when the source is a DFB laser and the injection current is sinusoidally modulated at a frequency of ω_m. As discussed in detail in Chapters 2 and 3, the current modulation produces both intensity and wavelength modulation of the output. Consequently, the heat generated as a function of time depends on both the time variation of the intensity and the time variation of the absorption, as described by (4.5). In Appendix 4.2 it is shown that the heat generation function in this case is:

$$H(r, \varphi, z, t) = (\alpha_0 C) I(r, \varphi, z) \left\{ a_0 + \frac{\kappa}{2} a_1 e^{j\psi_f} + s_1(\nu, t) + \sum\nolimits_{n=2}^{\infty} s_n(\nu, t) \right\} \tag{4.30}$$

where

$$s_1(\nu, t) = \left[\kappa a_0 + a_1 e^{j\psi_f} + \frac{\kappa}{2} a_2 e^{2j\psi_f} \right] e^{j\omega_m t} \tag{4.31}$$

$$s_n(\nu, t) = \left[\frac{\kappa}{2} a_{n-1} e^{j(n-1)\psi_f} + a_n e^{jn\psi_f} + \frac{\kappa}{2} a_{n+1} e^{j(n+1)\psi_f} \right] e^{jn\omega_m t} \tag{4.32}$$

Here ψ_f is the phase advance of the optical frequency modulation with respect to the current modulation and is related to the phase lag of the wavelength modulation by $\psi_f = (\pi - \psi_\lambda)$, as was explained in Chapters 2 and 3, α_0 is the gas absorption coefficient at line centre, and κ is the intensity modulation index as defined earlier. The Fourier coefficients a_n are defined in Chapter 3 – see (3.11) to (3.14) and Appendix 3.2. Although not explicitly shown for the sake of brevity, note that I, κ and a_n are all dependent on the optical frequency ν. Similar to Figures 3.5 and 3.7 in Chapter 3, Figures 4.3a and b show the phasor components of the first and second harmonic acoustic signals represented by (4.31) and (4.32), plotted in terms of the phase lag of the wavelength modulation. Note the absence of a background level in Figure 4.3 compared with Figures 3.5 and 3.7.

Equations (4.30) to (4.32) show that heat modulation occurs at the various harmonics of the modulation frequency, which in turn may excite several coincident or near-coincident resonances of the acoustic cavity. In general, the amplitude of the kth acoustic mode of the cavity excited by the nth harmonic of the modulation frequency will be:

$$A_{kn} = \frac{(\gamma_{sp} - 1)\alpha_0 C P_{\text{in}}}{A_c} \cdot \frac{Q_k}{\omega_k} \cdot \left\{ \frac{1}{1 - jQ_k \Delta_{kn}} \right\} F_k \cdot s_n(\nu, t) \tag{4.33}$$

where the overlap integral F_k is defined as before by (4.24) or (4.25), and $\Delta_{kn} = (\omega_k^2 - n^2\omega_m^2)/(n\omega_m\omega_k)$ is the deviation of the nth harmonic from the kth resonant frequency.

For pure longitudinal modes the acoustic frequencies are integer multiples of the fundamental and may be matched to the harmonics of the modulation frequency. Hence a series of odd-order acoustic longitudinal modes may be excited in an open-ended cell with a DFB laser source.

4.3 Design of Photoacoustic Cells

4.3.1 Open-Ended Resonators

A common form of photoacoustic cell is illustrated in Figure 4.4. It consists of an open-ended cylindrical tube forming the acoustic resonator with buffer volumes at each end, which suppress noise from absorption by the laser windows and from turbulence at the gas inlet and outlet. The laser beam is passed centrally along the resonator tube axis with a microphone placed at the centre.

The dimensions of the resonator tube may be chosen so as to allow only pure longitudinal modes to be excited. This may be seen by rewriting the resonant frequency condition of (4.15) in terms of the acoustic wavelength, $\lambda_k = 2\pi v/\omega_k$, as follows:

$$\left(\frac{2\pi}{\lambda_k}\right)^2 \left\{1 - \left(\frac{\alpha_{lm}\lambda_k}{2r_c}\right)^2\right\} = \left(\frac{q\pi}{l_c}\right)^2 \tag{4.34}$$

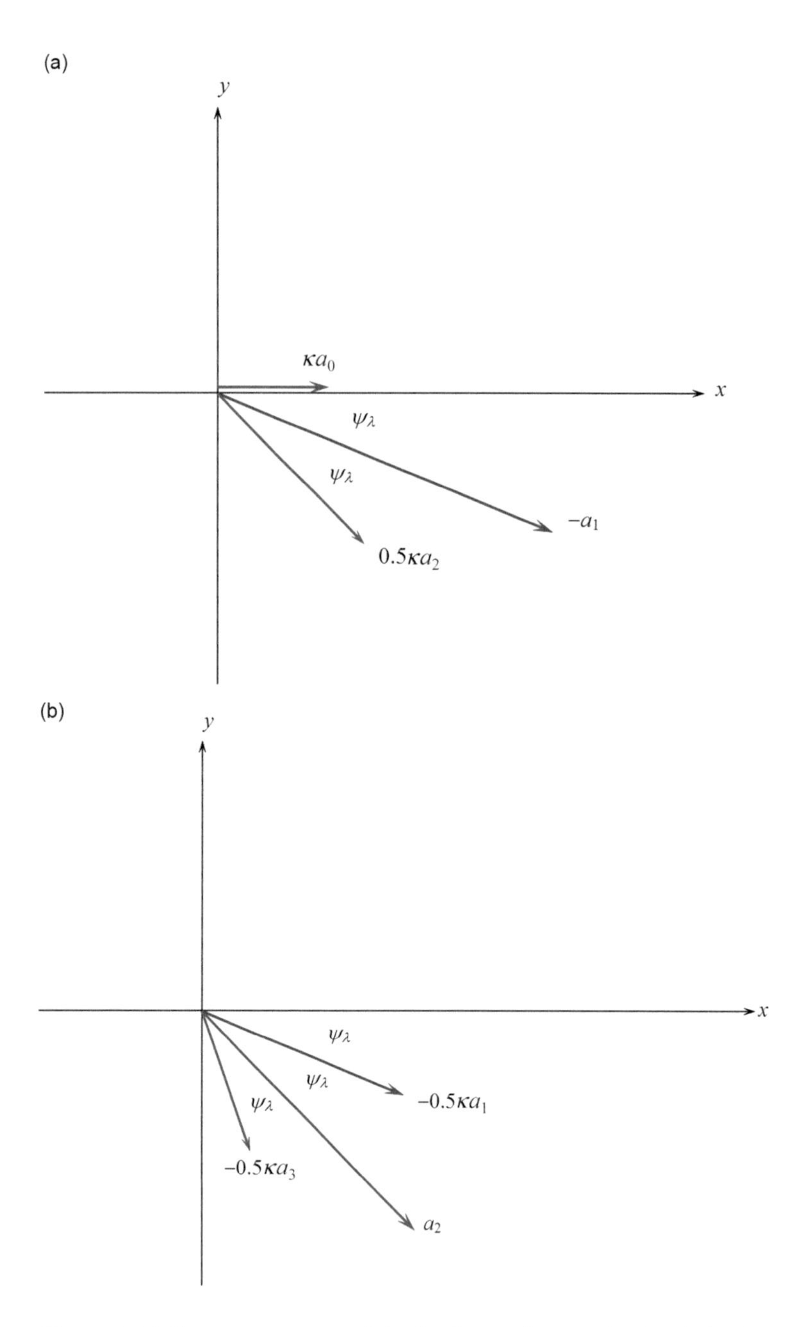

Figure 4.3 (a) Phasor components of the first harmonic acoustic signal. (b) Phasor components of the second harmonic acoustic signal.

Since the RHS of (4.34) is positive (or zero) we see that for the radial mode corresponding to a particular value of α_{lm} to exist, it is necessary for $r_c \geq \alpha_{lm}(\lambda_k/2)$. Hence with reference to the values of α_{lm} given in Table 4.1 and assuming that the beam profile

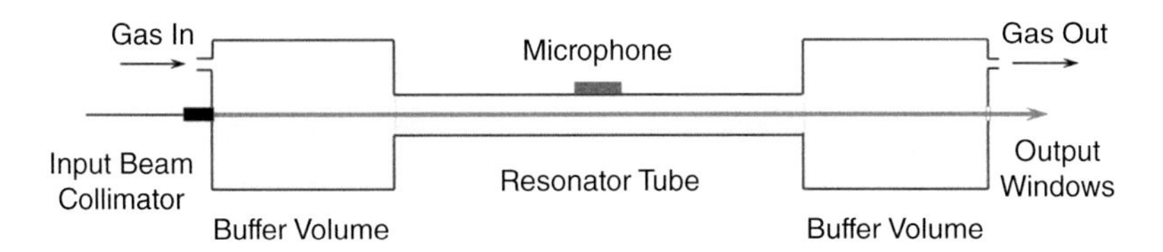

Figure 4.4 Cylindrical photoacoustic cell with open-ended resonator tube and buffer volumes for excitation of longitudinal modes.

has no azimuthal variation so that $m = 0$ is only possible, if the resonator tube radius is chosen so that $r_c < 1.22(\lambda_k/2)$, then no radial modes are possible for this value of λ_k. The overlap factor for these pure longitudinal modes is then given by (4.25) and their resonant frequencies are:

$$f_q = \frac{qv}{2l_{ce}} \tag{4.35}$$

where the effective cell length, $l_{ce} = l_c + \Delta l_c$ includes the 'end correction' for open-ended resonator cells to take into account the sound radiation into the buffer regions at the ends [7]. For flanged open ends, as shown in Figure 4.4, the correction to the cell length is: $\Delta l_c = \frac{16r_c}{3\pi}$.

Absorption of the intensity-modulated laser power by the cell windows also generates a modulated heat source giving rise to an undesirable background acoustic signal at the modulation frequency. The length of the buffer regions can be chosen to minimise the coupling of this background noise to the resonator [1, 8, 9]; the optimum buffer length for noise suppression is: $\lambda_k/4$.

For operation at pressures above 0.1atm, the Q-factor for the cell is dominated by surface losses arising from thermal and viscous loss. The thermal loss occurs in a thin region near the resonator walls where the gas expansion/contraction makes a transition from adiabatic in the bulk of the cell to isothermal at the walls, since the gas will have constant temperature at the walls where there is much higher thermal conductivity compared with the gas. The viscous loss also occurs over a thin region near the walls and arises from the component of the acoustic velocity parallel to a wall surface, which must vanish at the wall [1, 3, 8, 9]. The Q-factor can be expressed in terms of characteristic thermal and viscous skin depths of d_{th} and d_v, respectively, and, for longitudinal modes in an open-ended cell, is given by:

$$Q \cong \frac{r_c}{d_v + (\gamma_{sp} - 1)d_{th}} \tag{4.36}$$

The skin depths are given by $d_{th} = \sqrt{2\kappa/\rho\omega C_p}$ and $d_v = \sqrt{2\mu/\rho\omega}$, where ρ is the gas density, κ is its thermal conductivity, C_p is the specific heat at constant pressure and μ is its dynamic viscosity. Table 4.2 gives approximate values for these parameters for several gases.

Table 4.2 Approximate values of acoustic parameters for several gases at 20 °C and 1atm

Gas	Density kg m^{-3}	Acoustic velocity m s^{-1}	Cp J kg^{-1} K^{-1}	γ_{sp}	Thermal conductivity W m^{-1} K^{-1}	Dynamic viscosity kg m^{-1} s^{-1}
Air	1.29	343	1×10^3	1.4	2.5×10^{-2}	1.8×10^{-5}
O_2	1.43	326	0.9×10^3	1.4	2.7×10^{-2}	2×10^{-5}
N_2	1.25	349	1×10^3	1.4	2.6×10^{-2}	1.75×10^{-5}
CO_2	1.98	267	0.8×10^3	1.3	0.07×10^{-2}	1.38×10^{-5}

Hence the Q-factor for the qth longitudinal mode may be written as:

$$Q_q \cong \Upsilon_g r_c \sqrt{\omega_q} = \Upsilon_g r_c \sqrt{\frac{\pi q v}{l_{ce}}} \tag{4.37}$$

where Υ_g is purely a function of the gas parameters:

$$\Upsilon_g = \frac{\sqrt{\rho}}{\left\{\sqrt{2\mu} + (\gamma_{sp} - 1)\sqrt{\frac{2\kappa}{C_p}}\right\}} \tag{4.38}$$

and may be calculated using the values given in Table 4.2.

Note from (4.37) the dependence of the Q-factor on the cell dimensions according to $Q \propto r_c/\sqrt{l_{ce}}$ and on the mode number according to $Q \propto \sqrt{q}$. However, combining (4.33) for the signal amplitude with (4.37) for the Q-factor and (4.27) for the overlap factor, the acoustic signal amplitude for the odd modes q = 1, 3, 5,. . . is given by:

$$A_{qn} = \frac{2\sqrt{2}}{\sqrt{\pi^5}}\left\{\frac{1}{\sqrt{q^3}}\right\}\left\{\frac{(\gamma_{sp} - 1)\Upsilon_g}{\sqrt{v}}\right\}\left\{\frac{\sqrt{l_{ce}}}{r_c}\right\}\left\{\frac{1}{1 - jQ_q\Delta_{qn}}\right\}\alpha_0 CP_{\text{in}} \cdot s_n(\nu, t) \tag{4.39}$$

where $\Delta_{qn} = \left(\omega_q^2 - n^2\omega_m^2\right)/\left(n\omega_m\omega_q\right)$

Considering the dependence of the signal amplitude on the cell dimensions in (4.39), the best sensitivity is in fact achieved with a long thin cell operating at the resonance condition of its lowest-order longitudinal mode.

Let us consider an example of the design and performance of an acoustic cell of this form, excited by a DFB laser source. Typically, a bulk cell may be formed from brass or stainless steel with resonator dimensions [8] of 100 mm length and 3 mm radius. This gives the frequency of the first longitudinal mode from (4.35) as ~1.63 kHz, where the velocity of sound in air is taken as ~343 m s^{-1}. The acoustic wavelength is ~200 mm at this frequency, so an appropriate length for the buffer region would be ~50 mm with a radius of ~20 mm. The condition that $r_c < 1.22(\lambda_k/2)$ is well satisfied for this and higher-order modes, so a range of odd-order pure longitudinal modes may be excited by the harmonics of the modulation frequency, as described by (4.33) for a DFB laser source.

Using the values in Table 4.2 for air with the above cell dimensions in (4.37) gives Q_1 ~ 39 for the first longitudinal mode at ~1.63 kHz, Q_3~67 for the third longitudinal mode at ~4.9 kHz and Q_5 ~ 87 for the fifth longitudinal mode at ~8 kHz. Figure 4.5

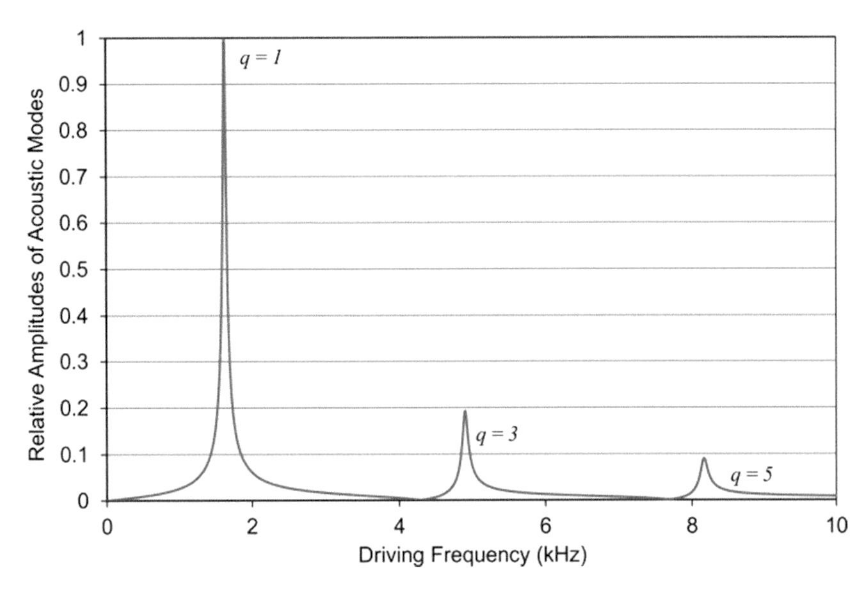

Figure 4.5 Relative amplitudes of the odd-order longitudinal acoustic modes as a function of the driving frequency of the laser source for the cell dimensions given in the text.

shows the relative amplitudes of these modes as a function of the driving frequency. This frequency dependence arises from the $q^{-3/2}(1 - jQ_q\Delta_{qn})^{-1}$factor in (4.39). Figure 4.5 clearly shows that the lowest-order mode gives the highest sensitivity, even though it has the lowest Q-factor, but to attain this sensitivity it is important that the modulation frequency (or the harmonic arising from the absorption line) is matched to the acoustic resonant frequency, with the need to avoid drift of the resonant frequency through mechanical or thermal disturbance of the photoacoustic cell.

Consider the case where the modulation frequency of the DFB laser source is precisely set at the resonant frequency of the lowest-order longitudinal mode (1.63 kHz) for the above photoacoustic cell, with air (or nitrogen) as the background gas. The total acoustic signal generated can then be calculated from (4.39) as:

$$A_T(\nu, t) \cong \{50s_1(\nu, t) + 10s_3(\nu, t) + 4.5s_5(\nu, t) + \cdots\}(\alpha_0 CP_{\text{in}}) \tag{4.40}$$

where $s_1(\nu, t)$, $s_3(\nu, t)$ and $s_5(\nu, t)$ are given by (4.31) and (4.32) and represent the harmonic components generated from modulation over the gas absorption line.

Equation (4.40) has the units of pressure (J m^{-3} or pascal, N m^{-2}) if the input power is expressed in watts and α_0 is in units of m^{-1}. The first term represents the excitation of the first longitudinal mode at the modulation frequency, the second term represents the excitation of the third longitudinal mode from the third harmonic of the modulation frequency, etc. The relative magnitudes of the terms reflect the $q^{-3/2}$ dependence on mode order, as shown by the resonance peaks in Figure 4.5. A similar formula may be derived for any given cell dimensions and gas properties, as given in Table 4.2.

Note that under the above conditions, only the odd-order harmonics generated from the gas absorption line are recovered from the acoustic signal. If it is desired to obtain the even-order components, in particular the second harmonic signal components, then the modulation frequency of the laser current would be chosen to correspond to half that of the resonant frequency of the lowest-order longitudinal mode, i.e. 0.815 kHz in this case. The total acoustic signal generated with the above cell is then:

$$A_T(\nu, t) \cong \{50 s_2(\nu, t) + 10 s_6(\nu, t) + 4.5 s_{10}(\nu, t) + \cdots\}(\alpha_0 C P_{in}) \quad (4.41)$$

where again the first term represents the excitation of the first longitudinal mode at the second harmonic of the modulation frequency, the second term represents the excitation of the third longitudinal mode at the sixth harmonic of the modulation frequency, etc.

4.3.2 Miniaturised Open-Ended Resonators

More recently, the desire to reduce the size of photoacoustic sensors has led to the development of miniature photoacoustic cells [10–12], which may be conveniently fabricated through 3-D printing techniques [11, 12]. Typical dimensions for a miniature 3-D printed cell would be a resonator tube of approximately 10 mm length and 1 mm radius with buffer regions of 5 mm length and 5 mm radius. Compared with the bulk cell discussed above, the shorter resonator length gives higher resonant frequencies of ~14.5 kHz for the first and ~43.5 kHz for the third longitudinal modes. The Q-factors and signal amplitudes are approximately the same for both bulk and miniature cells with the above dimensions, since a reduction in cell length by a factor of 10 is approximately compensated by a reduction in cell radius by a factor of 3 due to the $r_c/\sqrt{l_{ce}}$ dependence of the Q-factor in (4.37) and its inverse in (4.37) for the signal amplitude. However, the higher resonant frequencies give the advantage of lower noise in detection.

4.3.3 Azimuthal and Radial Modes in Closed Cells

Instead of using longitudinal modes, the excitation of azimuthal or radial modes has also been considered for the operation of closed photoacoustic cells, as shown in Figure 4.6. As noted earlier, only the $m = 0$ mode may be excited by a central passage of a uniform beam through the cell. However, azimuthal modes can be excited by an off-axis passage, as indicated in Figure 4.6a.

In particular, photoacoustic cells have been designed to make use of the lowest-order radial mode [4, 13, 14, 15], as shown in Figure 4.6b and c. The diagonal passage of the beam shown in Figure 4.6b allows the detector to be placed in the centre of the end face, on the central maximum of the radial mode. Additionally, noise from window heating and gas flow can be reduced by arranging the beam to enter and exit on the nodes of the radial mode.

As noted earlier, the viscous loss term in the Q-factor arises from the component of the acoustic velocity parallel to the resonator walls. For the lowest-order radial mode, the component parallel to the cylindrical walls of the cell is zero and hence viscous losses occur only at the end walls of a closed cell giving a higher Q-factor (see Figure 4.2 showing the direction of acoustic velocity for the different modes). For the lowest-order radial mode in a closed cylindrical cell, the Q-factor is given by [13]:

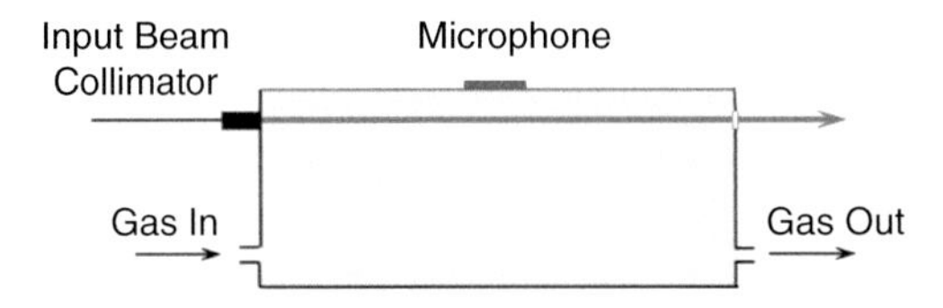

(a) Azimuthal Mode Excitation by Off-axis Beam

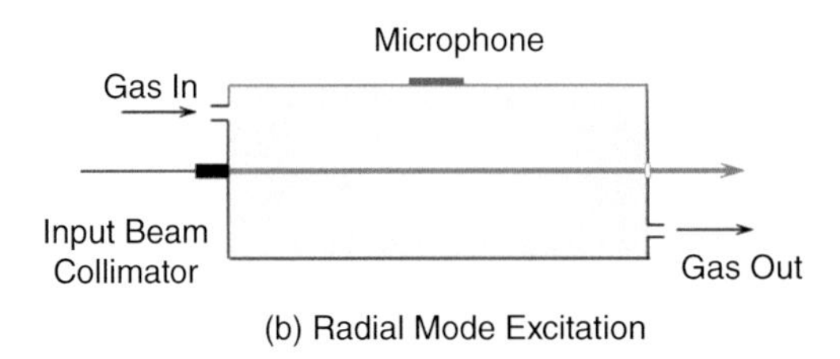

(b) Radial Mode Excitation

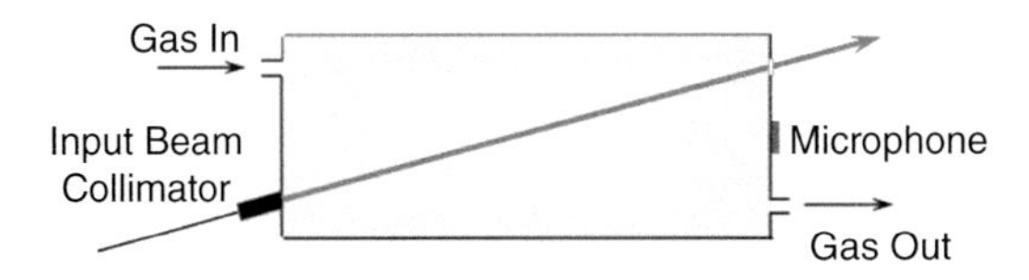

(c) Radial Mode Excitation by Diagonal Beam

Figure 4.6 (a) Excitation of azimuthal modes by off-axis beam passage. (b) Radial mode excitation by central passage of beam through cell. (c) Radial mode excitation by diagonal passage of beam with the detector in the centre.

$$Q_r \cong \frac{r_c}{(r_c/l_c)d_v + (\gamma_{sp} - 1)(1 + r_c/l_c)d_{th}} \tag{4.42}$$

Compared with (4.36) for the longitudinal modes in an open cell, note that the viscous loss contribution is reduced by the factor r_c/l_c (which is the ratio of the area of the two end faces to the cylindrical surface). The thermal loss factor is, however, increased due to the fact that it is now a closed cell. Using the expressions for the skin depths given earlier and (4.15) for the resonance frequency, $\omega_r = \pi\alpha_{lm}v/r_c$, of the lowest-order radial mode, the Q-factor may be written in a similar manner to (4.37) as:

$$Q_r \cong \Upsilon'_g r_c \sqrt{\omega_r} = \Upsilon'_g \sqrt{\pi\alpha_{lm} v r_c} \tag{4.43}$$

where $\alpha_{lm} = 1.22$ from Table 4.1 for the lowest-order radial mode. Note that, as well as the gas parameters, Υ'_g now contains the factor r_c/l_c and is given by:

$$\Upsilon'_g = \frac{\sqrt{\rho}}{\left\{(r_c/l_c)\sqrt{2\mu} + (\gamma_{sp} - 1)(1 + r_c/l_c)\sqrt{2\kappa/C_p}\right\}} \tag{4.44}$$

Consider now the design of a closed cylindrical cell operating in the lowest-order radial mode at a typical resonance frequency of ~10 kHz in air or nitrogen. The acoustic wavelength is λ_a ~ 35 mm and with α_{lm} = 1.22 in (4.15), the required cell radius is $r_c = 0.61\lambda_a$ or ~20 mm. If we assume that r_c/l_c~0.2 then the Q-factor is ~1200. As evident from (4.43) the increase in Q-factor compared with the longitudinal modes results from the increased cell radius as well as the reduction in the viscous losses.

The increased Q-factor does not, however, give a corresponding increase in sensitivity because the signal amplitude is also inversely proportional to the cross-sectional area of the cell. This can be seen from (4.33), where the amplitude of the lowest-order radial mode at resonance with the modulation frequency (or its nth harmonic) may be written as:

$$A_{rn} = \frac{(\gamma_{sp} - 1)\alpha_0 C P_{\text{in}}}{A_c} \cdot \frac{Q_r}{\omega_r} \cdot F_r \cdot s_n(\nu, t) \tag{4.45}$$

We can calculate the signal amplitude as follows. First, the overlap factor for the lowest-order radial mode distribution using (4.14) and (4.24) for small absorption and no azimuthal variation is given by:

$$F_r \cong \frac{B_r \int_0^{r_c} J_0(3.83r/r_c) I_{\text{in}}(r) r \mathrm{d}r}{\int_0^{r_c} I_{\text{in}}(r) r \mathrm{d}r} \tag{4.46}$$

To estimate the overlap factor for a central passage of the excitation beam through the resonator, as shown in Figure 4.6b, let us assume for simplicity that the beam has a uniform intensity over a radius of r_0 and zero elsewhere, with $r_0 \ll r_c$. Hence:

$$F_r \cong \frac{2B_r \int_0^{r_0} J_0(3.83r/r_c) r \mathrm{d}r}{r_0^2} \approx B_r \tag{4.47}$$

where we have used the relation: $\int x J_0(x) \mathrm{d}x = x J_1(x) \approx \frac{1}{2}x^2$ for x small [16, 17].

B_r may be obtained from the normalisation condition, (A4.1.15) in Appendix 4.1, to give:

$$2\pi l_c B_r^2 \int_0^{r_c} |J_0(3.83r/r_c)|^2 r \mathrm{d}r = V_c \tag{4.48}$$

Using the relation: $\int x J_0^2(x) \mathrm{d}x = \frac{1}{2}x^2 \{J_0^2(x) + J_1^2(x)\}$ [16, 17], we obtain B_r = 2.5.

For a diagonal passage of the excitation beam, as shown in Figure 4.6c, Schilt [14] plots the overlap factor as a function of the beam angle relative to the cell axis, with the overlap factor reducing to ~1.5 for an angle of 12°.

Hence, substituting the above expressions into (4.45) along with the Q-factor from (4.43) and the maximum overlap factor of 2.5, the amplitude of the lowest-order radial mode is:

$$A_{rn} = \frac{2.5}{\pi\sqrt{3.83}} \left\{ \frac{(\gamma_{sp} - 1) \Upsilon'_g}{\sqrt{\nu}} \right\} \left\{ \frac{1}{\sqrt{r_c}} \right\} \alpha_0 C P_{\text{in}} \cdot s_n(\nu, t) \tag{4.49}$$

Compared with (4.39) for the longitudinal modes, note the inverse dependence on the square root of the cell radius. With the values given above for the cell and gas parameters, the amplitude is:

$$A_{rn} \cong 15(\alpha_0 C P_{\text{in}}) \cdot s_n(\nu, t) \tag{4.50}$$

Hence, although the Q-factor for the radial mode is large, the overall sensitivity is reduced compared with that of the first-order longitudinal mode given by (4.40) due to the increase in cell radius. The sensitivity can of course be improved by using a smaller radius cell, but this requires a corresponding increase in operation frequency; for example, reducing the cell radius by a factor of 10 would increase the sensitivity by a factor of ~3, requiring operation at 100 kHz.

In general, the analytical models given in this section provide useful information for identifying the design criteria and key parameters for optimising the performance of acoustic cells. This may be complemented by the use of software modelling, such as finite element modelling with COMSOL Multiphysics [18], to simulate the operation of a practical cell, including the gas inlet/outlet and microphone cavities.

4.4 Detection, Calibration and Noise in PAS Systems

The DFB laser sources used for PAS are typically current-modulated at frequencies in the range of 1–100 kHz so the acoustic modes generated in the cell can be detected by standard commercial microphones [11, 12, 19, 20], such as condenser, electret or more recently MEMS microphones, with typical sensitivities in the range of 1–20 mV Pa^{-1}. Piezoelectric devices, which have the advantage of a wide frequency range, may also be used for acoustic detection, but they have low sensitivity due to the inherent low coupling efficiency of the acoustic vibrations in the gas to the solid piezoelectric surface. Microphones may be equipped with an integrated pre-amplifier circuit, but usually a lock-in amplifier is employed at the output with the reference input for the lock-in at the modulation frequency of the DFB laser source. Similar to the detailed description of the lock-in signals given in Chapter 3 and as specified by (4.31) and (4.32), the output harmonics from the lock-in contain a number of components related to the modulation parameters and the Fourier coefficients, and careful interpretation is required to extract gas absorption line profiles [21] and accurate values for gas concentration, pressure or temperature.

The magnitude of the theoretical raw signal obtained from the microphone using a DFB laser source for the measurement of gas absorption lines may be determined as follows. To illustrate, consider a PAS system for the measurement of CO_2 in air using a MEMS microphone such as the WM7331E from Cirrus Logic [22], where the microphone sensitivity is –38 dB (12.6 mV Pa^{-1}) over the frequency range of approximately 100 Hz to 15 kHz. From Table 1.2 in Chapter 1, the line centre absorbance for a CO_2 line at a wavelength of 1572.3 nm is $\alpha_0 \sim 0.14\ \text{m}^{-1}$ with half-linewidth of ~2.5 GHz. Let us assume that we use a miniaturised photoacoustic cell as described earlier, with a radius of 1 mm, resonator length, l_c of ~10 mm and first-order longitudinal mode resonance at ~15 kHz. The DFB laser characteristics and modulation parameters are assumed to be the same as those given in Chapter 3, Section 3.3.2 and Figures 3.6 and 3.8. Figure 4.7a shows the theoretical 15 kHz signal from the microphone, calculated from (4.39) with the above microphone sensitivity, for 100 ppm CO_2 in air as the laser wavelength is scanned through the absorption line with the laser current modulated at 15 kHz to excite the first longitudinal mode. Figure 4.7b shows the theoretical 15 kHz from the microphone when the modulation frequency of the laser current is halved to 7.5 GHz so that the second harmonic arising from the gas absorption line excites the first longitudinal mode. Similar results may be obtained for any gas using the absorption line parameters given in Table 1.2.

One important advantage of PAS over standard WMS is that large optical powers of a few watts can be used to improve the signal magnitude and hence the sensitivity of gas detection. Although the signal magnitude is proportional to the optical power in both standard WMS and PAS, unlike standard WMS, where the optical power falls on the detector and hence becomes saturated at high levels, with PAS it is the relatively small amount of absorbed power that generates the acoustic signal. The power from the modulated DFB laser source may be enhanced to watt levels by an erbium-doped fibre amplifier (EDFA) which, in standard form, provides gain over a wavelength range of approximately 1530–1560 nm or over different wavelength regions by other types of rare-earth-doped fibre amplifiers. For amplification of the output of DFB lasers in the 1600–1700 nm region, stimulated Raman scattering in silica fibres may be employed in combination with an EDFA, to form a fibre Raman amplifier system [12]. (Fibre amplifier systems will be discussed in detail in Chapters 6 and 7.) Of course the ultimate sensitivity that can be achieved with PAS is limited by noise and background levels. Although electronic and microphone noise will be present, it is normally the background noise which limits the sensitivity. Compared with standard WMS where there is a large first harmonic background level, there is, in principle, no background with PAS but, in practice, absorption of the beam at the cell windows or walls does generate a background level and background noise. Acoustic noise may also arise from the gas flow in the cell and from external disturbances. The background noise and signal level may be measured using a non-absorbing gas such as nitrogen in the cell. In standard WMS, the background level may be employed for calibration and referencing (such as in the *2f/1f* ratio technique explained in Chapter 3), but this is not generally possible with PAS, so referencing and calibration are more difficult. One way of reducing noise in the output signal is by use of a differential photoacoustic cell [4], which consists of two similar resonant acoustic cavities and microphones, only one of which is excited by the laser beam, so that coherent noise is cancelled in the differential output.

As an example of the sensitivity that can be achieved in the near-IR, Bauer [12] has demonstrated a minimum detection limit of a few tens of ppb for methane at 1651 nm with a normalised noise-equivalent absorption coefficient (NNEA) of 4.1×10^{-9} cm^{-1} W $Hz^{-1/2}$ using a miniaturised 3-D printed photoacoustic cell (resonator length of 10 mm, 0.9 mm radius with 5 mm long buffer regions) operating in the first longitudinal mode at 15.2 kHz, with a fibre Raman amplifier system to boost the input optical power to 1.1 W. Although a higher ultimate sensitivity can be

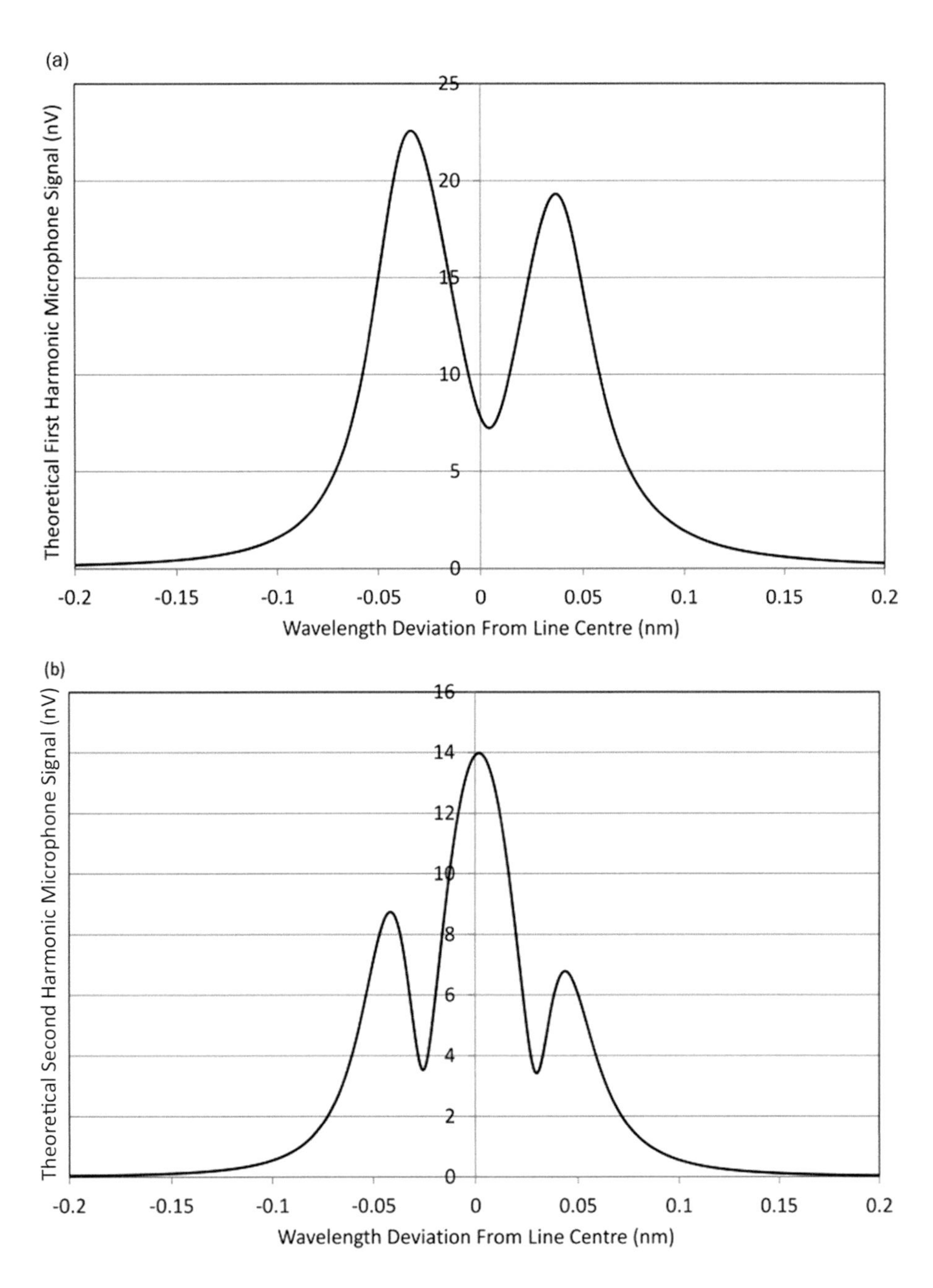

Figure 4.7 (a) Theoretical signal at 15 kHz from a MEMS microphone for the first harmonic of the absorption line for 100 ppm CO_2 gas with the laser modulation frequency set to the frequency of the first longitudinal acoustic mode. (b) Theoretical signal at 15 kHz from the microphone for the second harmonic of the absorption line for 100 ppm CO_2 with the laser modulation frequency set to half the frequency of the first longitudinal acoustic mode.

achieved in the mid-IR, the near-IR fibre-based system allows the potential for remote sensing and the multiplexing of a network of miniature sensors, and may be applied to a variety of gases with near-IR absorption lines.

4.5 Quartz-Enhanced Photoacoustic Spectroscopy

In quartz-enhanced photoacoustic spectroscopy (QEPAS), the acoustic cell and microphone are replaced by a quartz tuning fork (QTF), which detects an acoustic signal generated in the gas at the resonant frequency of the QTF, as illustrated in Figure 4.8a.

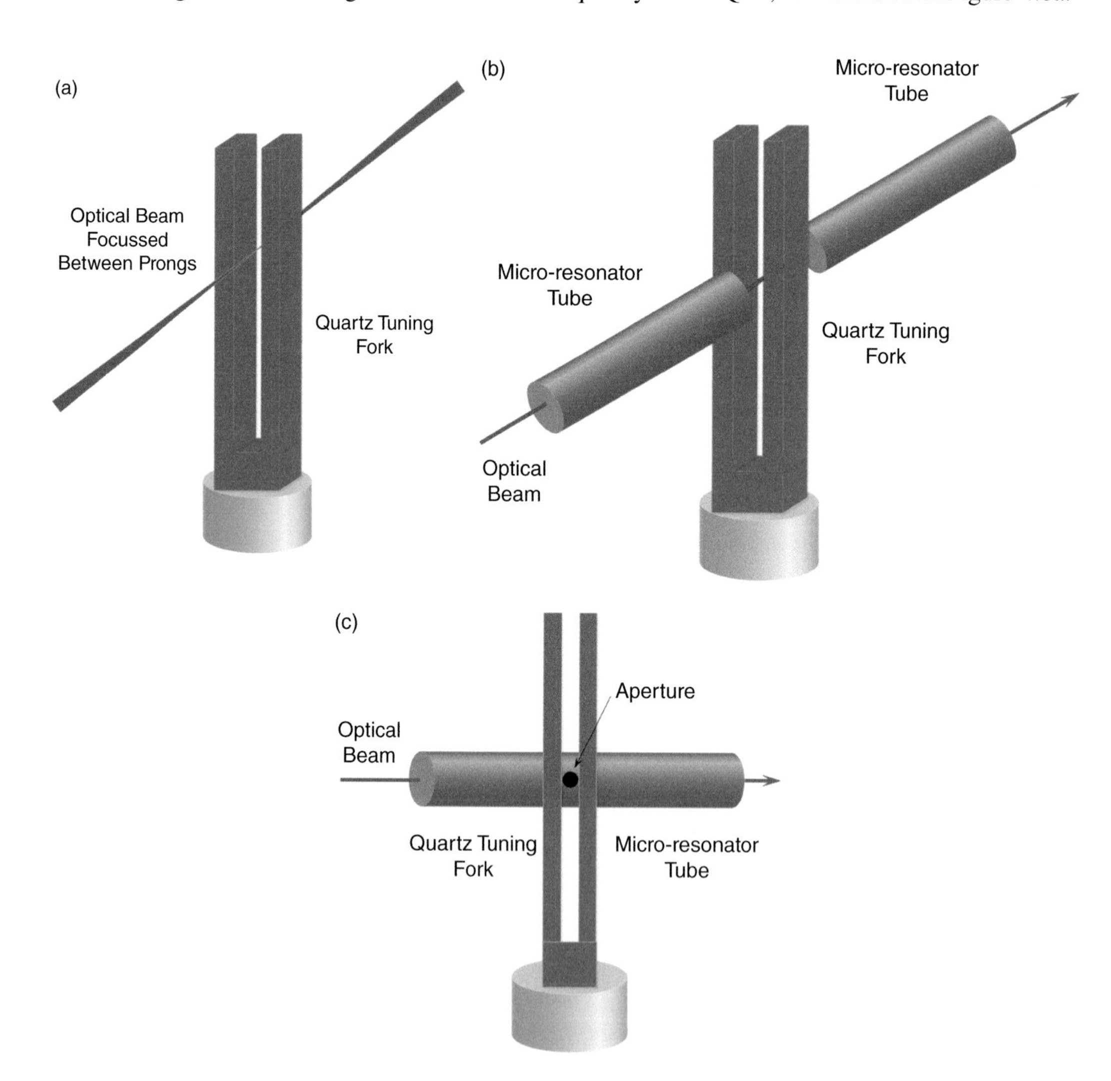

Figure 4.8 Quartz-enhanced photoacoustic spectroscopy: (a) beam focussed between prongs of tuning fork, (b) micro-resonator tubes before and after the QTF to enhance the acoustic signal, (c) micro-resonator tube with aperture on one side of the QTF.

A non-resonant cell may be used to separate the target gas from its surroundings and to control its pressure and temperature, etc. In order to excite the piezo-electrically active mode of the QTF, where the prongs move in opposite directions, the optical beam needs to be focussed between the prongs of the QTF. A comprehensive review of QEPAS is given in references [23, 24].

For low cost and convenience, a standard commercial QTF may be used, such as those used in clocks and timing devices. These have a fixed resonance frequency of 2^{15} Hertz (32.768 kHz) and a very high Q-factor, greater than 10^4. However, the spacing between the prongs is relatively small at ~300 μm and the volume between the prongs is ~300 μm × 300 μm × 3 mm. In situations where the optical beam cannot be adequately focussed between the narrowly spaced prongs, or where a lower resonance frequency is required to better match gas relaxation rates, a custom-made QTF may be designed to optimise the PAS signal.

As for standard PAS, the signal magnitude from QEPAS is proportional to the optical power absorbed and the Q-factor, and inversely proportional to the resonant frequency. The QEPAS signal may be enhanced by the use of micro-resonator tubes placed before and after the QTF, as shown in Figure 4.8b, or on one side of the QTF, as shown in Figure 4.8c, with a small aperture in the centre to allow the QTF to pick up the acoustic signal at the anti-node in the centre of the micro-resonator. With near-IR systems, a convenient way of generating the acoustic wave in the gas is to replace the free-space focussed beam of Figure 4.8a with a tapered region of an optical fibre placed between the prongs of the QTF. The evanescent field extending from the tapered region of the fibre is used to excite the acoustic wave and since the fibre is continuous, this allows the possibility of sensor multiplexing. Using this method with micro-resonator tubes, Li et al. [25] attained a minimum detection limit of ~20 ppm for CO sensing (210 s averaging time), corresponding to an NNEA coefficient of 1.44×10^{-8} cm^{-1} W Hz$^{-1/2}$ using a DFB laser at 2.3 μm wavelength.

4.6 Conclusion

This chapter has reviewed the basic principles of PAS using acoustic resonant cells and the harmonic signals that are generated with modulated DFB laser sources. The theoretical description and models presented allow the determination of the expected signal levels for different gases and assist in the design of suitable acoustic gas cells for efficient operation in the near-IR. However, the same principles apply for operation with mid-IR diode laser sources, as discussed in Chapter 8, since it is the source modulation frequency (and not the optical frequency) which determines the acoustic frequencies generated in the photoacoustic cell.

Appendix 4.1 Derivation of the Amplitudes of the Acoustic Eigenmodes

The lossless eigenmodes given by (4.13) to (4.15) are solutions of the acoustic wave equation:

$$\nabla^2 p_k - \frac{1}{v^2}\frac{\partial^2 p_k}{\partial t^2} = 0 \tag{A4.1.1}$$

However, in practice the modes are subject to thermal and viscous losses [3] and, in the absence of a driving heat source, the mode amplitudes will decay in time according to the factor $e^{-a_k t}$ where a_k is the decay constant of a mode and is related to the characteristic decay time by $a_k = 1/\tau_k$. To include the effect of these losses we need to include a damping term in the wave equation as follows:

$$\nabla^2 p_k - \frac{2a_k}{v^2}\frac{\partial p_k}{\partial t} - \frac{1}{v^2}\frac{\partial^2 p_k}{\partial t^2} = 0 \tag{A4.1.2}$$

The eigenmode solutions for this equation are:

$$p_k(r,\varphi,z,t) = \{p_k(r,\varphi,z)e^{-a_k t}\}e^{j\omega_k' t} = p_k(r,\varphi,z)e^{js_k t} \tag{A4.1.3}$$

where $s_k = \omega_k' + ja_k$ is the complex frequency incorporating the exponential amplitude decay and a (slightly) shifted frequency compared with (4.15) according to:

$$\omega_k' = \sqrt{\omega_k^2 - a_n^2} \tag{A4.1.4}$$

With intensity modulation of the laser input to the cell, the total acoustic field is driven at the modulation frequency and can be represented as a sum of the eigenmodes whose amplitudes are maintained by the driving heat source. Hence the acoustic field is the solution of the damped wave equation:

$$\nabla^2 p - \frac{2a}{v^2}\frac{\partial p}{\partial t} - \frac{1}{v^2}\frac{\partial^2 p}{\partial t^2} = -\frac{(\gamma_{sp}-1)}{v^2}\frac{\partial H}{\partial t} \tag{A4.1.5}$$

where $p(r,\varphi,z,t)$ is the eigenmode summation:

$$p(r,\varphi,z,t) = e^{j\omega_m t}\sum_k A_k p_k(r,\varphi,z) \tag{A4.1.6}$$

To solve for the amplitudes, A_k, of each eigenmode, we first obtain expressions for each term in (A4.1.5) as follows. Applying the Laplacian operator to (A4.1.6) gives:

$$\nabla^2 p = e^{j\omega_m t}\sum_k A_k \nabla^2 p_k(r,\varphi,z) \tag{A4.1.7}$$

Now since the individual eigenmode distributions, $p_k(r, \varphi, z)$ are solutions of the wave equation (A4.1.2) we can use (A4.1.2) and (A4.1.3) to write:

$$\sum\nolimits_k A_k \nabla^2 p_k(r, \varphi, z) = \frac{1}{v^2} \sum\nolimits_k \{2ja_k s_k - s_k^2\} A_k p_k(r, \varphi, z) \tag{A4.1.8}$$

Also from (A4.1.6) we can write the following expression for the time derivatives:

$$\frac{\partial p}{\partial t} = j\omega_m \mathrm{e}^{j\omega_m t} \sum\nolimits_n A_k p_k(r, \varphi, z) \tag{A4.1.9}$$

$$\frac{\partial^2 p}{\partial t^2} = -\omega_m^2 \mathrm{e}^{j\omega_m t} \sum\nolimits_n A_k p_k(r, \varphi, z) \tag{A4.1.10}$$

From (4.17) we obtain the time derivative of the heat source as:

$$\frac{\partial H}{\partial t} = j\omega_m H(r, \varphi, z) \mathrm{e}^{j\omega_m t} \tag{A4.1.11}$$

Combining all the above terms according to the wave equation (A4.1.5) gives:

$$\sum\nolimits_k \{(\omega_k^2 - \omega_m^2) + 2j\omega_m a\} A_k p_k(r, \varphi, z) = j\omega_m (\gamma_{sp} - 1) H(r, \varphi, z) \tag{A4.1.12}$$

To isolate each amplitude A_k we make use of the orthogonality of the eigenmodes, namely, $\int p_i p_j \mathrm{d}V = 0$ except for $i = j$. Hence by integration of (A4.1.12) we obtain:

$$A_k = \frac{j\omega_m (\gamma_{sp} - 1)}{\{(\omega_k^2 - \omega_m^2) + j\omega_k \omega_m / Q_k\}} H_k \tag{A4.1.13}$$

where Q_k is the Q-factor for the kth mode, $Q_k = \omega_k/2a$.

The quantity H_k is the coupling or spatial overlap of the heat distribution with the acoustic mode distribution and is given by:

$$H_k = \frac{\iiint p_k(r, \varphi, z) H(r, \varphi, z) \mathrm{d}V}{\iiint |p_k(r, \varphi, z)|^2 \mathrm{d}V} \tag{A4.1.14}$$

The dimensionless constant B_k in (4.14) may be obtained by normalisation of the eigenfunctions through setting:

$$\iiint |p_k(r, \varphi, z)|^2 \mathrm{d}V = V_c \tag{A4.1.15}$$

with the integrations performed over the cell volume, V_c.

Appendix 4.2 Derivation of the Heat Generation Function with a Modulated DFB Laser

With combined intensity and wavelength modulation, the heat generated as a function of space and time from (4.5) is:

$$H(r, \varphi, z, t) = \alpha(\nu, t)CI(r, \varphi, z, t) \tag{A4.2.1}$$

For the intensity modulation, the time variation of the intensity is the real part of (4.8) which can be written in the form:

$$I(t) = I\left\{1 + \frac{\kappa}{2}\left(e^{j\omega_m t} + e^{-j\omega_m t}\right)\right\} \tag{A4.2.2}$$

For the wavelength modulation, the optical frequency of the laser is a function of time according to (3.5) from Chapter 3:

$$\nu_l = \nu + \delta\nu \cos(\omega_m t + \psi_f) \tag{A4.2.3}$$

Hence the absorption as a function of time from (3.7) is:

$$\alpha(\nu, t) = \alpha_0 f(\nu_l) \tag{A4.2.4}$$

where the Fourier expansion of $f(\nu_l)$ can be expressed from (3.10) as the real part of the following function:

$$f(\nu_l) = a_0 + \sum_{n=1}^{\infty} a_n e^{j(n\omega_m t + n\psi_f)} \tag{A4.2.5}$$

The Fourier coefficients, a_n, are defined in Chapter 3 by (3.11) to (3.14) and in Appendix 3.2.

Hence we can write the heat generation function with a modulated DFB laser as:

$$H(r, \varphi, z, t) = (\alpha_0 C)I(r, \varphi, z)\left\{1 + \frac{\kappa}{2}\left(e^{j\omega_m t} + e^{-j\omega_m t}\right)\right\}\left\{a_0 + \sum_{n=1}^{\infty} a_n e^{j(n\omega_m t + n\psi_f)}\right\} \tag{A4.2.6}$$

Equation (A4.2.6) may be re-cast into the series:

$$H(r, \varphi, z, t) = (\alpha_0 C)I(r, \varphi, z)\left\{a_0 + \frac{\kappa}{2}a_1 e^{j\psi_f} + s_1(\nu, t) + \sum_{n=2}^{\infty} s_n(\nu, t)\right\} \tag{A4.2.7}$$

where:

$$s_1(\nu, t) = \left[\kappa a_0 + a_1 e^{j\psi_f} + \frac{\kappa}{2}a_2 e^{2j\psi_f}\right]e^{j\omega_m t} \tag{A4.2.8}$$

$$s_n(\nu, t) = \left[\frac{\kappa}{2}a_{n-1}e^{j(n-1)\psi_f} + a_n e^{jn\psi_f} + \frac{\kappa}{2}a_{n+1}e^{j(n+1)\psi_f}\right]e^{jn\omega_m t} \tag{A4.2.9}$$

References

1. J.-P. Besson, S. Schilt and L. Thevenaz, Multi-gas sensing based on photoacoustic spectroscopy using tunable diode lasers, *Spectrochim. Acta A*, 60, 3449–3456, 2004.
2. S. Schilt and L. Thevenaz, Wavelength modulation photoacoustic spectroscopy: theoretical description and experimental results, *Infrared Phys. Techn.*, 48, 154–162, 2006.
3. L.B. Kreuzer, The physics of signal generation and detection in *Optoacoustic Spectroscopy and Detection*, Y.-H. Pao, Ed., New York, Academic Press, ch. 1, 1977.
4. A. Miklos and P. Hess, Application of acoustic resonators in photoacoustic trace gas analysis and metrology, *Rev Sci Instrum.*, 24, (4), 1937–1955, 2001.
5. S. Schafer, A. Miklos and P. Hess, Quantitative signal analysis in pulsed resonant photoacoustics, *Appl. Opt.*, 36, (15), 3202–3211, 1997.
6. J. Li, X. Gao, W. Li, et al., Near-infrared diode laser wavelength modulation-based photoacoustic spectrometer, *Spectrochim. Acta A*, 64, 338–342, 2006.
7. L. E. Kinsler, A. R. Frey, A. B. Coppens, J. V. Sanders, Pipes, resonators and filters in *Fundamentals of Acoustics*, New York, John Wiley & Sons Inc., ch. 10, 2000.
8. F. G. C. Bijnen, J. Reuss and F. J. M. Harren, Geometrical optimization of a longitudinal resonant photoacoustic cell for sensitive and fast trace gas detection, *Rev. Sci. Instrum.*, 67, (8), 2914–2923, 1996.
9. M. Tavakoli, A. Tavakoli, M. Taheri and H. Saghafifar, Design, simulation and structural optimization of a longitudinal acoustic resonator for trace gas detection using laser photoacoustic spectroscopy, *Opt. Laser Technol.*, 42, (5), 828–838, 2009.
10. E. L. Holthoff, D. A. Heaps and P. M. Pellegrino, Development of a MEMS-scale photoacoustic chemical sensor using a quantum cascade laser, *IEEE Sens. J.*, 10, (3), 572–577, 2010.
11. R. Bauer, G. Stewart, W. Johnstone, E. Boyd and M. Lengden, A 3D-printed miniature gas cell for photoacoustic spectroscopy of trace gases, *Opt. Lett.*, 39, (16), 4796–4799, 2014.
12. R. Bauer, T. Legg, D. Mitchell, et al., Miniaturized photoacoustic trace gas sensing using a Raman fiber amplifier, *IEEE J. Lightwave Technol.*, 33, (18), 3773–3780, 2015.
13. A. Karbach and P. Hess, High precision acoustic spectroscopy by laser excitation of resonator modes, *J. Chem. Phys.*, 83, (3), 1075–1084, 1985.
14. S. Schilt, L. Thevenaz, M Nikles, L. Emmenegger and C. Huglin, Ammonia monitoring at trace level using photoacoustic spectroscopy in industrial and environmental applications, *Spectrochim. Acta A*, 60, 3259–3268, 2004.
15. L.-Y. Hao, J.-X. Han, Q. Shi, et al., A highly sensitive photoacoustic spectrometer for near infrared overtone, *Rev. Sci. Instrum.*, 71, (5), 2000.
16. W. Rosenheinrich, Tables of some indefinite integrals of Bessel functions of integer order, 2019. [Online]. Available: http://web.eah-jena.de/~rsh/Forschung/Stoer/besint.pdf (accessed April 2020)
17. J. R. Culham, Bessel functions of the first and second kind, 2004. [Online]. Available: www.mhtlab.uwaterloo.ca/courses/me755/web_chap4.pdf (accessed April 2020)
18. Comsol Inc, COMSOL Multiphysics, 2019. [Online]. Available: www.comsol.com/products (accessed April 2020)
19. C. Haisch, Photoacoustic spectroscopy for analytical measurements, *Meas. Sci. Technol.*, 23, 1–16, 2012.
20. www.analog.com/en/analog-dialogue/articles/understanding-microphone-sensitivity.html (accessed April 2020)

21. L. Li, N. Arsad, G. Stewart, et al., Absorption line profile recovery based on residual amplitude modulation and first harmonic integration methods in photoacoustic gas sensing, *Opt. Comm.*, 284, (1), 312–316, 2011.
22. Cirrus Logic, Inc., 2019. Analog MEMS microphones [Online]. Available: https://www.cirrus.com/products/ (accessed April 2020)
23. P. Patimisco, A. Sampaolo, L. Dong and F. K. Tittel, Recent advances in quartz enhanced photoacoustic spectroscopy, *Appl. Phys. Rev.*, 5, (011106), 1–20, 2018.
24. P. Patimisco, G. Scamarcio, F. K. Tittel and V. Spagnolo, Quartz-enhanced photoacoustic spectroscopy: a review, *Sensors*, 14, 6165–6206, 2014.
25. Z. Li, Z. Wang, Y. Qi, W. Jin and W. Ren, Improved evanescent-wave quartz-enhanced photoacoustic CO sensor using an optical fiber taper, *Sens. Actuators B: Chem.*, 248, 1023–1028, 2017.

5 Design and Application of DFB Laser Systems and Optical Fibre Networks for Near-IR Gas Spectroscopy

5.1 Introduction

The high performance, wide availability, relatively low cost and maturity of near-IR DFB lasers make them ideally suited for a wide range of applications in near-IR gas spectroscopy. Application areas include atmospheric, environmental and safety monitoring, combustion analysis and medical diagnostics [1–4]. Additionally, operation in the near-IR region opens up the opportunity for increased flexibility in system design with the use of standard optical fibres, fibre amplifiers, fibre networks and a wide range of optical components readily available from the data communications industry. Compared with traditional gas sensor technologies [5], such as those based on catalytic sensors (pellistors), optical fibre systems have the advantages of safety in a hazardous environment due to their non-electrical nature, and the ability to remotely and simultaneously address a large number of locations through a fibre optic network. Also, as was discussed in detail in Chapter 3, the use of various source modulation formats combined with good signal-to-noise ratio detection, made possible with mature receiver technology and data processing algorithms, means that sufficient sensitivity can be attained in many practical applications, despite the inherently weak strength of near-IR gas absorption lines. This chapter will review a number of techniques that can be applied in near-IR spectroscopy to enhance the sensitivity, including the use of multi-pass cells, ring-down and cavity-enhanced spectroscopy, as well as non-enhanced systems employing evanescent-wave cells, micro-optic cells for sensing over a wide area with optical fibre networks and open-path free space propagation. A number of examples will be presented of near-IR experimental and commercial systems for the detection of gas leaks and for combustion, emission and atmospheric monitoring.

5.2 Gas Cells for DFB Lasers and Optical Fibre Systems

There are a variety of gas cell types that can be employed in the near-IR with DFB laser and fibre optic systems. Examples range from relatively large, bulk and multi-pass cells to enhance the sensitivity, to relatively small micro-optic cells that are readily compatible with optical fibres and suitable for the creation of multi-point systems. In practice, the type of cell deployed will depend on whether the application is in a controlled or field environment and on trade-offs between sensitivity, cost and low maintenance for field deployment. Note that photoacoustic gas cells were discussed in detail in Chapter 4 and will not be considered further here.

5.2.1 Bulk and Multi-Pass Cells

With appropriate collimation optics, the output of a DFB laser source may be directed through a bulk gas cell ranging in length from tens of centimetres to several metres. However, to maximise sensitivity, especially with weak near-IR absorption lines, it is desirable to have long interaction path lengths with a compact cell configuration and for this reason, multi-pass absorption cells are often used, the most popular being the White cell [6] or the Herriott cell [7, 8]. These can give effective interaction lengths of several metres from physical cell lengths of only ~10 cm.

The White cell, illustrated in Figure 5.1a, is formed from three spherical concave mirrors, M_1, M_2 and M, all with the same radius of curvature.

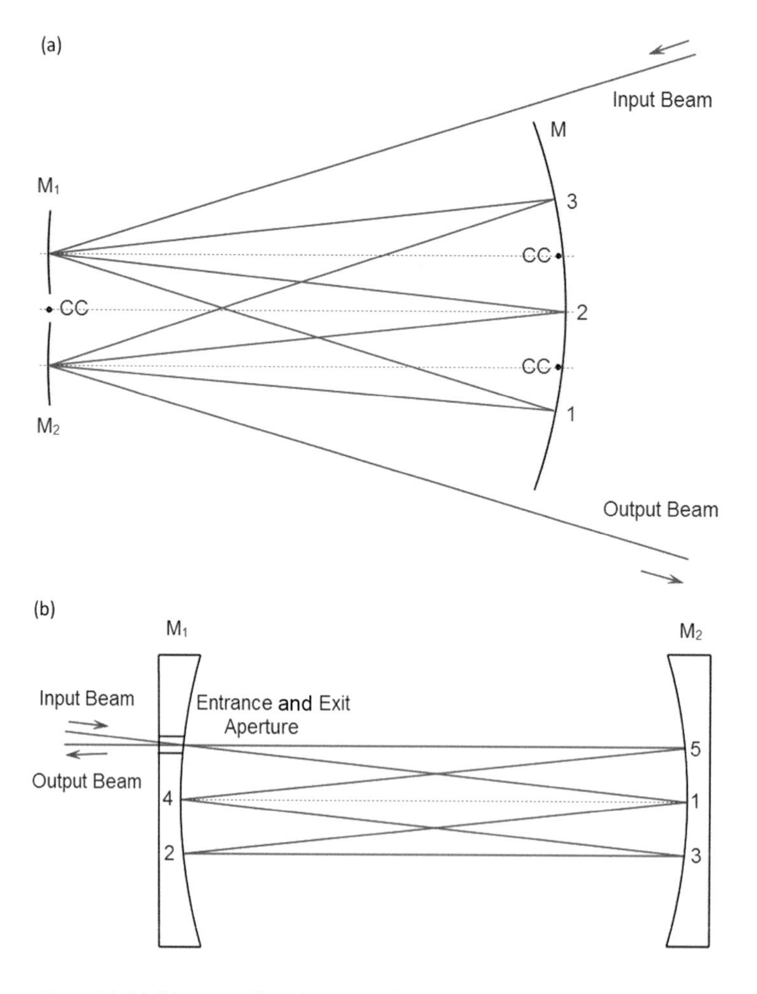

Figure 5.1 Multi-pass cell designs: (a) eight-pass White cell configuration (after [6]), (b) six-pass Herriott cell (after [9]).

As shown in Figure 5.1a, the mirrors are arranged so that the centre of curvature (CC) of mirrors M_1 and M_2 are on the surface of M and the centre of curvature of M is midway between M_1 and M_2. This configuration ensures that light emerging from any point on M_1 and reflecting from M is brought to a focus at the corresponding point on M_2 and is then returned to the original point on M_1 on the next refection from M. If M_1 and M_2 are positioned symmetrically with respect to M, then the ratio L_M/S_C determines the number of passes through the cell, in multiples of four, where L_M is the length of M and S_C is the separation between the centres of curvature of mirrors M_1 and M_2, which can be adjusted as required.

The basic Herriott cell, illustrated in Figure 5.1b, is formed from two spherical mirrors with the same radius of curvature, f, separated by a distance d where $f < d < 2f$. The beam enters through a hole in one of the mirrors and, after the prescribed number of passes that satisfies the closure condition [9], the beam exits from the same hole, but at a different angle from the input beam. Adjustment of the mirror separation changes the number of passes that satisfy the closure condition and also the elliptical pattern of the beam spots on the mirrors. In the astigmatic version of the Herriott cell [10], the mirrors have different radii of curvature, giving the advantage that the beam spot patterns on the mirrors are more evenly distributed over the mirror surfaces with longer path lengths possible for a given cell volume. Custom-designed Herriot cells are available from a number of commercial sources [11, 12].

5.2.2 Micro-Optic Cells

For fibre optic systems, especially for multi-point sensors where, as discussed later, a fibre optic network is employed to monitor gas concentrations at a large number of locations, it is convenient to use micro-optic gas cells which can readily be incorporated into the network through standard fibre optic connectors. The typical construction of such a micro-optic cell is illustrated in Figure 5.2.

As shown in Figure 5.2a, an optical fibre is located and aligned to a GRIN lens by the use of a commercially available capillary tube matched to standard fibre dimensions and the GRIN lens diameter. The input GRIN lens collimates the divergent output beam from the fibre through an open pathlength of typically 5 cm, which is then focussed into the output fibre by the output GRIN lens. As illustrated in Figure 5.2b, the capillaries and GRIN lenses are mounted in a V-groove which ensures simple alignment of the components, with fine adjustment by rotation of the GRIN lenses. Although not shown in Figure 5.2 for clarity, the interfaces between the capillaries and GRIN lenses are in physical contact or index-matched bonded to avoid introducing interference effects at the interfaces. For 5 cm path lengths, insertion losses of <1 dB are readily obtained. In practice, the micro-optic cell is accommodated within a protective housing arrangement with suitable filters which allow gas access, but protect the optics from water and dust ingress. Fibre-coupled reference cells with standard connectors are available commercially [13] with path lengths ranging from 3–16 cm and with multi-pass folded optics for long path lengths up to 80 cm.

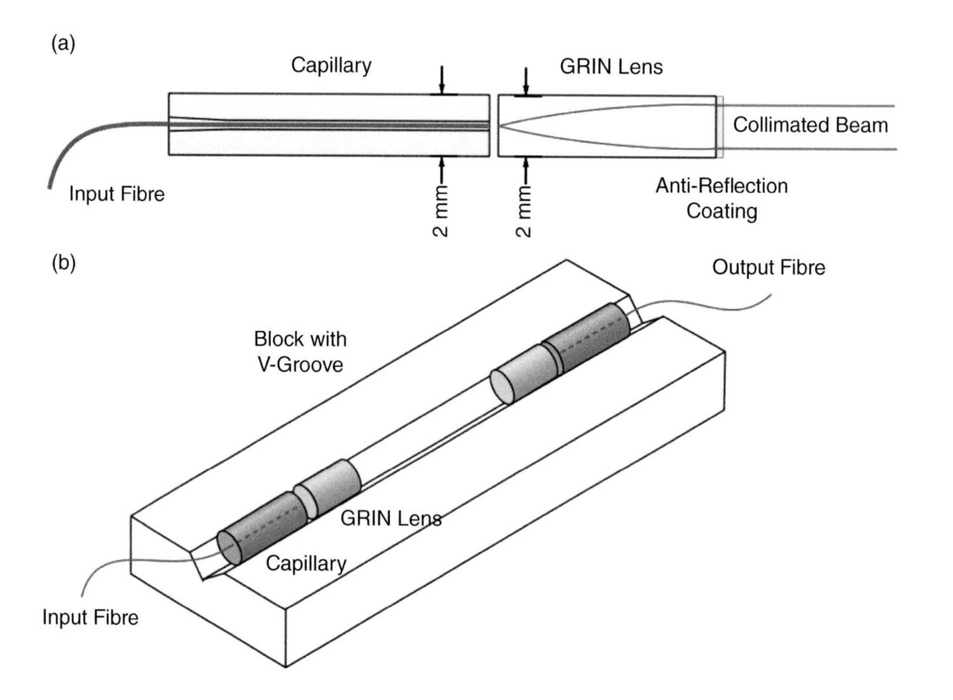

Figure 5.2 Typical construction of a micro-optic gas cell: (a) capillary for mounting and aligning the fibre to the GRIN lens, showing typical dimensions, (b) use of a V-groove block to align and form the open path cell.

5.2.3 Etalon Fringe Reduction

In system operation with the above gas cells, etalon fringes may arise due to interference from reflections within the cell or from connectors or other components within a fibre optic network. These fringes may completely swamp the desired gas absorption signal, so it is very important to take appropriate steps to minimise their magnitude. Consider the example of the micro-optic cell [14] where reflections can occur from the front surfaces of the GRIN lenses, as illustrated in Figure 5.3a.

Considering Figure 5.3a, we assume that, with anti-reflection coatings on the GRIN lenses, the reflection coefficients are small, $r_1 \simeq r_2 = r \ll 1$, and the transmission coefficient at the surfaces is approximately unity. Hence we only need to consider the first reflections from the GRIN lens surfaces. We also assume that the gas absorption in the cell is small and has a negligible effect on the reflected component. Hence the output electric field from the cell that gives rise to the etalon fringes is described by:

$$E_T(t) \cong E(t) + E_R(t) = E\mathrm{e}^{j2\pi\nu t} + (\eta r)^2 E\mathrm{e}^{j2\pi\nu(t-\tau)} \tag{5.1}$$

where η represents the (single-pass) attenuation of the electric field from the less than perfect collimation of the cell, ν is the optical frequency and τ is the round trip delay of the cell, $\tau = 2l/c$.

The total output power can be calculated from the time-average $\langle E_T \cdot E_T^* \rangle$ giving:

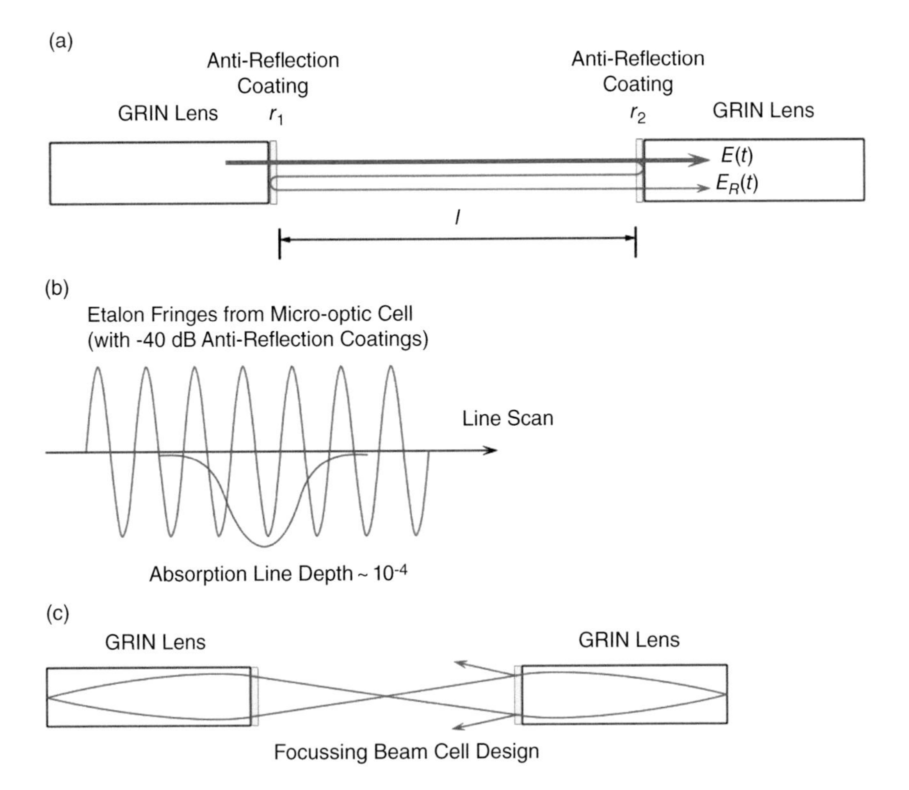

Figure 5.3 (a) Reflections in a gas cell which give rise to etalon fringes, (b) fringe pattern with –40 dB anti-reflection coatings compared with an absorption line with line centre absorbance of 10^{-4}, (c) focussing cell design to reduce the magnitude of the etalon effects.

$$P_T(\nu) \cong P\left[1 + 2\eta^2 r^2 \cos\left(\frac{2\pi\nu}{\Delta\nu_c}\right)\right] \tag{5.2}$$

where $\Delta\nu_c = 1/\tau = c/2l$ is the etalon fringe frequency spacing (or FSR) of the cell.

For a typical micro-optic cell with an open-space path length of ~5 cm, the fringe spacing is 3 GHz, which is similar to the linewidth of pressure-broadened gas absorption lines at atmospheric pressure. Thermal or mechanical disturbance of the cell will cause the fringes to drift across the field of view. Scanning across a gas absorption line will generate a fringe pattern superimposed on the absorption line profile, as illustrated in Figure 5.3b. Comparing the magnitude of the fringes to a gas absorption line with a line centre absorbance of $\sim 10^{-4}$ we see from (5.2) that they are comparable when $2\eta^2 r^2 \approx 10^{-4}$. Hence, with a typical 1 dB collimation transmission loss for the cell ($\eta^2 \approx 0.8$), the anti-reflection coatings on the GRIN lenses must reduce the reflectance (r^2) to better than –40 dB. Alternatively, the fringe magnitude can be reduced by reduction of η^2 for the reflected light using the focussing cell design illustrated in Figure 5.3c, where the back-reflected light from the GRIN lenses is now divergent.

When the method of wavelength modulation spectroscopy (WMS) is used, then harmonics of the modulation frequency are generated from the etalon fringes, similar to the process of generating harmonics from the absorption line, as explained in Chapters 3 and 4. From (3.4) and (3.5) in Chapter 3 and (A4.2.2) in Appendix 4.2 of Chapter 4, with WMS, the DFB laser power and optical frequency are modulated according to:

$$P_l(\nu,t) = P(\nu) + \Delta P(\nu)\cos(\omega_m t) = P(\nu)\left\{1 + \frac{\kappa}{2}\left(e^{j\omega_m t} + e^{-j\omega_m t}\right)\right\} \tag{5.3}$$

$$\nu_l = \nu + \delta\nu \cos(\omega_m t + \psi_f) \tag{5.4}$$

where κ is the intensity modulation index and $\delta\nu$ is the amplitude of the optical frequency modulation.

Substituting these expressions in (5.2) gives:

$$P_T(\nu,t) \cong P(\nu)\left\{1 + \frac{\kappa}{2}\left(e^{j\omega_m t} + e^{-j\omega_m t}\right)\right\}\left[1 + 2\eta^2 r^2 \cos\left(\theta + \beta\cos\left(\omega_m t + \psi_f\right)\right)\right] \tag{5.5}$$

where $\theta = 2\pi\nu/\Delta\nu_c$ and $\beta = 2\pi\delta\nu/\Delta\nu_c$.

Following the same procedures as the analysis given in Chapters 3 and 4, we can expand (5.5) into its harmonic components using the Bessel function expansion:

$$e^{j\{\theta + \beta\cos(\omega t + \psi)\}} = e^{j\theta}\sum_{n=-\infty}^{\infty} J_n(\beta) e^{jn\left(\omega t + \psi + \frac{\pi}{2}\right)} \tag{5.6}$$

with $J_{-n}(\beta) = (-1)^n J_n(\beta)$.

For example, the second harmonic arising from the etalon fringes can then be obtained in phasor notation, after some algebraic manipulation, as:

$$h_2(\nu,t) = -4\eta^2 r^2 P(\nu)\left\{\frac{\kappa}{2} J_1(\beta)\sin(\theta)e^{j\psi_f} + J_2(\beta)\cos(\theta)e^{j2\psi_f} - \frac{\kappa}{2} J_3(\beta)\sin(\theta)e^{j3\psi_f}\right\}\cdot e^{j2\omega_m t} \tag{5.7}$$

This expression is analogous to that obtained from gas absorption lines, see (3.20) in Chapter 3 and (4.32) in Chapter 4. Note that the sine and cosine terms in (5.7) describe the periodic fringe ripple across a wavelength scan, while the fringe magnitude is determined by the Bessel functions, which lie in the range $0.6 > J_n(x) > -0.4$ for $n > 0$. The Bessel functions are zero at certain values of β so the fringe magnitude can be reduced (but not eliminated) by choice of the amplitude, $\delta\nu$, of the optical frequency modulation.

Although we have specifically considered the case of a micro-optic cell in the above analysis, the results apply to any source of interference arising from reflections within a fibre optic system. There may in fact be several sources of reflections giving rise to multiple fringe patterns of different magnitudes and periodicities. If necessary, data processing algorithms and adaptive filtering may be used to remove residual ripples.

5.3 High-Finesse Cells for Sensitivity Enhancement

In contrast to the previous discussion where it is desirable to minimise interference effects, there are various techniques which make use of high-finesse gas cells with very high reflectance mirrors, close to unity, to enhance the effective path length of the cell [4, 15–37]. These methods include ring-down spectroscopy and the various forms of cavity-enhanced spectroscopy, which we shall review in the next sections.

5.3.1 Ring-Down Spectroscopy

The basic principles behind cavity ring-down spectroscopy are illustrated in Figure 5.4. An input pulse, whose width is much less than the round-trip cavity time, τ, is launched into a high-finesse cell, as shown in Figure 5.4.

The directly transmitted pulse is $P_0 = (T_1T_2)P_{in}e^{-\alpha Cl}$ where T_1 and T_2 are the transmittances of the mirrors and the exponential function represents the attenuation from the gas absorption in the cell, assuming the gas occupies the full cell length. The next pulse, arising from reflections from the mirrors is given by $P_1 = T_2R_1R_2(T_1P_{in}e^{-\alpha Cl})e^{-2\alpha Cl} = R_1R_2P_0e^{-2\alpha Cl}$

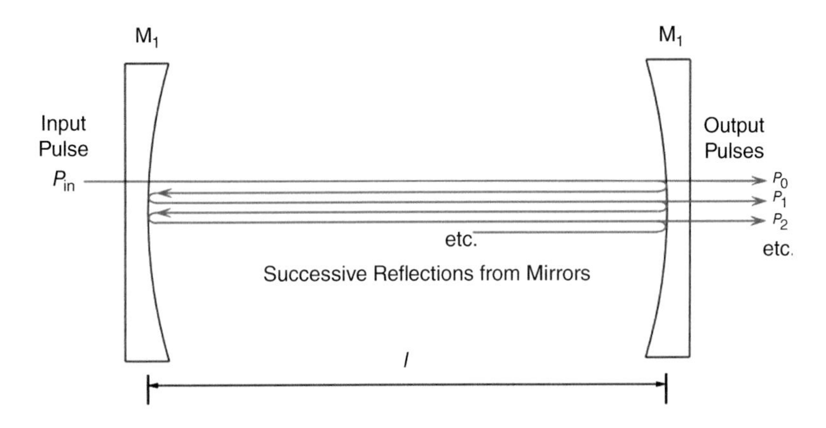

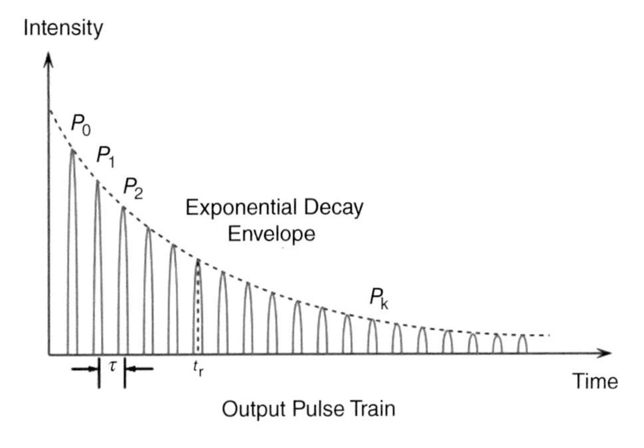

Figure 5.4 Ring-down of pulses launched into a high-finesse cell.

where R_1 and R_2 are the reflectances of the mirrors. Similarly, the next pulse is given by $P_2 = (R_1R_2)^2P_0e^{-4\alpha Cl}$ so that we can write an expression for the kth output pulse as:

$$P_k = (R_1R_2)^kP_0e^{-2k\alpha Cl} \tag{5.8}$$

The output pulse train is illustrated in Figure 5.4, where each output pulse is separated in time by the cavity round-trip time of $\tau = 2l/c$. The time taken to reach the kth pulse is $t = k\tau$, so replacing the discrete variable k in (5.8) by the continuous variable t/τ gives the envelope of the pulse train as:

$$P(t) = P_0 \exp\left\{-\frac{t}{\tau}(2\alpha Cl - \ln R_1R_2)\right\} \tag{5.9}$$

Hence the exponential decay has a characteristic decay time or 'ring-down time' of:

$$t_r = \frac{\tau}{(2\alpha Cl - \ln R_1R_2)} = \frac{l}{c\left(\alpha Cl - \ln\sqrt{R_1R_2}\right)} \tag{5.10}$$

Note that only the mirror reflectance has been taken into account in (5.10), but if there is background transmission loss or scattering loss in the cell, this must also be included in the denominator of (5.10). From (5.10) we see that the gas absorption causes a change in the ring-down time and measurement of this change is the basic principle behind cavity ring-down spectroscopy.

Viewed as a multi-pass cell, a length enhancement factor may be defined for the ring-down cell as the number of (one-way) passes in the cell in reaching the ring-down time, namely:

$$N = 2\left(\frac{t_r}{\tau}\right) = \frac{1}{\left(\alpha Cl - \ln\sqrt{R_1R_2}\right)} \tag{5.11}$$

For a large enhancement factor, the mirror reflectance must be close to unity and any background losses minimised. For $R_1 = R_2$ and close to unity, we can make the approximation $\ln\sqrt{R_1R_2} \cong -(1-R)$ in (5.11), so the cell length enhancement factor in the absence of gas is $N = 1/(1 - R)$, which is 1000 for a mirror reflectance of 99.9%. A cell of length 1 m has then an effective path length of 1 km with a ring-down time of 3.3 μs.

The fractional change in ring-down time as a result of a small change in gas absorbance, $\delta A = \alpha(\delta C)l$, can be obtained by differentiation of (5.10) as:

$$\frac{\delta t_r}{t_r} = N\delta A = (Nl)\alpha\delta C \tag{5.12}$$

The sensitivity, which may be defined as the minimum value of $\alpha\delta C$ that can be measured over a 1 s interval with a 1σ certainty and expressed as a noise-equivalent absorption coefficient (NEA), is [15, 17, 18]:

$$NEA = \frac{1}{Nl}\left(\frac{\delta t_r}{t_r}\right)\sqrt{\frac{2}{f_{rep}}}_{\min} \tag{5.13}$$

where f_{rep} is the data collection rate.

If, for example, with a data collection rate of ~200 MHz, a 1% change in ring-down can be reliably measured (through an exponential fit to the pulse decay envelope), then from (5.12) and (5.13) a resolution of $\delta A \sim 10^{-5}$ and a NEA of ~10^{-11} cm^{-1} $Hz^{-1/2}$ can be attained with a mirror reflectance of 99.9% and cell length of 1 m. In practice, other factors such as system drift and cavity mode coupling may limit the sensitivity. It is important to note that the enhancement factor and the sensitivity are dependent on maintaining the high mirror reflectance with freedom from contamination and condensation. Also, when the level of absorbance becomes significant compared with $(1 - R)$, the enhancement factor and resolution are reduced as indicated by (5.11) and (5.12).

Instead of pulsed excitation, there are alternative ways for interrogation of the cavity to determine the ring-down time. For example, the CW output of a laser may be coupled to a cavity mode and the laser output rapidly switched off to observe the exponential decay of power from the cavity. The decay may also be observed by arranging intermittent coupling to the mode(s) through a sweep of either the laser frequency or the cavity length by a piezo-electric transducer (PZT) to shift the cavity mode position [19, 20]. In phase-shift cavity ring-down spectroscopy [21], the CW laser input to the cavity is intensity-modulated at an angular frequency of ω_m. The ring-down time may then be measured from the phase shift, ϕ_{mr}, observed in the modulated output from the cavity. The phase shift is related to the ring-down time through the relationship, $\tan\phi_{mr} = \omega_m t_r$.

5.3.2 Cavity-Enhanced Spectroscopy

Cavity-enhanced absorption spectroscopy (CEAS) [16] uses a similar high-finesse absorption cell to that in ring-down spectroscopy, but differs by measuring the integrated intensity from the cell excited by a CW input, as illustrated in Figure 5.5 rather than the ring-down time with a pulsed input. The term integrating cavity output spectroscopy (ICOS) has also been used, particularly where integration is performed over a large number of cavity modes which are sequentially excited.

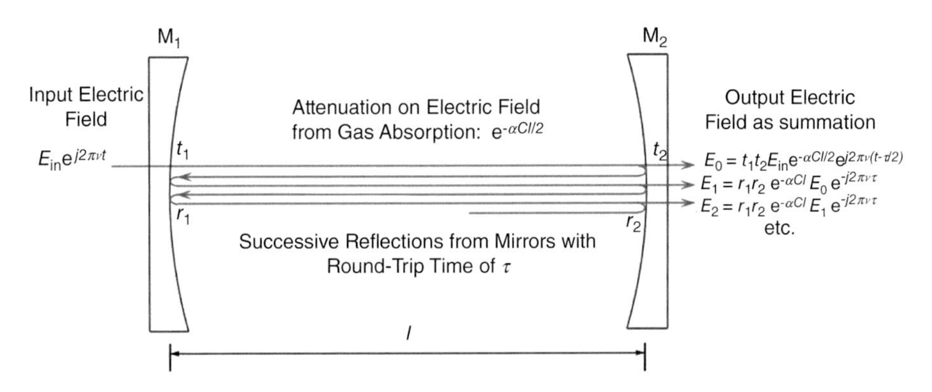

Figure 5.5 Cavity-enhanced absorption spectroscopy (CEAS).

With reference to Figure 5.5, the basic principle behind CEAS can be explained by considering the CW output electric field from the cell as a sum of delayed and attenuated waves as follows:

$$E_0(t) = t_1 t_2 E_{in} e^{-\frac{\alpha Cl}{2}} e^{j2\pi\nu\left(t-\frac{\tau}{2}\right)} \sum_{k=0}^{\infty} e^{-\frac{k\tau}{2t_r}} e^{-j2\pi\nu k\tau} \tag{5.14}$$

where t_1 and t_2 are the transmission coefficients at the mirrors. The attenuation of successive waves, delayed by the round-trip time, τ, is represented in terms of the ring-down time (note that since we are considering the electric field at this point and not the intensity, the attenuation terms include a factor of ½).

The geometric series of (5.14) has a sum to infinity given by:

$$E_0(t) = \frac{t_1 t_2 E_{in} e^{-\frac{\alpha Cl}{2}} e^{j2\pi\nu\left(t-\frac{\tau}{2}\right)}}{1 - e^{-\frac{\tau}{2t_r}} e^{-j2\pi\nu\tau}} \tag{5.15}$$

The output intensity or power can be obtained from the time average $\langle E \cdot E^* \rangle$ so the output power in terms of the input power is:

$$\frac{P_{out}}{P_{in}} = \frac{T_1 T_2 e^{-\alpha Cl}}{1 + e^{-\frac{\tau}{t_r}} - 2e^{-\frac{\tau}{2t_r}} \cos(2\pi\nu\tau)} \tag{5.16}$$

where $T_1 = t_1^2$ and $T_2 = t_2^2$ are the transmittances of the mirrors.

By using the trigonometric identity, $\cos 2\theta = 1 - 2\sin^2\theta$, this may be recast in the form:

$$\frac{P_{out}}{P_{in}} = \frac{T_1 T_2 e^{-\alpha Cl}}{\left(1 - e^{-\frac{\tau}{2t_r}}\right)^2} \left\{ \frac{1}{1 + F\sin^2(\pi\nu\tau)} \right\} \tag{5.17}$$

The expression in the curly brackets is the familiar Airy function for a Fabry–Perot type resonator [22], where F is the coefficient of finesse given, in this case, by:

$$F = \frac{4e^{-\frac{\tau}{2t_r}}}{\left(1 - e^{-\frac{\tau}{2t_r}}\right)^2} \cong 16\left(\frac{t_r}{\tau}\right)^2 \tag{5.18}$$

The linewidth of the resonance is obtained from (5.17) by noting the shift in frequency from the peak position where the response falls by a factor of 2, that is, when $F\sin^2\left(\pi\Delta\nu_{\frac{1}{2}}\tau\right) \cong F\left(\pi\Delta\nu_{\frac{1}{2}}\tau\right)^2 = 1$ where $\Delta\nu_{\frac{1}{2}}$ is the half-width-half-maximum (HWHM) linewidth.

An important parameter of the cavity is the finesse, which is defined as the ratio of the free spectral range (FSR $= c/2l = 1/\tau$) to the full-width-half-maximum (FWHM) linewidth of the resonance. Making use of (5.18), the finesse is:

$$\mathcal{F} = \frac{\text{FSR}}{\text{FWHM}} = \frac{1}{\tau(2\nu_{1/2})} \cong \frac{\pi\sqrt{F}}{2} \cong \frac{2\pi t_r}{\tau} \tag{5.19}$$

The transmission peaks of the cell occur when $\nu\tau = 0, 1, 2, \ldots$, that is, on the resonances when the cell length is an integer number of wavelengths.

Using the expression (5.10) for the ring-down time, the output power on the transmission peaks from (5.17) and (5.18) is given by:

$$\frac{P_{out}}{P_{in}} = \frac{T_1 T_2 e^{-\alpha Cl}}{\left(1 - \sqrt{R_1 R_2} e^{-\alpha Cl}\right)^2} \tag{5.20}$$

The effective path length at a transmission peak can be deduced from (5.20) by deriving the change in output power as a result of the absorption. If we assume that the absorption is small, $\alpha Cl \ll 1$, and use the approximation that $e^{-x} \cong 1 - x$, we obtain:

$$\frac{P_{in} - P_{out}}{P_{in}} \cong \frac{\alpha Cl(1 - R_1 R_2) + R_1 \left(1 - \sqrt{\frac{R_2}{R_1}}\right)^2}{\left(1 - \sqrt{R_1 R_2}\right)^2} \tag{5.21}$$

where we have also assumed loss-less mirrors so that $T = 1 - R$ for both mirrors.

If the mirror reflectances are similar, so that $R_1 \approx R_2 = R$ and close to unity, (5.21) further simplifies to:

$$\frac{P_{in} - P_{out}}{P_{in}} \cong \frac{\alpha Cl\left(1 + \sqrt{R_1 R_2}\right)}{\left(1 - \sqrt{R_1 R_2}\right)} \approx \frac{2\alpha Cl}{(1 - R)} \tag{5.22}$$

The approximation in (5.22) shows that the effective cell length is $2l/(1 - R)$ which can be very large when R is close to unity.

Of course the above analysis is based on the idealised situation where the laser wavelength is perfectly coupled or locked to a single cavity mode and for a sufficiently long time (i.e. longer than the ring-down time) so as to allow the build-up of successive reflections in the cavity to produce the Airy function distribution. Electronic locking of the laser frequency to a cavity mode may be performed by the Pound–Drever–Hall method [23], but there are, however, a number of issues in using this method in practice, particularly where absorption line scanning and WMS techniques are required. We shall return to these issues in Section 5.3.5, but first we consider two important alternative techniques to avoid the complexities of external locking.

5.3.3 Off-Axis Cavity-Enhanced Spectroscopy

Instead of locking the laser wavelength to a cavity mode, a more practical approach is to rapidly sweep the laser wavelength or cavity modes (or both) with respect to each other, adding successive scans to improve the SNR. However, the rapid sweep means that an individual mode response no longer follows an Airy function because there is insufficient time for the summation described by (5.14) to be fully completed at a given wavelength point. For example, sweeping the laser output over a wavelength range of 0.2 nm (~25 GHz) at a typical rate of 500 GHz per second gives a sweep time of 50 ms. The FSR of a 1 m cavity is 150 MHz, so ~170 modes are scanned in 50 ms. For the cavity round-trip time of 7 ns and a typical ring-down time of 3 μs, the cavity finesse from (5.19) is ~2700, so the mode linewidth is ~55 kHz and hence the time spent at each mode is ~0.1 μs, much less than the ring-down time or build-up time for

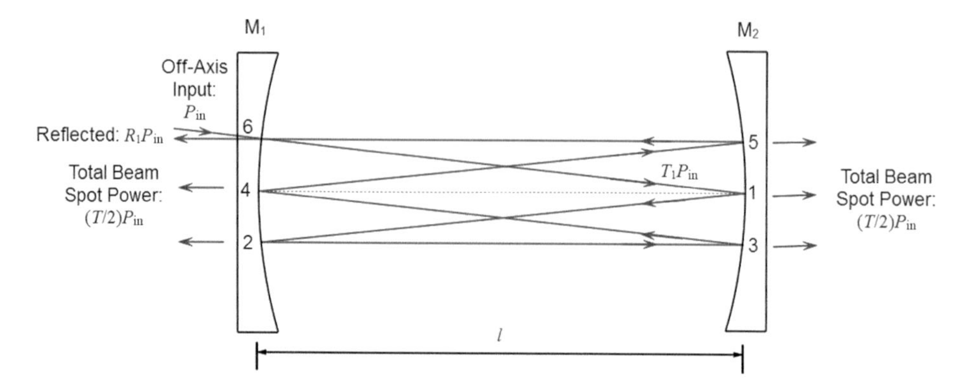

Figure 5.6 Off-axis cavity-enhanced cell for increasing mode density.

the resonance condition. Indeed, for this situation of sweeping across multiple modes, it is desirable to simply 'wash out' the mode structure or modal noise that arises in the output intensity. A convenient way of achieving this is to greatly increase the mode density through the method of off-axis cavity-enhanced absorption spectroscopy (OA-CEAS) [24]. Here, instead of the laser input beam being directed along the cell axis, it is aligned off-axis to create a multi-pass beam in the cell (equivalent to the excitation of transverse modes of the resonator), as illustrated in Figure 5.6, similar to the Herriott cell of Figure 5.1b but without the entrance/exit aperture.

The beam retraces its original path after the prescribed number of passes that satisfies the closure or re-entrant condition (six passes shown in Figure 5.6). As noted earlier, the number of passes that satisfy the re-entrant condition is dependent on the mirror separation and can be increased to more than 1000 by use of astigmatic mirrors. Hence the round-trip time of the cavity, i.e. the time taken for the beam to retrace its path, can be made very large and consequently the FSR is greatly reduced, creating a very dense mode spectrum. When the FSR becomes comparable to the mode width, the mode spectrum is effectively suppressed, with the cavity exhibiting a virtually uniform optical frequency response to laser modulation and scanning, as performed with WMS techniques.

For this case, where interference from successive beams can be ignored, the output from the cavity can be derived as an intensity sum without the need for the phase terms in (5.14). For the example of the six-pass off-axis cavity configuration shown in Figure 5.6, the output intensity from spot 1 on mirror 2 can be written as the intensity sum of the first transmitted beam and the subsequent contributions from the reflected beams (which make six passes in the cell before retracing their path):

$$\begin{aligned} P_1 &= T_1 T_2 P_{\text{in}} \mathrm{e}^{-\alpha Cl} + T_1 T_2 P_{\text{in}} \mathrm{e}^{-\alpha Cl} (R_1 R_2)^3 \mathrm{e}^{-6\alpha Cl} + \cdots \\ &= T_1 T_2 P_{\text{in}} \mathrm{e}^{-\alpha Cl} \sum_{k=0}^{\infty} (R_1 R_2)^{3k} \mathrm{e}^{-6k\alpha Cl} \end{aligned} \tag{5.23}$$

This may be generalised to an N-pass cell:

$$\frac{P_1}{P_{\text{in}}} = T_1 T_2 \mathrm{e}^{-\alpha Cl} \sum_{k=0}^{\infty} (R_1 R_2)^{\frac{Nk}{2}} \mathrm{e}^{-Nk\alpha Cl} = \frac{T_1 T_2 \mathrm{e}^{-\alpha Cl}}{1 - (R_1 R_2)^{\frac{N}{2}} \mathrm{e}^{-N\alpha Cl}} \tag{5.24}$$

For the mirror reflectance close to unity, the same expression applies for each of the $N/2$ spots on mirror 2. Hence if the beam spot pattern (which may be a circle, ellipse or Lissajous-type figure with astigmatic mirrors) is focussed onto a detector, the total output power from mirror 2 is:

$$\frac{P_{out}}{P_{in}} = \frac{NT_1T_2e^{-\alpha Cl}}{2\left\{1-(R_1R_2)^{N/2}e^{-N\alpha Cl}\right\}} \tag{5.25}$$

Consider the case of lossless mirrors with similar reflectance, $R_1 \cong R_2 = R$ and $R = 1 - T$. If the reflectance is close to unity, we can make the approximations $R^N = (1 - T)^N \simeq 1 - NT$ and $(1 - R^N) \simeq NT = N(1 - R)$. With these approximations and if the absorbance is small so that $e^{-N\alpha Cl} \approx (1 - N\alpha Cl)$, we can write (5.25) as:

$$\frac{P_{out}}{P_{in}} \simeq \left(\frac{T}{2}\right)\frac{T(1-\alpha Cl)}{\{T+R^N(\alpha Cl)\}} \tag{5.26}$$

Considering first the case of no absorbance, it is evident from (5.26) that the power collected from mirror 2 is $(T/2)P_{in}$. Compared with (5.20) for on-axis CEAS where $P_{out} \cong P_{in}$ for no absorbance when locked to a cavity mode, we see that the power output with OA-CEAS is much reduced, which is the key disadvantage of this approach. Note that (5.26) also applies to the power collected from the beam spots on mirror 1, but there is also a strong back-reflected beam of RP_{in} at the incident point on mirror 1. Hence, as expected, the total power emanating from the lossless cavity is equal to the input power.

We can now calculate the sensitivity of the output to small changes in the absorbance by differentiating (5.26) with respect to (αCl). This gives:

$$\frac{\delta P_{out}}{P_{out}} \simeq \frac{T(T+R^N)}{\{T+R^N(\alpha Cl)\}^2}\cdot\delta(\alpha Cl) \tag{5.27}$$

The sensitivity is dependent on the level of absorbance, but at small absorbance $\alpha Cl \to 0$ this becomes:

$$\frac{\delta P_{out}}{P_{out}} \simeq \frac{(T+R^N)}{T}\cdot\delta(\alpha Cl) \approx \frac{\alpha Cl}{(1-R)} \tag{5.28}$$

Comparing (5.28) with (5.22), we see that the cell length enhancement factor is only reduced by a factor of 2 compared with on-axis CEAS. However, OA-CEAS has the key advantages that WMS techniques can be readily used and it is more suitable for field deployment since there is no need to lock the laser wavelength to a mode of the cell. For example, Engel et al. [25] have demonstrated a sensitivity of 1.9×10^{-12} cm^{-1} Hz$^{1/2}$ using this technique for CO monitoring at a wavelength of 1.57 μm with a 1.1 m cavity and a 30 mW DFB laser source. The main disadvantages of OA-CEAS are the much reduced power output from the cavity and the limitations on the data collection rate as a result of the long round-trip time of the cavity.

5.3.4 Optical Feedback Cavity-Enhanced Spectroscopy

Another very important way of avoiding the complexities associated with external locking of the laser wavelength to a cavity mode, but also avoiding the problem of the low output power that occurs with OA-CEAS, is the method known as 'optical feedback cavity-enhanced spectroscopy' or OF-CEAS for short [26–28]. The technique is based on the principle of self-locking the laser frequency to a cavity mode by controlled optical feedback from the cavity. Typically a V-shaped optical cavity is used, as illustrated in Figure 5.7, which ensures that the direct reflection at mirror M_1 from the laser input is not returned to perturb the laser while a controlled part of the resonant field in the cavity is fed back to the laser. The V-shaped cavity also has the practical advantage that a long baseline cavity can be accommodated within a relatively small volume.

Important parameters for effective self-locking [26–28] are the feedback phase, controlled by the PZT as shown in Figure 5.7, and the power fraction returned, controlled by a variable attenuator or isolator. Self-locking with a high-finesse cavity also has the desirable effect of narrowing the linewidth of the laser emission since the optical feedback acts as injection seeding of the laser from the very narrow linewidth cavity resonance.

When the DFB laser wavelength is tuned by a current ramp over a range of several cavity modes, the laser output self-locks at the peak position of each mode in turn, giving a mode-by-mode spectrum at the precisely defined optical frequencies of the modes.

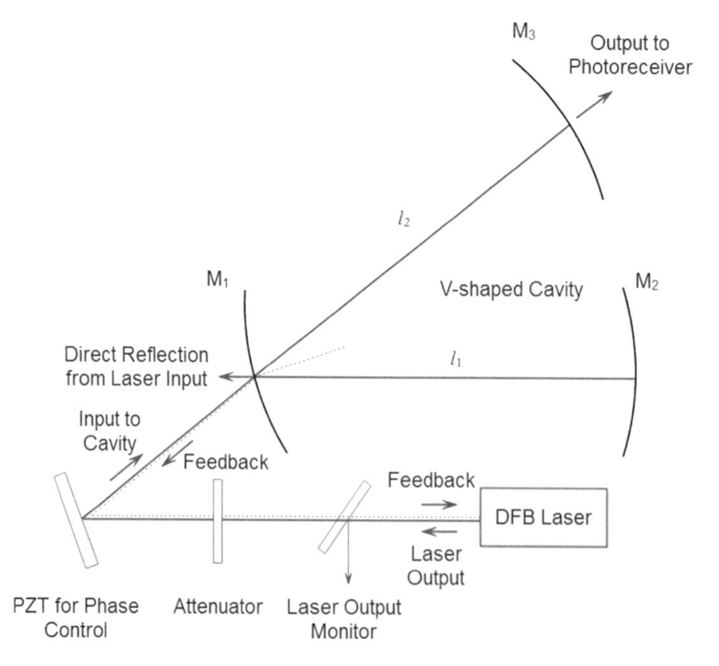

Figure 5.7 V-shaped cavity for optical feedback cavity-enhanced spectroscopy (OF-CEAS) (after [27]).

Note that the laser wavelength remains locked to a particular mode over a range of currents from the ramp before jumping to the next mode. This interval, over which the laser wavelength remains locked to a mode even as the ramp current is increased, is known as the 'locking range' and can be increased by increasing the feedback fraction. This effect can be understood by noting that, with the introduction of optical feedback, the external V-shaped cavity now forms part of the laser cavity so that the thermal and electronic current tuning mechanisms within the laser itself have less of an influence on the extended laser cavity, effectively retarding the tuning rate. Interestingly, this phenomenon was observed in 1985 by Uttam [29] in research on frequency-modulated continuous-wave ranging (FMCW) for optical time domain reflectometry, but was an undesirable effect in this context. Of course, the rate of the current ramp (tuning speed) needs to be sufficiently slow so as to allow build-up of the cavity resonance and hence provide sufficient feedback for self-locking at each mode position.

A number of demonstrations using OF-CEAS for trace gas monitoring in the near-IR (and mid-IR) have been performed in the literature and reviewed by Morville [30], with noise-equivalent absorption sensitivities (NEAS) down to ~10^{-10} cm^{-1} $Hz^{-1/2}$. The company ap2e [31] has exploited OF-CEAS for the development of commercial gas sensors.

5.3.5 Noise-Immune Cavity-Enhanced Optical Heterodyne Molecular Spectroscopy

Returning to the situation discussed in Section 5.3.2, where the laser wavelength is externally locked to a cavity mode by, for example, the Pound–Drever–Hall method [23], a way of achieving very high sensitivity, with noise levels approaching the intrinsic AM laser noise, is the method of 'noise-immune cavity-enhanced optical heterodyne molecular spectroscopy' or NICE-OHMS [32–37]. This method combines the enhancement from a high-finesse cavity with the advantages of frequency modulation spectroscopy, while achieving immunity to laser frequency noise. The principles behind the method are illustrated schematically in Figure 5.8 and can be explained as follows.

First the laser frequency is tightly locked to a cavity mode by the Pound–Drever–Hall method [23] where low-frequency modulation of the optical frequency (or phase) of the laser around the peak of the cavity mode generates an AM error signal for an electronic feedback loop to hold the laser wavelength at the peak of the mode. With appropriate PZT actuators on the cavity, the position of the mode can be scanned over a limited spectral range (for example over an absorption line), with the laser wavelength tracking the mode if it is tightly locked to it [36]. There are, however, two potential problems with this basic set-up. First, a very high-finesse optical cavity, where the mode linewidth is extremely narrow, is a very efficient converter of laser frequency noise to amplitude noise, so any residual frequency-locking noise from the laser-cavity lock can severely degrade the noise performance. Second, with use of WMS where the modulation frequency is relatively low, the sidebands are coupled to the same mode and the coupling is greatly reduced if the WMS modulation frequency is comparable or greater than the linewidth of the mode.

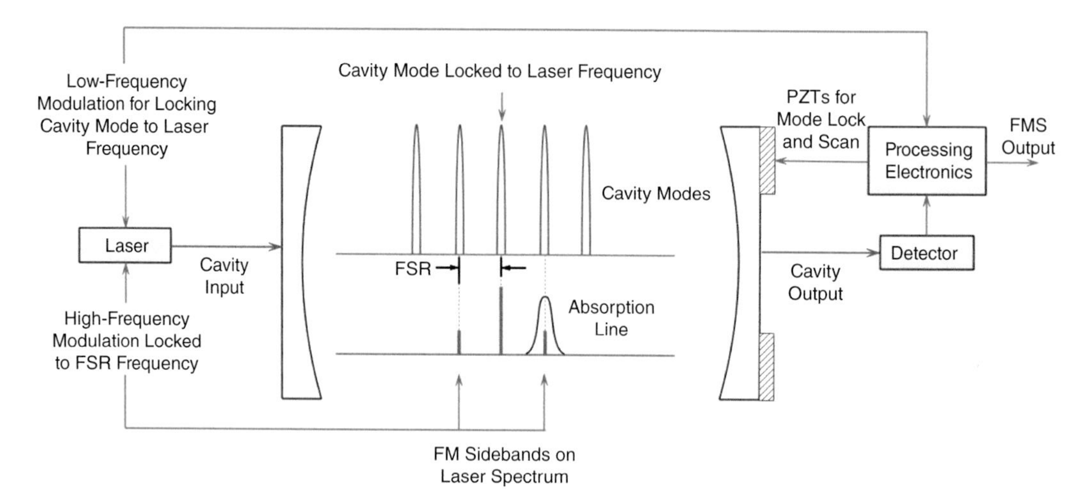

Figure 5.8 Schematic of NICE-OHMS method.

The NICE-OHMS method provides a solution to both of these problems by using frequency modulation spectroscopy (FMS), where the high modulation frequency is exactly matched to the FSR (or a multiple of the FSR) of the cavity. In this way, the two sidebands (only two sidebands are significant for FMS where the modulation amplitude is low) are efficiently coupled to cavity modes in the same way as the central carrier frequency is coupled to the central mode, as shown in Figure 5.8. Also, all three components (carrier plus two sidebands) experience the same residual frequency shifts/noise from the laser-cavity lock and hence there is no contribution to AM noise from this source on the de-modulated signal from the detector at the FMS modulation frequency (detected in the usual way by phase-sensitive demodulation). When an absorption line of a gas is present within the cavity, only one component (either the carrier or one of the sidebands) is affected, which is then attenuated by the gas absorption (or slightly frequency-shifted by the change in refractive index by gas dispersion) giving a resultant FMS signal at the detector representing the gas absorption (or dispersion). With this method, sensitivities close to the shot noise limits have been demonstrated, down to ~10^{-13} cm^{-1} over a 1 s averaging time and an NEA of ~10^{-11} cm^{-1} $Hz^{-1/2}$ [32–37].

5.4 Optical Fibre and Waveguide Gas Cells

In this section we will consider some special types and designs of optical fibre or waveguide which can be used as a gas cell without the need to connect to an external cell.

5.4.1 Evanescent-Wave Cells

The evanescent field refers to the exponentially decaying electric field distribution from the higher-index core region into the lower-index cladding region of a waveguide.

Early work [38–43] considered the use of the evanescent field of standard optical fibres or waveguides as a way of directly interacting with a gaseous species. For an optical fibre, the evanescent field can be accessed by polishing away the cladding until the core is reached, but this is generally limited to an interaction length of a few centimetres [44]. Specialist D-shaped fibre, as illustrated in Figure 5.9a, can be made at the fibre fabrication stage, potentially giving interaction lengths of tens or hundreds of metres

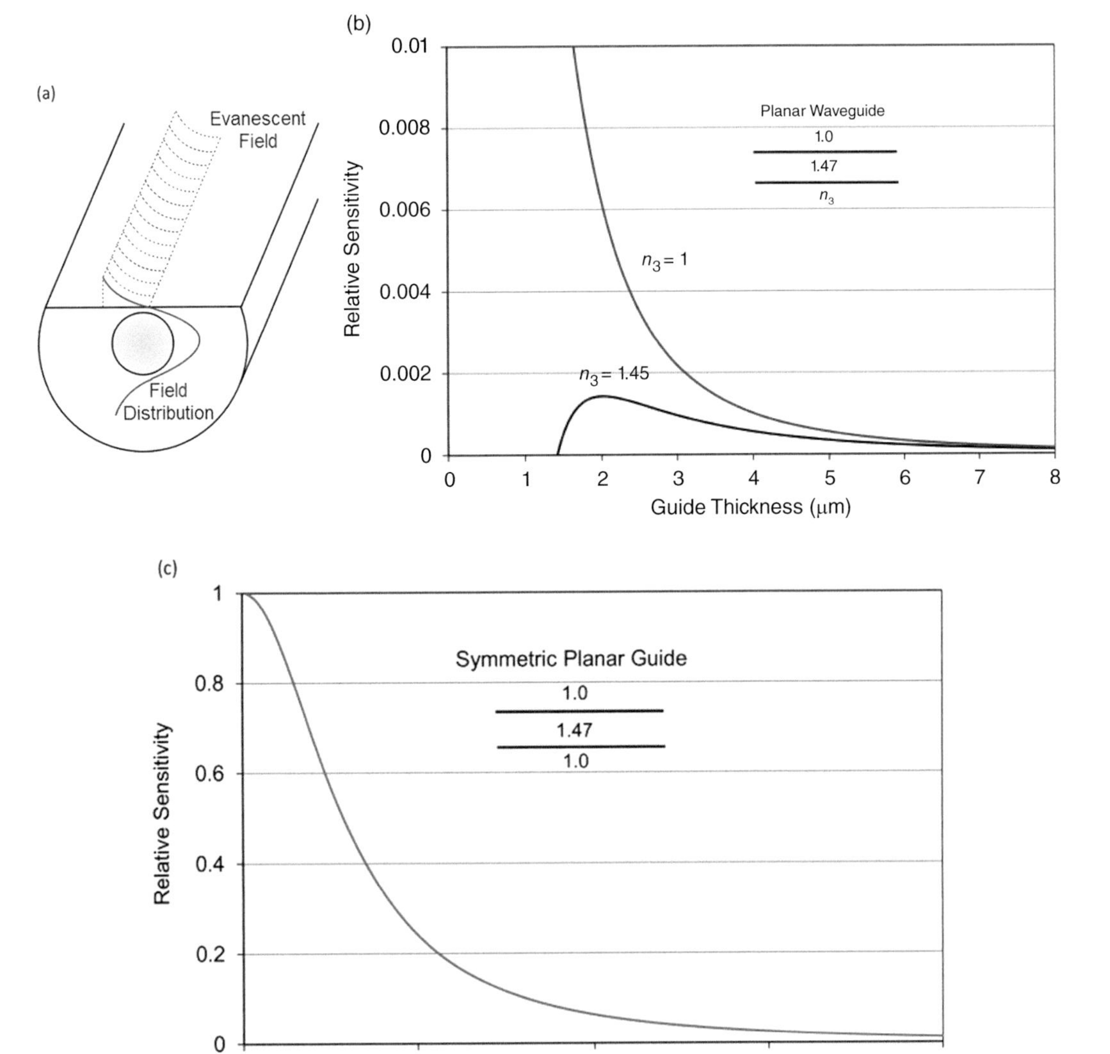

Figure 5.9 (a) Polished or D-shaped fibre allowing access to the evanescent field. (b) Theoretical relative sensitivity of a planar silica guide to gas absorption as a function of guide thickness for symmetric and asymmetric cases. (c) Theoretical sensitivity of a symmetric planar guide as the thickness is reduced to zero (assuming gaseous absorber on both sides).

in a compact fibre coil. However, apart from practical considerations of availability, cost and non-standard splicing, there are a number of challenges with using D-shaped or polished fibre for evanescent-wave interaction with gas absorbers.

First, the sensitivity is directly proportional to the fraction of the optical power carried in the evanescent field (see Appendix 5.1). This fraction is small in an asymmetric silica waveguide structure due, in part, to the significant index difference between the core region ($n \sim 1.45$–1.47) and the gas ($n \sim 1$), which results in a low penetration depth of the evanescent field into the gaseous region. The trends in the relative sensitivity, Γ_e, of evanescent field interaction compared with a direct absorption cell can be simply deduced from an asymmetric planar waveguide model, where an analytical expression can be obtained for TE modes as (see Appendix 5.1):

$$\Gamma_e = \left(\frac{n_1}{n_e}\right)\frac{1}{k_0 d_e\sqrt{n_e^2 - n_1^2}}\left(\frac{n_2^2 - n_e^2}{n_2^2 - n_1^2}\right) \tag{5.29}$$

where n_1, n_2 and n_3 are the superstrate, guide and substrate indices, respectively, n_e is the effective index of the guided mode ($n_1 < n_e < n_2$), $k_0 = 2\pi/\lambda_0$ and the effective depth, d_e, of the guided mode is obtained by adding the penetration depths of the evanescent fields into the superstrate, and substrate layers to the guide layer thickness:

$$d_e = d + d_{p1} + d_{p3} \tag{5.30}$$

The penetration depths (or $1/e$ depths) of the evanescent field into the superstrate and substrate layers are given by:

$$d_{p1} = \frac{1}{k_0\sqrt{n_e^2 - n_1^2}};\quad d_{p3} = \frac{1}{k_0\sqrt{n_e^2 - n_3^2}} \tag{5.31}$$

The guide thickness and effective index are related through the eigenvalue equation for TE modes in a planar guide as:

$$k_0 d\sqrt{n_2^2 - n_e^2} = m\pi + \tan^{-1}\sqrt{\frac{n_e^2 - n_1^2}{n_2^2 - n_e^2}} + \tan^{-1}\sqrt{\frac{n_e^2 - n_3^2}{n_2^2 - n_e^2}} \tag{5.32}$$

Figure 5.9b shows the relative sensitivity of the fundamental mode at 1600 nm wavelength in an asymmetric guide as a function of guide thickness using (5.29) to (5.32) for a gas absorber ($n_1 = 1$) with a core index of $n_2 = 1.47$ and substrate index $n_3 = 1.45$ compared with a symmetric guide of substrate index $n_3 = 1$. Note that, while the relative sensitivity of the symmetric guide continues to increase as the thickness is reduced, for the asymmetric structure it reaches a peak and then falls to zero. This is a consequence of the fact that as the guide thickness is reduced, the mode approaches cut-off into the substrate region, that is, $n_e \to n_3$ (or equivalently, from a ray optics viewpoint, the ray approaches the critical angle at the core/substrate interface). As this condition is approached, the evanescent field penetration into the substrate grows very large, as indicated by (5.31), while the power fraction in the superstrate falls to zero. Figure 5.9b indicates that the best relative sensitivity of an asymmetric silica guide is only ~0.1–0.2%. Numerical modelling of D-shaped fibre [39] gives a similar result of

0.1–0.2% relative sensitivity, so a 5–10 m interaction length would be required to equal that of a 1 cm open-path length.

Steps can be taken to increase the relative sensitivity of silica fibres or guides. The power fraction in the evanescent field of a D-shaped or polished fibre can be increased by coating the surface with a high-index film [38, 42], which has the effect of 'pulling' the electric field distribution toward the higher-index region and hence increasing the field strength at the surface. Dip-coating with a TiO_2-SiO_2 sol-gel film during fibre fabrication has been considered for this purpose [45], typically giving an improvement factor of up to ~10 with a ~200 nm thickness and index of 1.8, but this introduces additional fabrication challenges, especially if long lengths are to be uniformly coated. Another method is to greatly reduce the diameter of a standard clad optical fibre (by heat pulling) until the evanescent field extends sufficiently into the surrounding region (see the example in Chapter 7, Section 7.3.5 on evanescent-wave spectroscopy with fibre laser combs). The sensitivity improvement follows from the simple symmetric planar model of Figure 5.9c, where the relative sensitivity approaches 100% as the thickness is reduced to zero (assuming the gas surrounds the guide or fibre). However, only short lengths are practical and such a small diameter fibre is extremely fragile.

A second major problem with evanescent field sensors is contamination on the surface. The penetration depth (5.31) into an air or gas region (n_1 ~ 1) is typically only ~250 nm in the near-IR and so contamination or condensation on the surface can readily reduce or prevent interaction of the evanescent field with the gas. In practice, careful protection of the sensor in the packaging is required, such as by the use of filters, gas-permeable membranes and hydrophobic coatings. Some forms of contamination may make their presence known by their effect on the overall transmission, but liquid condensation on the surface may be virtually invisible. For this latter case, a method for compensation with D-fibre has been investigated, based on the principle that water condensation on the surface causes a change in the birefringence of the fibre [40, 41], which in turn changes the optical path length difference (OPD) between the two orthogonal polarisation modes if launched into the fibre. With an input and output polariser, a polarimetric interferometer can be formed whose OPD may be monitored by measuring the fringe spacing through, for example, white light interferometry [41]. However, this adds a degree of complexity to the system and the method is limited in its scope and range of operation – clearly the sensor becomes ineffective if the evanescent field is totally obscured. Note also that the significant birefringence of D-fibre (typically $B \sim 3 \times 10^{-4}$, because of its asymmetry) may give rise to background signals with WMS since there will be a small degree of coupling between the orthogonal polarisation modes of the fibre (for example at fibre splices) and some degree of polarisation-dependent loss. This means that there is a weak polarimetric interferometer present, with a fringe spacing of c/BL ~ 5 GHz for a 200 m length, similar to the linewidths of pressure-broadened gas absorption lines. Hence, as discussed in Section 5.2.3, this may give rise to significant background signals with WMS, unless steps are taken to minimise polarisation coupling and polarisation-dependent losses.

Recently, the use of the evanescent field for gas sensors has been revisited using silicon photonics technology [46, 47]. Silicon photonics fabrication can be done through the existing commercial CMOS manufacturing infrastructure and has provided a new platform for the design and integration of a range of photonic components. For gas sensors, the evanescent field from a silicon photonic ridge waveguide, as illustrated in Figure 5.10a, has been used to interact with the R(4) absorption line of methane in the near-IR at 1650.96 nm [46].

Compared with standard silica-type fibres or waveguides with a core index of n~1.45, the high index of the silicon guide, n ~ 3.47 at 1600 nm, formed on a SiO_2 substrate (n ~ 1.45), means that a much larger fraction of the power can be contained in the superstrate evanescent field, as illustrated in the planar model of Figure 5.10b. Although the penetration depth into the gaseous region is similar to silica guides, it can be seen by comparing Figures 5.9b and 5.10b that the high core index allows the guide thickness to be reduced to a much lower value before the cut-off condition takes effect, thus increasing the power fraction in the evanescent field. A relative sensitivity of ~25%

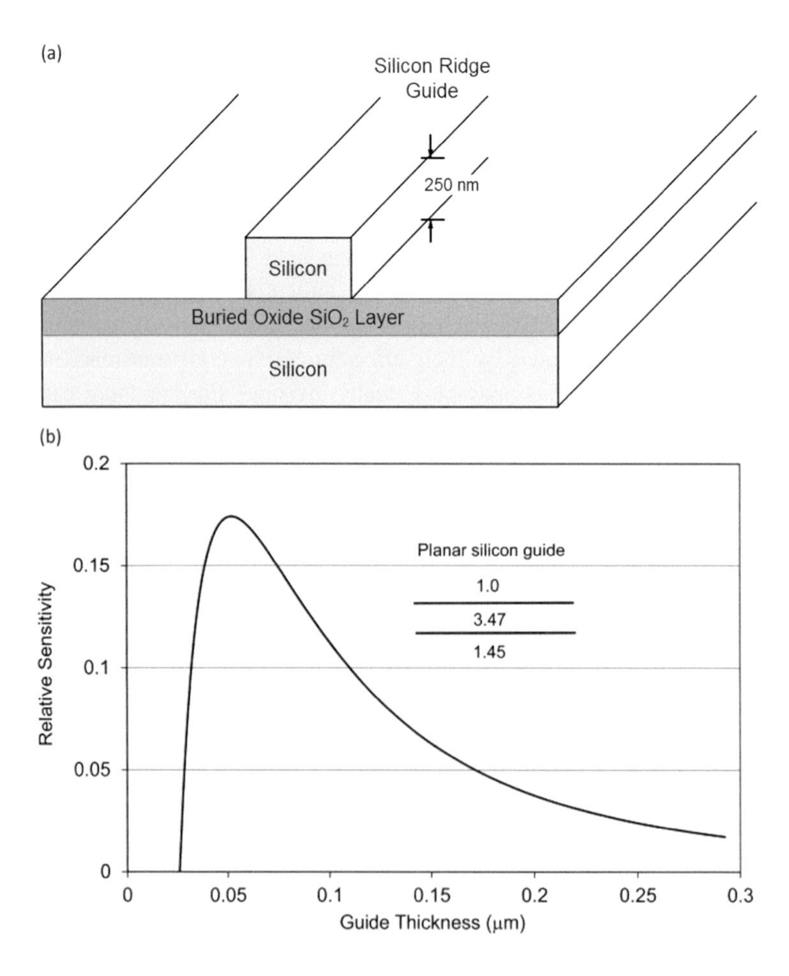

Figure 5.10 (a) Silicon ridge waveguide for evanescent field interaction (after [46]). (b) Theoretical relative sensitivity of a planar silicon guide to gas absorption (core and substrate indices of 3.47 and 1.45, respectively) as a function of guide thickness.

has been experimentally demonstrated [46] for the TM_{00} mode in the ridge waveguide structure of Figure 5.10a. The high index also allows the waveguide to be tightly folded so that a 10 cm interaction length can be folded into a footprint of 16 mm^2, making possible a high level of integration on a single chip.

Using this technology, an integrated, miniature photonic chip for methane detection at 1650 nm has been demonstrated [47], with a chip area of only ~30 mm^2. The chip incorporates a III-V semiconductor optical amplifier chip coupled to a silicon cavity to form the tuneable laser source, with signal and reference photodetectors formed on the same III-V chip. Two coiled silicon waveguides for evanescent-wave interaction are included on the chip, one acting as a sealed reference methane cell and the other for ambient methane sensing. The reference cell is used for real-time active wavelength control, calibration and normalisation purposes. Preliminary work has indicated a methane sensitivity of ~90 ppmv $Hz^{-1/2}$ or a noise-equivalent absorbance of 1.9×10^{-4} $Hz^{-1/2}$. It is envisioned that the data from a network of these low-cost photonic sensor chips, deployed over oil and gas production sites, could be wirelessly streamed to a central location for continuous monitoring of flow rates and location of methane leakages. Furthermore, the technology can in principle be extended to a range of other gases with near-IR absorption lines, with potential applications in areas such as environmental monitoring, the energy industry and security.

5.4.2 Micro-Structured Optical Fibre Gas Cells

Over the last two decades, various forms of micro-structured fibre [48, 49] have been investigated for potential application in areas such as fibre data communication, delivery of ultra-short high-power pulses and as low-loss fibres for the mid-IR region. In 'holey' fibre, the core is solid silica with air holes in the cladding region to effectively lower its refractive index for attaining guiding in the core or in a suspended core structure. In photonic band-gap fibre, the core is hollow and guidance is achieved through photonic band-gap effects from periodic arrays. Anti-resonant hollow core fibres also have air holes running down the fibre length but, unlike photonic band-gap fibres, do not require a 2-D periodic array of holes.

Similar to the use of evanescent waves, as discussed in the previous section, micro-structured fibre has also been investigated as a way of interacting with a gaseous species [4, 50–53]. In photonic band-gap fibres, over 98% of the guided optical power may be contained in the hollow core and air holes of the fibre so that if these are filled with a gas, a high relative sensitivity is possible. However, there are a number of serious practical issues in attempting to use micro-structured fibres as gas cells, such as the efficient filling of the air holes with a gas along the full length of the fibre, slow response time, water or dirt blocking the holes, multi-mode interference effects, attenuation with long fibre lengths and connection to standard fibres. To solve the problems of gas filling and slow response time, one solution is to drill periodic side holes along the length of the fibre to allow the gas to access the holes. This has been done in a 7 cm length of hollow core photonic band-gap fibre using a femtosecond Ti-sapphire laser to drill holes at 1 cm intervals. A response time of 3 s was demonstrated for methane sensing at

1665.5 nm with a sensitivity of ~650 ppm [51]. However, drilling a periodic series of holes along a long length of fibre is still a major challenge.

5.5 Fibre Optic Gas Sensor Networks

One major advantage of near-IR gas absorption spectroscopy is the availability of low-cost and low-loss fibre optics and components to form networks for multi-point sensing over large areas. Several techniques have been investigated for the creation and interrogation of multi-point sensor networks, including spatial division multiplexing (SDM), time division multiplexing (TDM), sequential multiplexing and frequency-modulated continuous wave (FMCW) methods.

5.5.1 Multi-Point Gas Sensor Network with Spatial-Division Multiplexing

One of the most practical multi-point systems for methane sensing employing SDM has been developed and commercialised by OptoSci Ltd (trademark 'OptoSniff') [54–56] and is illustrated schematically in Figure 5.11.

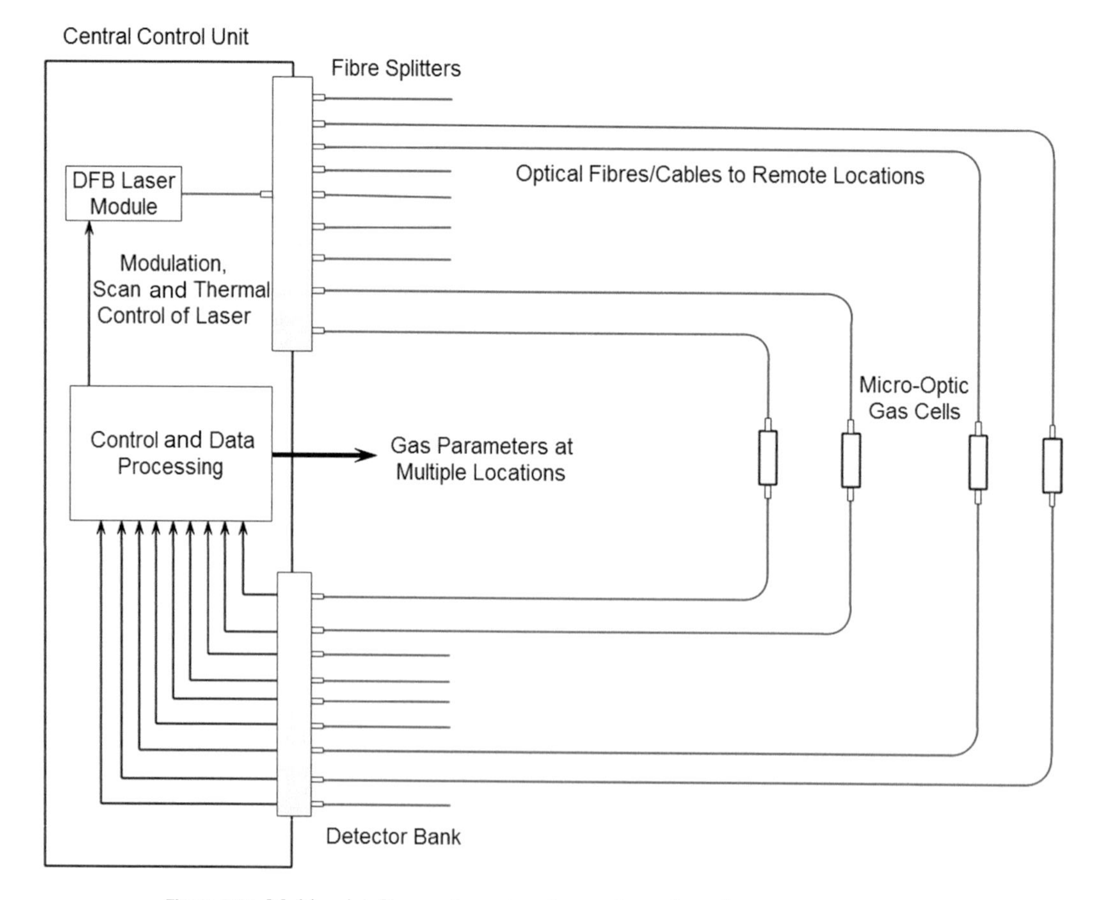

Figure 5.11 Multi-point fibre optic system for methane detection.

The central control unit, sited in a control room remote from the gas-sensing area, contains a single DFB laser operating around the near-IR absorption lines of methane at 1650 or 1665 nm (see Figure 1.9 in Chapter 1). WMS techniques are applied by modulation, scan and thermal control of the laser wavelength, as discussed in detail in Chapter 3. By cascading 2×2 fibre couplers (see Figure 5.12), the fibre output from the single laser is split into multiple fibre lines using standard fibre cables which are led to the sensing region for remotely interrogating up to 256 points at distances of up to 20 km over a wide area. Low-power optical signals carried over a fibre optic network make the system inherently safe with no spark or EMI risk. The sensor points make use of the micro-optic cells discussed in Section 5.2.2, which are readily connected into the fibre optic network using standard SC/APC fibre connectors. In practice, the sensor units are ruggedised to make them resistant to water, humidity, temperature changes, impact, etc. The fibres returning from the sensor locations are connected to a bank of photodiode detectors, also located in the central control unit, which contains a data processing unit, giving a readout of the gas parameters for the multiple locations across the site.

An analysis of receiver noise, as given in Appendix 5.2, shows that sensitivities of less than 100 ppm (0.2% LEL methane) for each point may readily be achieved, depending on the spatial extent of the system. This sensitivity is more than adequate for many industrial, safety and environment applications in methane monitoring. Consider, for example, the case of a 128-point system with a 5–10 mW DFB laser source, so that the power to each cell is reduced by a factor of 21 dB by the fibre splitters with, say, an extra 2 dB for fibre and cell losses. The laser power level returned from each sensor point to the photodiode receiver is therefore ~30 μW. As shown from (A5.2.8) in Appendix 5.2, the minimum detectable absorbance is $\sim 2\times10^{-9}$ $W^{1/2}$ $Hz^{-1/2}$ at an SNR

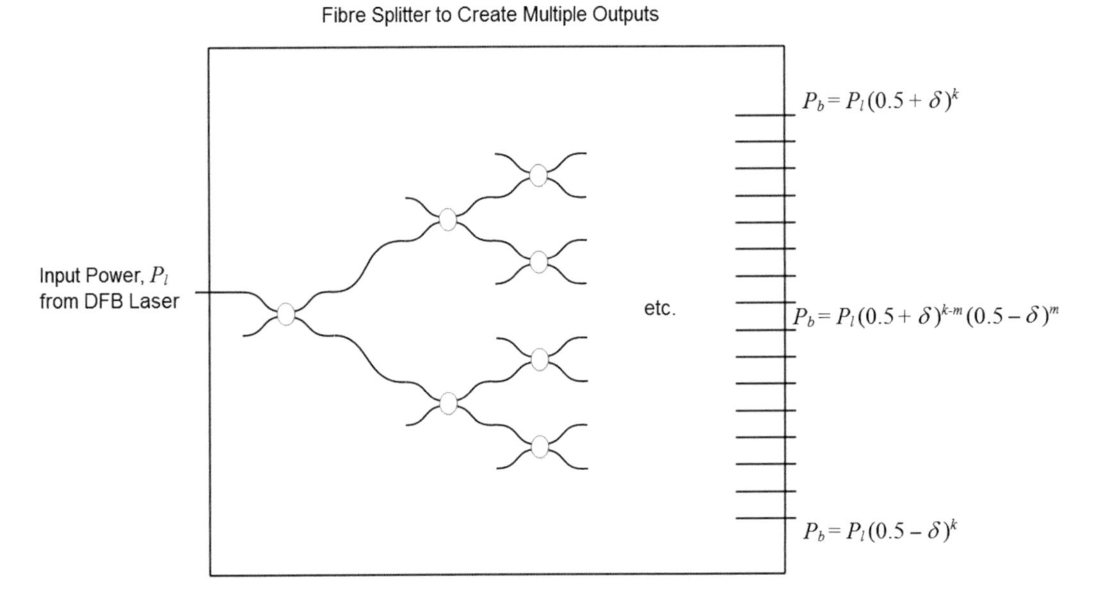

Figure 5.12 Branched network formed from a series of 2×2 fibre couplers.

of unity for the peak value of the WMS second harmonic, under receiver shot noise-limited conditions. Hence for the power level 30 μW and a typical lock-in noise bandwidth of 1 kHz at a 10 Hz scan rate, the minimum detectable absorbance is ~10^{-5}. This absorbance translates to a methane concentration of ~5 ppm with a 5 cm pathlength micro-optic cell (using the line centre absorption coefficient of 0.36 cm^{-1} for methane at 1651 nm from Table 1.2 in Chapter 1). This is, of course, the limit in sensitivity for the given noise bandwidth and to approach this level of performance requires that other noise sources are minimised, with appropriate calibration and referencing procedures, such as use of the *2f/1f* method, as described in Chapter 3.

As noted, 2×2 fibre couplers are used to form the branched network for the interrogation of multiple points, as illustrated in Figure 5.12. In order to form 2^k outputs, $(2^k - 1)$ couplers are required.

Of course, individual couplers may not have an exact 50:50 split ratio, especially if the operation wavelength is different from the standard 1550 nm coupler wavelength. Hence there will in practice be a degree of variation in power levels to each cell, as illustrated in Figure 5.12 for the case where all the couplers are assumed to have the same split ratio of $(0.5 + \delta)$:$(0.5 - \delta)$ and δ represents the deviation from the ideal 50:50 coupler. The fractional power level to individual branches can then be written as:

$$\frac{P_b}{P_l} = (0.5 + \delta)^{k-m}(0.5 - \delta)^m \tag{5.33}$$

where m goes from 0 to k. Note that, while there are 2^k outputs, there are only $(k + 1)$ different levels from (5.33), so a number of outputs have the same level.

With the assumption that $\delta \ll 0.5$, then (5.33) may be approximated, in units of dB, as:

$$\frac{P_b}{P_l} \simeq -3k + 8.7(k - 2m)\delta \tag{5.34}$$

Hence all outputs fall within the range: $(-3k \pm 8.7k\delta)$ dB. For example, for a 128-point system (k = 7) with δ = 0.01, all the output levels can be approximately specified by $-(21 \pm 0.6)$ dB.

It is worth noting here that a single-point fibre optic system utilising a single micro-optic cell has important applications, for example, in the determination of gas distribution profiles. Such a system (with the trademark 'OptoMole') has been developed by OptoSci [56] for application in leak detection. An escape of methane from a leak in a gas main may often find its way to the surface via underground utility ducts that are used for telephone or other services. Locating the actual point of leakage may involve considerable cost due to the need for excavations, as well as significant disruption to local services. A single-point fibre optic system may easily be fed along an underground duct to generate a real-time profile of the methane distribution, rapidly identifying regions of high ingress and thus assisting in the process of identifying the escape point from a gas main.

Similar single- or multi-point fibre optic systems may be used for a range of other gases with near-IR absorption lines, provided the near-IR lines are sufficiently strong to obtain useful sensitivities. As noted above, the multi-point system can detect an absorbance of $<10^{-4}$ at each point with a 5 cm micro-optic cell, so the minimum concentration $C_{\min}$ that can be measured for various gases may then be estimated using the parameters listed in Tables 1.1 and 1.2 in Chapter 1. The pathlength of the fibre-coupled micro-optic cell may of course be increased beyond 5 cm (at the expense of cell insertion loss) to reduce the value of $C_{\min}$, and versions up to 80 cm are available commercially [13]. In other configurations, several short micro-optic cells may be connected in series along the same fibre path to either improve the sensitivity or to obtain an averaged concentration if the cells are spaced out over a wide area [57].

5.5.2 Multi-Point Gas Sensor Network with Time-Division Multiplexing

As an alternative to SDM, time-division multiplexing (TDM) with a pulsed laser source has been experimentally demonstrated for the interrogation of several gas sensors [58], as illustrated in Figure 5.13.

The individual gas sensors are arranged in a ladder network with fibre delay lines of ~60 m between sensors to delay the pulse output from successive sensors by ~300 ns. The coupler ratios are chosen to equalise the power from the sensors, namely, C1 = 66:33 and C2 = 50:50. The pulse width used is 200 ns at a repetition rate of 500 kHz with wavelength modulation of the laser source for WMS at a

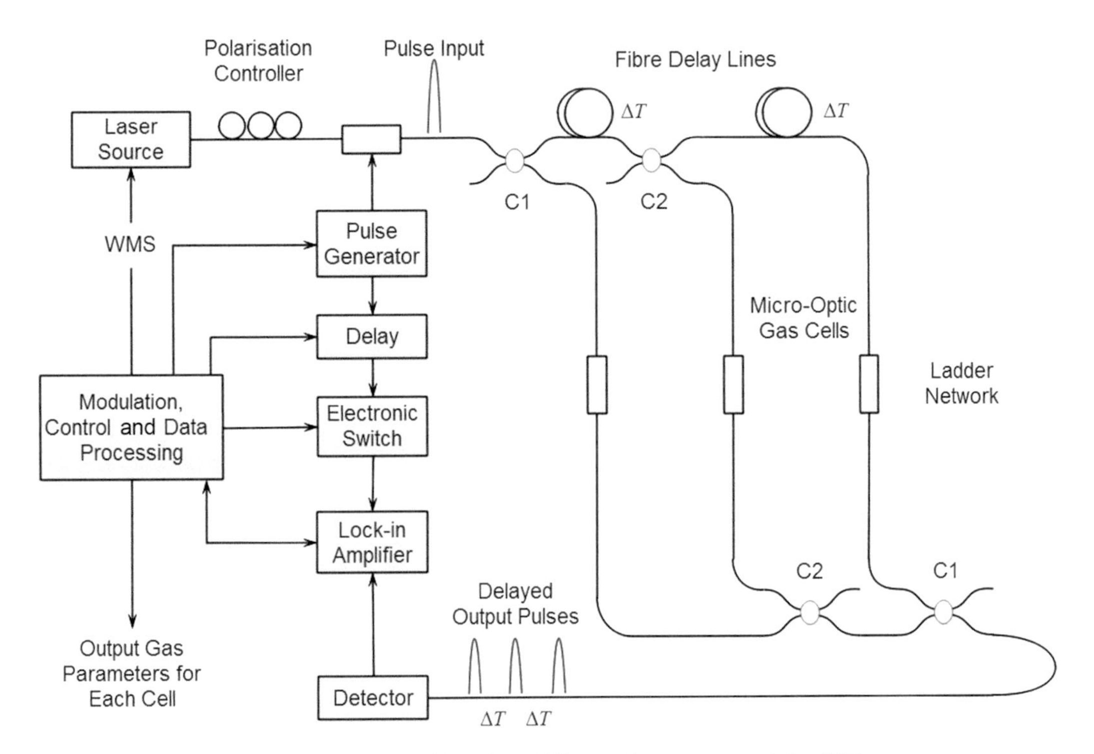

Figure 5.13 Time-division multiplexing of fibre optic gas sensors (after [58]).

relatively slow sinusoidal modulation rate of 400 Hz. From measurements made on the three-point system, the noise-equivalent detection sensitivity was ~150 ppm acetylene on its near-IR line at 1530.2 nm for a time constant of 300 ms and 25 mm path-length gas cells. The cross-talk between cells was estimated to be ~–30 dB. Compared with SDM, TDM has the advantage of needing only a single detector, but in practice the number of cells that can be multiplexed by TDM is limited due to cross-talk issues.

5.5.3 Multi-Point Gas Sensor Network with FMCW Multiplexing

Another technique for multiplexing fibre optic sensors is to use a ranging method derived from radar and applied to fibre sensors [29, 59], known as frequency-modulated continuous-wave (FMCW) ranging. Two forms of FMCW have been investigated for fibre optic gas sensors, namely: (i) optical coherent FMCW, where the optical frequency is ramped over time [60] and the coherent mixing of two mutually delayed signals gives rise to beat signals on the detector and (ii) sub-carrier RF FMCW, where a swept RF frequency modulation is applied to the source and the beat frequencies generated by mixing the delayed signals with a local RF reference after detection.

Figure 5.14 illustrates the possible use of optical FMCW to address individual micro-optic cells. Normally, as was explained in Section 5.2.3, the reflections from the surfaces of the GRIN lenses are minimised, but here they are deliberately used to create individual beat signals for each cell by interferometric mixing [61–63].

As shown in Figure 5.14, several micro-optic cells are connected in series along a single fibre optic line, with each cell chosen to have a different path-length, l_i. The two back-reflections from a cell, one from the input GRIN lens and the other from the output

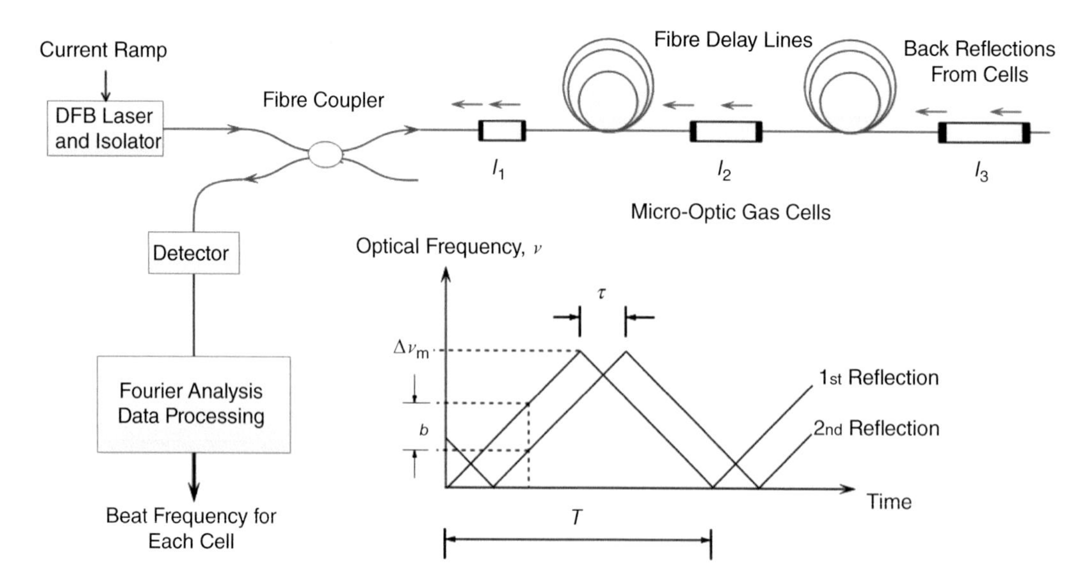

Figure 5.14 Optical FMCW method for addressing multiple cells along a single fibre optic line.

lens will have a relative time delay on arrival at the detector of $\tau_i = 2l_i/c$. A current ramp, applied to the DFB laser source, produces a linear time ramp of the output optical frequency so the instantaneous optical frequencies of the two reflections will differ on arrival at the detector, creating a difference or beat frequency given by:

$$f_{bi} = 2f_r\tau_i(\Delta\nu_m) \tag{5.35}$$

where $f_r = 1/T$ is the frequency of the (triangular) ramp and $\Delta\nu_m$ is the maximum optical frequency shift produced by the ramp, as shown in Figure 5.14.

Fibre delay lines inserted between successive cells ensure that reflections from neighbouring cells do not interfere and hence each cell is identified by its individual beat signal, dependent on the cell length. Typically, with a 100 Hz ramp frequency and $\Delta\nu_m$ ~50 GHz, the beat frequencies are in the range of 3–6 kHz for cell lengths of 5–10 cm. While the physical construction of the system is relatively simple, using only a single fibre line with serial cells, there are several practical issues for effective operation. The linearity of the optical frequency ramp must be of a high quality and application of WMS methods and the associated signal extraction is more complex [63]. Also, the gas absorption for each individual cell needs to be de-convolved from all the beat signals since reflected signals from more distant cells return though the nearer cells (with reduced signal-to-noise ratios). Simulations indicate that a 20-point system should be feasible [61] although a practical gas sensor system based on optical FMCW has not been demonstrated.

As noted earlier, instead of ramping the optical frequency, the alternative way of using FMCW for multiplexing is to ramp the frequency of an RF signal used for intensity modulation of the DFB laser output and extract the beat frequencies by mixing a local RF reference with the delayed signals from the cells after detection. Figure 5.15 illustrates such a system [64, 65] with the same construction as the TDM multi-point fibre network of Figure 5.13, but now with intensity modulation on the source instead of a pulsed output.

The beat frequencies are given by a similar expression to (5.35), but now in terms of the maximum frequency shift of the swept RF modulation:

$$f_{bi} = 2f_r\tau_i(\Delta f_m) \tag{5.36}$$

Note that the maximum frequency shift, Δf_m, is typically ~10 MHz, which is around three orders of magnitude smaller that the equivalent optical frequency shift in (5.35). However, in this case, transmitted (rather than reflected) optical signals from the cells are employed and the delay times determined by the optical delay in the path from the source modulation to the receiver (set for each cell by the choice of the delay line) in relation to the electrical delay of the reference, giving delay times of the order of a few hundred nanoseconds. A three-point sensor system with 25 mm pathlength cells and beat frequencies of 20, 40 and 50 kHz has been demonstrated [64] using WMS on the 1530.2 nm absorption line of acetylene with a sensitivity of 270 ppm $Hz^{-1/2}$ (corresponding to an absorbance of ~3×10^{-4}) and cross-talk between cells better than –22 dB. Similar to TDM, FMCW has the advantage over SDM of requiring only a single detector, but in practice the number of cells that can be multiplexed and the sensitivity

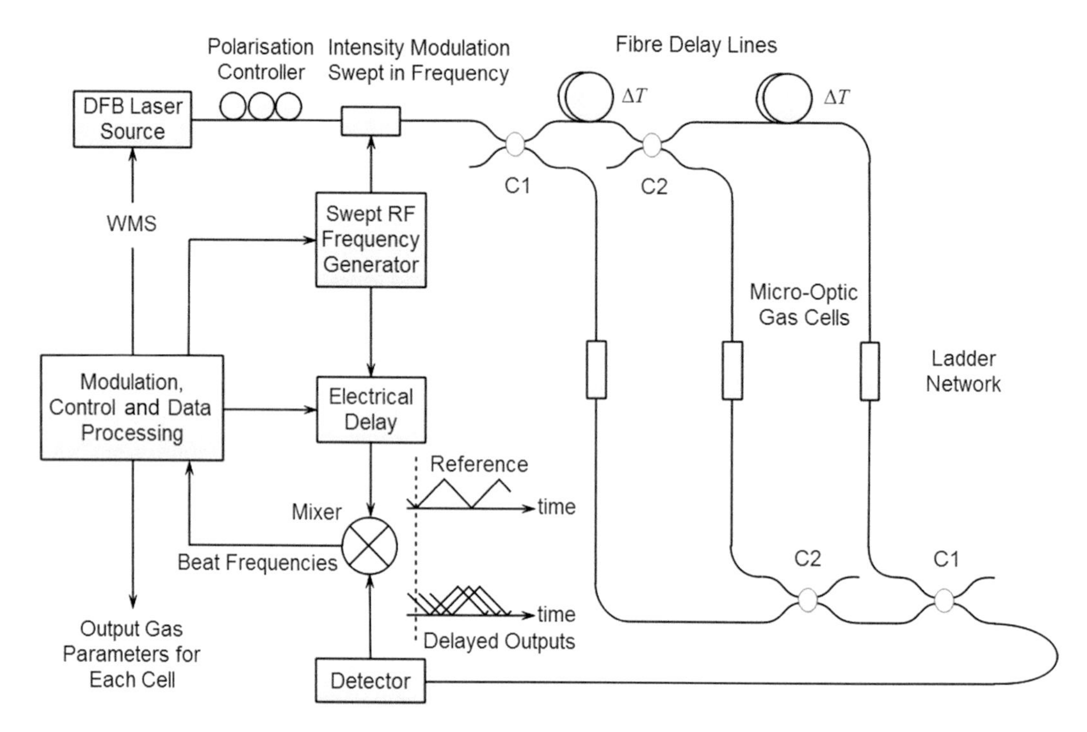

Figure 5.15 Multiplexing of fibre optic gas sensors with FMCW and swept RF intensity modulation (after [64]).

are limited by coherent mixing and cross-talk between sensor paths. However, with appropriate steps taken to minimise these effects, it may be possible to multiplex ~30 cells or more with a sensitivity of $\sim 10^{-3}$ in absorbance [65].

5.6 Open-Path and Free-Space Systems

In a number of important applications, no gas cell as such is used, but the beam from a DFB laser is directed along an open path, such as in the detection of gas leaks, analysis of combustion processes, emission measurements on exhaust plumes and in atmospheric monitoring. We consider here some examples using DFB lasers operating on near-IR gas absorption lines using the methods of tuneable diode laser absorption spectroscopy (TDLS), as discussed in Chapters 2 and 3. Examples of the use of fibre laser combs in open-path systems are considered later in Chapter 7.

5.6.1 Detection and Imaging of Gas Leaks

The ability to quickly detect and locate gas leaks from pipelines or storage facilities is very important for environmental and safety reasons. An open-path continuous methane detection system based on TDLS for gas leak detection has been field tested by the US energy provider, Pacific Gas and Electric, at one of its natural gas storage facilities in northern California [66]. The sensor was developed by the company Acutect as part of the Methane Detectors Challenge [67]. The company Physical Science Inc. (PSI) [68] has developed a number of products [69, 70] for detecting, locating and quantifying methane leakage based on the methods of TDLAS with the near-IR absorption lines of methane. The remote methane leak detector (RMLD) is a portable, hand-held instrument used for surveying leaks from gas pipelines. It directs the near-IR beam along an open path over the area of interest and collects the scattered light from a natural surface. Analysis of the returned scattered light by TDLAS allows the determination of the integrated methane concentration over the beam path. Processing of collected data over a contaminated area or around the vicinity of a leak can identify the location of a leak and quantify the emission rate through temporal and spatial changes in methane distribution. In other developments, the instrument is being mounted on vehicles for surveying leaks in municipal areas and on drones to fly over gas pipelines and storage facilities. In this latter application, the unmanned aerial vehicle (UAV) flies in a search pattern around the facility and if a leak is detected, it homes in on the location and determines the leak rate by flying in circular patterns around the leak.

Imaging gas leaks at video rates enables the direction of the gas flow to be determined and hence fast location of the leak. Such an imaging system has been experimentally demonstrated for methane by Gibson et al. [71] using a 1651 nm DFB laser, driven by a square-wave current for switching the wavelength on and off the near-IR methane absorption line. Rather than raster scanning the laser output over the area to be monitored, a digital micro-mirror device projects a patterned output from the laser through the gas-affected area and the backscattered light from the scene is collected by a lens and focussed onto a single photodiode detector. The patterned output projected onto the scene consists of a sequence of mask pattern pairs (corresponding to Hadamard matrices) which are displayed at a rate of 20 kHz. The signal capture from the detector is synchronised to the displayed pattern, providing information on the correlation between the projected pattern and the gas distribution. This information, along with the known projected patterns, enables an image of the gas distribution to be created. Low-resolution methane gas imaging at a rate of 25 frames per second and for a gas leak of ~0.2 litres per minute from a distance of ~1 m was demonstrated, with the images overlaid on a high-resolution colour image of the scene taken by a standard CMOS camera in order to assist in leak location. The whole system, including the laser, digital micro-mirror, CMOS camera, lens and detector is packaged as a single 3-D printed unit.

5.6.2 Combustion Analysis and Emissions Monitoring

An important application area for TDLAS is in the characterisation of combustion processes and emissions monitoring. A range of key parameters involved in combustion processes can be extracted with TDLAS, including gas temperature, pressure, concentration and gas velocity, as discussed in Chapter 1. TDLAS with near-IR DFB lasers has been demonstrated for application in a diverse range of areas [1, 2], including fundamental combustion studies, and in natural-gas and coal-fired power plants, gas turbines, internal combustion engines, high-speed combustion flows, detonation combustors, etc. Companies such as Physical Science Inc. [72] and Zolo Technologies [73] market a range of combustion sensors based on TDLAS for gases, such as O_2, H_2O, CO, CO_2, NO, NO_2, suitable for a number of these application areas.

5.6.3 Tomographic Imaging of Emissions and Combustion Processes

The drive for reduction in harmful emissions, cleaner fuels and improved efficiency from jet and combustion engines has led to the need for real-time, in situ imaging of gas distributions in exhaust plumes or directly within the chambers where combustion processes are taking place [74]. One technique currently under development for aero-engine exhaust emissions [75–78] is the creation of real-time 2-D tomographic images of the CO_2 concentration in the exhaust plume by simultaneously passing ~126 beams, distributed uniformly around its circumference, through the plume, as illustrated in Figure 5.16.

To perform the measurements, WMS based on the *2f/1f* ratio, as described in Chapter 3 is used with a DFB laser operating on the CO_2 line at 1997.2 nm. After amplification by a thulium-doped fibre amplifier [78], the DFB output is split and fibre-coupled to the 126 paths through the exhaust plume. Measurements of the integrated absorption along each ~1 m pathlength through the plume are then processed to generate real-time 2-D tomographic images of the CO_2 concentration during testbed operation of the engine.

For direct analysis of combustion processes, the creation of tomographic images with TDLAS has been demonstrated for the analysis of a swirling flame [79]. Swirl injectors are widely employed to achieve lean burn combustion, which gives lower flame temperatures, reducing thermal NO_x emissions and improving engine reliability. An understanding of the reaction processes within a swirling flame is important for optimising the design of the swirling injector and real-time, in situ monitoring can be used to prevent the risk of a lean blowout, which is a significant safety hazard for an aero-engine. In the experimental demonstration, the outputs from two time-division-multiplexed DFB lasers, operating on

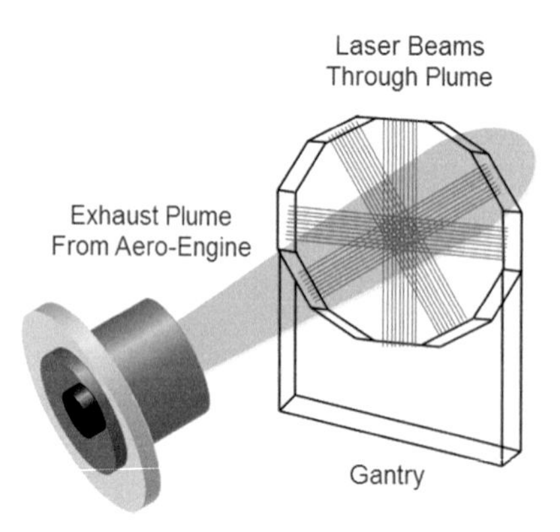

Figure 5.16 System for creating tomographic images of CO_2 emission from aero-engines.

H_2O lines at 1343.3 nm and 1391.7 nm, were used to produce an array of five fanned beams through the swirling flame, with each fanned beam sampled by a 12-photodiode array. In this way, real-time 2-D images of temperature and H_2O concentration over the flame cross-section were reconstructed with a spatial resolution of ~8 mm and dynamic temperature profiles were captured during a lean blowout event. Similar tomographic techniques with TDLAS have been proposed and modelled for the reconstruction of two-dimensional velocity distributions in a scramjet combustor [80].

The mixing and combustion processes in conventional compression-ignition engines, such as in diesel engines, lead to high levels of soot and NO_x emissions. Various techniques have been explored to mitigate these problems, such as the use of pre-mixed charge compression-ignition, homogeneous charge compression-ignition or use of fuels less prone to auto-ignition. To assist research of this type, it is desirable to image the concentration distribution of the fuel vapour during the mixing process under different engine operating conditions. This has been demonstrated [81] using a 31-beam array in a co-planar grid situated below the injector on a Volvo D5 engine, modified for single cylinder operation. A diode laser source at 1700 nm for hydrocarbon sensing and a reference source at 1651 nm were used for dual-wavelength ratiometric measurements on each path. In this way, tomographic images were obtained during the compression stroke at a rate of 13 frames per crank angle degree within the same engine cycle at 1200 rev min^{-1}, allowing observation of the fuel evaporation rate and mixing evolution under different operating conditions, such as injection timing, intake pressure and temperature.

5.6.4 Atmospheric Sensing and Monitoring

An area where laser absorption spectroscopy is playing an increasingly key role is in the measurement of various gases important for atmospheric monitoring, such as greenhouse gases, trace gas exchange between the biosphere and the atmosphere, or molecules related to ozone chemistry [82]. Of great concern in recent years is the rise of atmospheric greenhouse gases such as CO_2 from fossil fuel consumption, CH_4 from human activities and from natural sources (wetlands, oceans, rice paddy fields and the Arctic and Siberian permafrost regions) and N_2O from agricultural fertilisers. Several systems based on near-IR DFB lasers have been demonstrated for atmospheric monitoring.

An example of interest is the monitoring of CO_2 and CH_4 in the near-surface atmosphere, at the interplay between microbial activity and photosynthesis. An open-path system using a DFB laser operating around the R(16) line of CO_2 at 1572.33 nm has been tested for this purpose at a research site near Fairbanks in Alaska for measurement of trace-gas species above the thawing permafrost [83]. In the system design, the output from the DFB laser is transported via single-mode fibre to the launch optics and co-aligned with a visible laser for alignment purposes. The outgoing beam is directed through a 3.2 mm hole in a parabolic mirror and travels to a corner cube retroreflector placed downfield at a distance of 55 m, giving a round-trip open pathlength of 110 m. The parabolic mirror collects the returning beam, which is then focussed into a multi-mode fibre for transmission to a standard InGaAs photodetector. Direct absorption measurements of the CO_2 absorption line were made by sweeping the laser wavelength over ~0.4 nm at a rate of 500 Hz and the data collected were stored as 10 s sweep averages with field data averaged for 2.5 minutes. The instrument was characterised in the laboratory using a White cell (with up to a 40 m pathlength) and CO_2 concentrations were obtained by spectral fitting of measured data to known spectra. Diurnal cycles of CO_2 concentration were successfully observed where the daytime CO_2 levels remained relatively low near 400 ppmv with early morning increases to 500 ppmv or higher. The authors noted a number of challenges in achieving the required sensitivity for this application, notably, the weak line strength of CO_2 at 1572 nm and the required pointing accuracy of 0.05° (at a distance of 55 m from the 50 mm diameter retroreflector), exacerbated by the large diurnal temperature swings and an unstable ground base. However, an order of magnitude improvement could be attained by using the stronger CO_2 lines around 2 μm.

Another example concerns the remote monitoring of ambient methane around industrial areas such as oil and gas processing plants. Unintended or irregular methane leaks from these plants, known as fugitive emissions, are difficult to monitor and can be a significant contributor to greenhouse gas emissions. Detection of these emissions demands a measurement precision and detection limit of less than 2 ppmv in air. This level of detection has been demonstrated using a fibre optic networked system operating with a single DFB laser on the R(3) methane absorption line at 1653.73 nm [84] with a single detector for the sensor cells. The fibre optic system includes a fibre-coupled methane reference cell, laser power reference detector, a fibre ring resonator for wavelength calibration and an optical switch for interrogation of several remotely located sensors. In the fibre-coupled sensor heads, the beam is routed through an open pathlength of ~5.8 m in eight passes using an arrangement of planar reflective surfaces. Gas concentrations were measured from the *2f* signal from WMS with a polarisation scrambler and thermal stabilisation of optical components to improve measurement accuracy.

TDLAS on near-IR absorption lines is also finding important applications in atmospheric water vapour sensing [85–87] on board commercial jet liners. The company SpectraSensors [87] has developed the Water Vapor Sensing System (WWVS-II) based on a 1.37 μm laser for the near-IR absorption lines of water vapour using the *2f* signal from WMS with a scan rate of 4 Hz. An air sampler, attached to the external surface of an aircraft, collects and directs a sample to the internal measurement unit, which has an optical path length of 22.7 cm. WVSS-II has been designed specifically to provide meteorological measurements of in situ water vapour of the upper atmosphere as part of the Aircraft Based Observations Programme (ABOP). Under ABOP, coordinated by the World Meteorological Organization, real-time measurements of air temperature, pressure, wind speed and direction are collected from commercial aircraft worldwide and, with currently more than 5000 aircraft reporting into the global ABOP network, a wealth of data is available for meteorological purposes. The addition, through WWVS-II, of atmospheric water vapour measurements to the range of global data collected by in-flight aircraft creates a complete vertical profile of the atmosphere for meteorological applications, with clear benefits to weather forecasting, aviation and society at large. Interestingly, TDLAS for atmospheric sensing on board aircraft operates in a very challenging environment where, for example, during ascent the water vapour concentration can typically change from ~40 000 ppmv at ground level to ~50 ppmv at high altitude, air pressure from 1 bar to ~0.2 bar and air temperature from +20 °C to –50 °C, all of which affect the absorption line profile and have to be taken into account in interpreting the *2f* WMS signal as discussed in Chapters 1 and 3. The aviation environment also demands stringent requirements in terms of low weight, drag, power and icing risk, as well as automated operation with low maintenance. These requirements have been met and the WVSS-II has been certified for use on commercial flights with a number of different aircraft types.

5.7 Further Information on Near-IR Gas Sensing and Applications

A number of examples have been described in this chapter on the use of near-IR spectroscopy for a range of application areas and a very extensive literature on this topic has been published over the last two decades. In particular, the reader is referred to the review by Hodgkinson [4], which contains a very useful table of performance indicators from the literature for a range of gases, namely, NH_3, C_6H_6, CO_2, CO, C_2H_6, H_2CO, H_2S, CH_4, NO, N_2O, NO_2, SO_2 and H_2O. The table provides information on both near-IR and mid-IR absorption lines for these gases and the minimum detectable absorbance or concentration that can be achieved with the optical techniques employed. More recently and for additional application areas, the reader is referred to the series of conferences on Field Laser Applications in Industry and Research (FLAIR) and the special issues of *Applied Physics B, Lasers and Optics,* vol. 110, no. 2, February 2013, vol. 119, no. 1, April 2015 and vol. 123, 2017, where a number of full-length papers from the conference series are published. There are also a considerable number of companies, such as those listed in references [11–13, 31, 44, 56, 69, 72, 73, 87–91] that market a range of TDLAS gas sensors or optical components for gas sensing applications.

5.8 Conclusion

This chapter has reviewed a number of methods that can be applied in near-IR spectroscopy, such as the use of multi-pass cells, ring-down and cavity-enhanced spectroscopy, evanescent-wave cells, open-path systems and the multiplexing of micro-optic cells for wide-area optical fibre networks. A number of examples have been presented of experimental and commercial systems for the detection of gas leaks and for combustion, emission and atmospheric monitoring.

Not unexpectedly, the best-developed commercial near-IR systems are for gases such as CH_4, C_2H_2 and H_2O, where the near-IR lines are relatively strong. Water vapour sensing in the near-IR for atmospheric and combustion monitoring is particularly noteworthy. Methane has also received much attention due to safety and environmental concerns and several commercial or near-commercial systems have been developed. The fibre optic multi-point system based on spatial-division multiplexing has the ability to detect low levels of methane at more than 100 locations spread across a wide area. An interesting alternative strategy for methane sensing at multiple locations is the methane photonic sensor chip based on evanescent-wave interaction in silicon waveguides with the potential of being IoT-enabled with wireless streaming of data to a central location. However, compared with the fibre optic system, it is not electrically passive and EMI immune. Another strategy which has proved to be commercially viable is the use of long pathlength, open-path-type sensors based on the collection of scattered light from a surface or from a retroreflector. These types of devices are finding important application in the detection and imaging of methane leaks and in atmospheric monitoring.

For gases with weak near-IR lines, commercial development has been more challenging, but in areas such as combustion and emission analysis, gas concentrations may be relatively high and near-IR systems are providing new insights through tomographic imaging of gas concentrations, particularly for CO_2, which has stronger lines around the 2 μm wavelength region. We also discussed in this chapter a number of ways of enhancing the sensitivity through, for example, the methods of cavity-enhanced spectroscopy. However, these techniques are more costly, complex and less amenable to field deployment, so for any particular application, their use must be weighed against the issues and costs associated with the direct use of the very much stronger mid-IR absorption lines. We shall return to these issues in Chapter 8.

Appendix 5.1 Evanescent-Wave Interaction

In order to model the interaction of the evanescent wave of an optical waveguide with a gaseous absorber, a complex refractive index is used to describe the optical properties of the gas. To define the complex index, consider first the electric field of a simple plane wave as it propagates through a bulk gaseous medium in the z-direction:

$$E = E_0 \cdot \exp j\{2\pi \nu t - \beta z\} \tag{A5.1.1}$$

where β is the propagation constant.

The attenuation experienced as a result of absorption by the gas can be modelled as an imaginary component in the refractive index so that $\beta = k_0(n_1 - jk)$ with $k_0 = 2\pi/\lambda_0$. Hence (A5.1.1) may be expanded as:

$$E = E_0 e^{-k_0 k z} \cdot \exp j\{2\pi \nu t - k_0 n_1 z\} \tag{A5.1.2}$$

Noting that the intensity is proportional to the square of the electric field and comparing (A5.1.2) with the Beer–Lambert law (1.8) in Chapter 1 for a direct absorption cell of length l we can write:

$$P_{\text{out}} = P_{\text{in}} e^{-2k_0 k l} = P_{\text{in}} e^{-\alpha C l} \tag{A5.1.3}$$

Hence we obtain, $k = \alpha C/2k_0$, relating the imaginary component of the refractive index to the gas concentration and the absorption coefficient.

For gases, k is several orders of magnitude smaller that the real part of the refractive index, so for evanescent interaction with a waveguide, the gaseous absorption acts as a small perturbation on the waveguide properties. Hence the relative sensitivity, Γ_e, of evanescent field interaction as compared with a direct absorption cell is given by the perturbation expression [39]:

$$\Gamma_e = \frac{n_1}{n_e} \frac{\iint_{A_e} |E(x,y)|^2 \mathrm{d}A}{\iint_{\infty} |E(x,y)|^2 \mathrm{d}A} \tag{A5.1.4}$$

where $E(x, y)$ is the transverse electric field distribution of the guided mode, A_e is the cross-sectional area over which the gas is present in the evanescent field and n_e is the effective index of the guided mode.

The relative sensitivity can thus be calculated knowing the transverse electric field distribution of the guided mode, either analytically or numerically, by computation of the integrals in (A5.1.4).

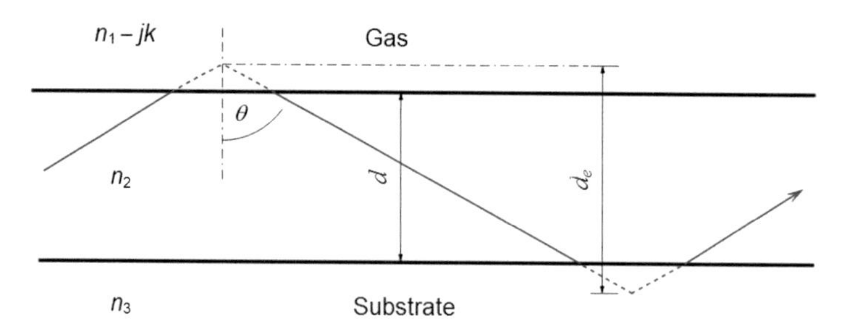

Figure A5.1.1 Ray optics model of a planar guide with attenuated reflections at the gaseous boundary due to gas absorption.

For a planar guide, the modal field distributions are well known and (A5.1.4) may be evaluated analytically. Alternatively a simple ray optics model may be used for the planar guide, as illustrated in Figure A5.1.1.

From the ray trajectory shown in Figure A5.1.1, the number of reflections per unit length that the ray makes in contact with the gaseous species is given by:

$$\xi = \frac{1}{2d_e \tan\theta} = \frac{\sqrt{n_2^2 - n_e^2}}{2n_e d_e} \tag{A5.1.5}$$

where we have used the expression $n_e = n_2 \sin\theta$ for the effective index of the guide mode.

Hence the attenuation per unit length experienced by the guided mode as a result of the gas absorption is:

$$\frac{P_0}{P_\mathrm{i}} = \left(|r|^2\right)^\xi = \exp\left\{\xi \ln|r|^2\right\} \tag{A5.1.6}$$

where r is the reflection coefficient of the ray at the gaseous boundary, $|r|^2 < 1$ due to the gas absorption.

For s-polarised and p-polarised light rays, corresponding respectively to TE and TM modes of the waveguide, the reflection coefficients are given by [39]:

$$r_s = \frac{j\gamma - k_x}{j\gamma + k_x}, \quad r_p = \frac{j\zeta\gamma - k_x}{j\zeta\gamma + k_x} \tag{A5.1.7}$$

where $\gamma = k_0\sqrt{n_e^2 - (n_1 - jk)^2}$, $k_x = k_0\sqrt{n_2^2 - n_e^2}$ and $\zeta = \{n_2/(n_1 - jk)\}^2$.

For k small and $|r|^2$ close to unity we can make the approximation:

$$\ln\left\{|r|^2\right\} \simeq \ln\left\{\frac{1-\delta}{1+\delta}\right\} \simeq -2\delta \tag{A5.1.8}$$

An expression for δ_s (and similarly for δ_p) may be obtained from (A5.1.7) as:

$$\delta_s = k \cdot \frac{2n_1\sqrt{n_2^2 - n_e^2}}{\left(n_2^2 - n_1^2\right)\sqrt{n_e^2 - n_1^2}} \tag{A5.1.9}$$

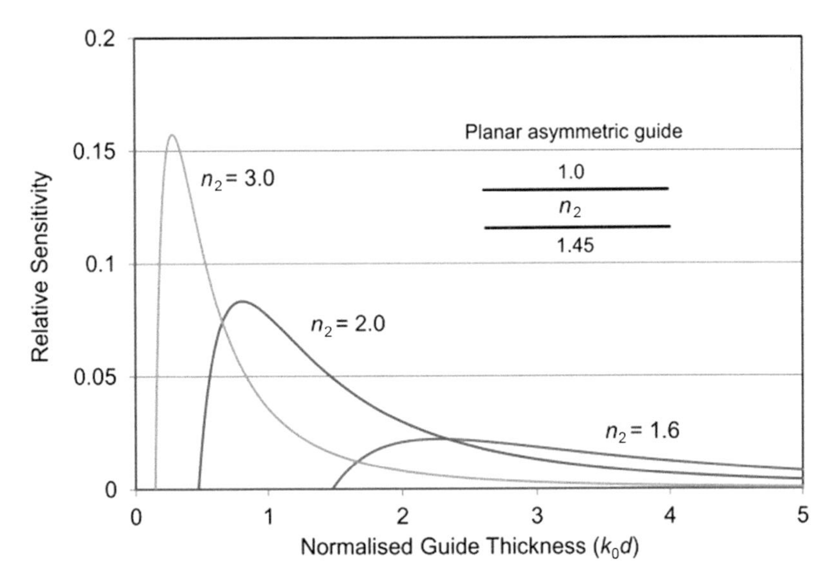

Figure A5.1.2 Relative sensitivity of the evanescent field of the TE_0 mode in an asymmetric planar guide to gas absorption for different guide indices.

Comparing (A5.1.6) for the evanescent case with (A5.1.3) for a direct absorption cell and using (A5.1.8) gives the relative sensitivity of evanescent-wave interaction for a planar waveguide as:

$$\Gamma_e = \frac{-\xi \ln |r|^2}{2k_0 k} = \frac{\xi\delta}{k_0 k} \tag{A5.1.10}$$

Substitution of the expressions for ξ and δ from (A5.1.5) and (A5.1.9) into (A5.1.10) gives the expression (5.29) in the main text. Figure A5.1.2 shows universal curves of the relative sensitivity plotted against the normalised thickness for the TE_0 mode in an asymmetric planar guide with a substrate index of 1.45.

Note from Figure A5.1.2 that the peak value of the relative sensitivity increases significantly with increase in the guide index. Since the x-axis represents normalised thickness, k_0d, the peak value can theoretically be attained at any wavelength of operation by the appropriate choice of the waveguide thickness, although tolerances are more relaxed at longer wavelengths where the guide is thicker. As also shown in Figure A5.1.2, a rapid drop-off in sensitivity occurs as the mode approaches the cut-off condition where the power drains into the substrate.

Appendix 5.2 Photodiode Receiver Circuit and Signal-to-Noise Ratios

The most common photodiode detectors for the near-IR are based on silicon (Si) for the wavelength range of 0.2–1.1 μm and indium-gallium-arsenide (InGaAs) for 0.5–2.6 μm. [92, 93]. There are several parameters that are commonly used to describe the detector performance. The responsivity, $\mathcal{R}$, in units of A W^{-1} defines the photodiode output current per watt of optical power falling on the detector. The noise-equivalent power (NEP) per unit bandwidth, with units of W $Hz^{-1/2}$, defines the signal level for a signal-to-noise ratio of unity and is typically of the order of 10^{-12} W $Hz^{-1/2}$. The detectivity, D, is the inverse of the NEP with D^* defined as $D^* = D\sqrt{A_d}$, in units of cm $Hz^{1/2}$ W^{-1}, where A_d is the detector area. The photodiode is normally connected to some form of receiver circuit, such as the transimpedance front end [94, 95] illustrated in Figure A5.2.1.

The photodiode may be reversed biased for operation in the photoconductive mode, as shown in Figure A5.2.1a, or grounded, as shown in b for operation in the photovoltaic mode. Biasing of the photodiode reduces the junction capacitance and hence improves the bandwidth but, on the other hand, the zero bias photovoltaic mode has better noise characteristics due to the almost complete suppression of surface leakage currents. Also with the transimpedance amplifier, a linear response is obtained in the

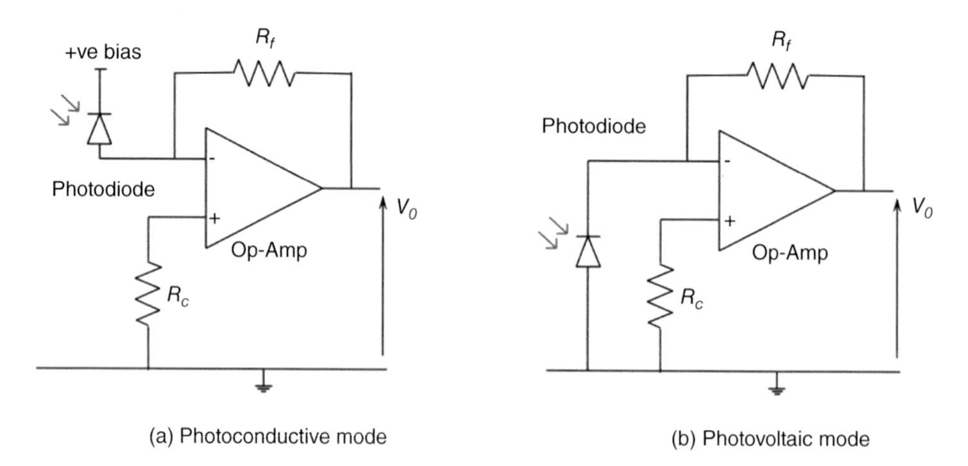

Figure A5.2.1 Transimpedance photodiode receiver circuit: (a) photoconductive mode with DC bias, (b) photovoltaic mode with zero bias.

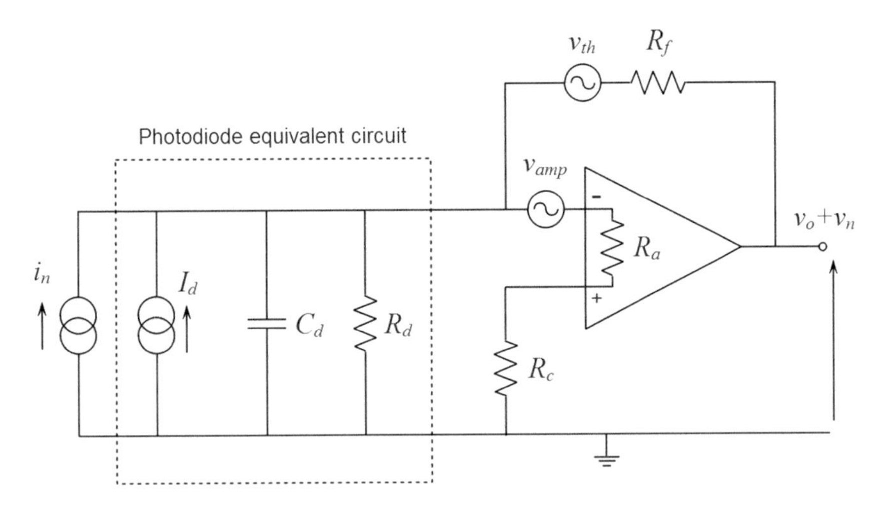

Figure A5.2.2 Equivalent circuit model of the transimpedance photodiode receiver showing the various noise sources.

photovoltaic mode due to the low load resistance on the photodiode from the virtual earth at the inverting input of the operational amplifier.

The bandwidth and noise characteristics of the transimpedance photodiode receiver can be derived from its equivalent circuit model, shown in Figure A5.2.2. The signal-to-noise ratio is determined using the superposition principle, where each source in Figure A5.2.2 is considered independently, while the other sources are disabled (voltage source set to zero and current source open circuited).

Consider first the signal output voltage v_o as a result of the optical power P_d on the detector, which generates a photodiode current of $I_d = \mathcal{R}\, P_d$, where $\mathcal{R}$ is the photodiode responsivity, as noted above. Adding currents at the inverting input gives:

$$I_d = -\left(\frac{v_o - v^-}{R_f}\right) + v^-\left(\frac{1}{R_d} + j\omega C_d\right) \qquad \text{(A5.2.1)}$$

where v^- is the voltage at the inverting input and we have assumed that no current flows into the inverting input due to the high input impedance of the op-amp.

Since for the op-amp, $v_o = A(v^+ - v^-)$ where A is the gain and $v^+ \cong 0$ in Figure A5.2.2, then from (A5.2.1) we can write the output signal voltage as:

$$v_o \simeq \frac{-R_f I_d}{1 + j\left(\frac{f}{f_c}\right)} \qquad \text{(A5.2.2)}$$

where the 3 dB bandwidth, $f_c = A/(2\pi R_d C_d)$, and we have assumed that $A \gg 1 + (R_f R_d)$.

Due to the relatively low frequencies employed in WMS, bandwidth limitations are not usually a problem and the output signal is then simply: $v_o \simeq -R_f I_d$.

The noise sources in the transimpedance receiver consist of thermal noise, amplifier voltage and current noise, shot and dark current noise. Consider each of these noise sources in turn where they are represented in Figure A5.2.2 by noise voltage sources, v_{th} and v_{amp} and a noise current source, i_n.

(i) The thermal noise of the feedback resistor is $v_{th}^2 = 4kTR_f$, where k is Boltzmann's constant and T is the temperature. Since the inverting input of the op-amp is at a virtual earth, this noise is directly transferred to the amplifier output, $v_{nf} = v_{th}$.

(ii) The amplifier voltage noise v_{amp} translates to output voltage noise v_{na}, which may be derived by summing currents at node (a) in Figure A5.2.2 to give:

$$\frac{v_{amp} - v_{na}}{R_f} + v_{amp}\left(\frac{1}{R_d} + \frac{1}{(R_a + R_c)} + j\omega C_d\right) = 0 \tag{A5.2.3}$$

Hence the voltage noise from this source is:

$$v_{na} = v_{amp}R_f\left(\frac{1}{R} + j\omega C_d\right) \tag{A5.2.4}$$

where R represents the parallel combination of resistors, $R = R_f//R_d//(R_a + R_c)$

(iii) The current noise source shown in Figure A5.2.2 is used to represent the amplifier current noise, the thermal noise of the parallel combination of resistors, $R_p = R_d//(R_a + R_c)$, the shot noise and the dark current noise as follows:

$$i_n^2 = i_{amp}^2 + \frac{4kT}{R_p} + 2e(I_d + i_d) \tag{A5.2.5}$$

where e is the electronic charge and i_d is the dark current. The voltage noise on the output from this source can be obtained in the same way as (A5.2.2) so $v_{ni} \simeq -R_f i_n$.

Since each of the above noise sources are mutually independent and uncorrelated, the total spectral density of the voltage noise on the output is obtained by adding their mean square values, $v_n^2 = v_{nf}^2 + v_{na}^2 + v_{ni}^2$. Integrating with respect to frequency gives the voltage noise over a frequency bandwidth B as:

$$v_n = R_f\sqrt{\left\{\frac{4kT}{R} + v_{amp}^2\left\{\frac{1}{R^2} + \frac{4\pi^2B^2C_d^2}{3}\right\} + i_{amp}^2 + 2e(I_d + i_d)\right\}B} \tag{A5.2.6}$$

Hence the voltage signal-to-noise ratio of the receiver is:

$$\text{SNR} = \frac{\mathcal{R}P_d}{\sqrt{\left\{\frac{4kT}{R} + v_{amp}^2\left\{\frac{1}{R^2} + \frac{4\pi^2B^2C_d^2}{3}\right\} + i_{amp}^2 + 2e(\mathcal{R}P_d + i_d)\right\}B}} \tag{A5.2.7}$$

With a sufficiently large value of feedback resistor to reduce the thermal noise and with low amplifier and dark current noise (in the photovoltaic mode), the receiver SNR is dominated by shot noise at modest power levels on the detector. With WMS as discussed in Chapter 3, the optical power on the detector consists of a DC level plus the harmonics of the modulation frequency that contain the desired signals to be measured. For example, considering for simplicity just the primary component of the second harmonic signal, we write the power falling on the detector as: $P_d \approx P_{dc}\{1 + a_2 A_0 \cos(2\omega_m t + 2\psi_f) + \cdots\}$ where A_0 is the line centre absorbance and a_2 is the Fourier coefficient shown in Figure 3.4, with a maximum value of 0.343 at line centre for $m = 2.2$. The signal-to-noise ratio for detecting the second harmonic under the shot noise limit at the receiver is then:

$$(SNR)_{shot} \approx a_2 A_0 \sqrt{\frac{\mathcal{R} P_{dc}}{2eB}} \tag{A5.2.8}$$

Hence with a typical value of 0.6 A W^{-1} for the responsivity, the minimum detectable absorbance is $\sim 2 \times 10^{-9}$ W$^{1/2}$ Hz$^{-1/2}$ at an SNR of unity for the peak value of the second harmonic, or a normalised noise-equivalent absorption (NNEA) of $\sim 2 \times 10^{-9}$ cm^{-1} W$^{1/2}$ Hz$^{-1/2}$.

In systems using WMS, the receiver output is usually fed to a lock-in amplifier [96] to extract the various harmonics from the signal. The lock-in may be in either analogue form, as illustrated in Figure A5.2.3, or in digital form where the signal is first fed to an analogue-to-digital converter (ADC) and all subsequent operations, as described below, are performed by digital signal processing. In either case, the lock-in amplifier receives a reference input at the modulation frequency, which is used to extract the magnitude and phase of the desired harmonic component by multiplying the signal by the reference and also by a $\pi/2$-shifted version of the reference, as illustrated in Figure A5.2.3. This multiplication creates sum and difference frequencies and, in particular, the difference frequency is at DC for the harmonic component of the signal at the reference frequency and is extracted by a low-pass filter.

The characteristics of the low-pass filter (time constant, filter order, roll-off, bandwidth and settling time) are important, since these determine the noise bandwidth in the above relationships for the SNR and also set a limit in how fast the output can track changes in the input signal. In practice, an appropriate time constant for the lock-in measurement is selected by the user, taking into account the trade-off

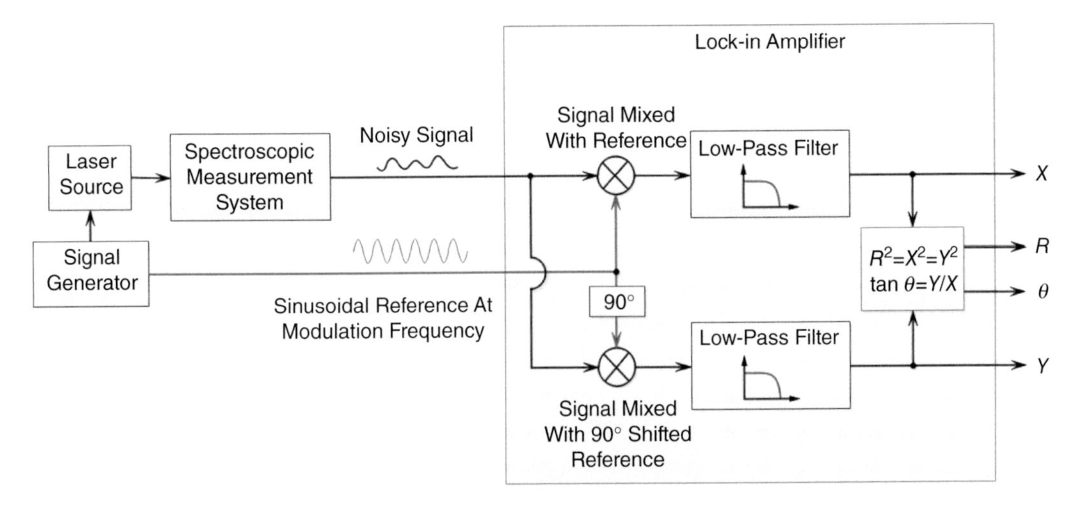

Figure A5.2.3 Schematic of a lock-in amplifier for spectroscopic measurements.

between a long time constant for low noise and a short time constant for a fast response. For example, with WMS, as described in Chapter 3, if the wavelength of the DFB laser is scanned through an absorption line at a typical rate of 10 Hz, then the bandwidth of the low-pass filter must be sufficiently large to capture the rate of change of the second harmonic component across the scan while minimising the noise bandwidth. Typically, a bandwidth of up to 1 kHz (time constant of ~150 μs) may be required to accurately reproduce the second harmonic signal shape for absorption line fitting.

Instead of improving the SNR by reducing the bandwidth, an alternative (and equivalent) method is that of signal averaging. Here, the data collected from repetitive scans over an absorption line is averaged over a specified time interval or number of scans, k. Since the noise from each individual scan is uncorrelated, the noise signal adds according to: $v_t^2 = v_1^2 + v_2^2 + v_3^2 + \cdots$ and the averaged noise over k scans (with $v_1^2 = v_2^2 = v_3^2 = \cdots = v_n^2$) is:

$$v_{av} = \frac{1}{k}\sqrt{\sum_{i=1}^{k} v_i^2} = \frac{\left(\sqrt{k}\right) v_n}{k} = \frac{v_n}{\sqrt{k}} \tag{A5.2.9}$$

Hence averaging reduces the noise by the factor $\sqrt{k}$, which is equivalent to reducing the detection bandwidth by k for the non-averaged case.

There is, however, a limit in the sensitivity improvement that can be achieved through the improvement in the SNR from signal averaging [97]. The above analysis has centered on receiver (white) noise, but in practice there are a number of other noise sources in spectroscopic systems, such as from drift, jitter, etalon fringes, source noise or transmission fluctuation through thermal and mechanical disturbances, etc. For optimum sensitivity, these noise sources must be minimised through techniques such as referencing, as discussed in Chapter 3 and adaptive signal processing [98, 99]. A useful way to characterise the sensitivity of a spectroscopic system and to identify the optimum averaging time is through the Allan–Werle plot [97, 100], which graphically illustrates how the sensitivity improves with integration time before reaching an optimum value after which system drift takes over.

To illustrate, consider making repetitive scans over an absorption line at a rate of 10 Hz (scan time of 0.1 s) and extracting the gas concentration from the peak value of the second harmonic signal, as explained in Chapter 3, assuming that the gas concentration and other gas parameters are not changed over the course of the experiment. Let us say that we perform the scans over a period of 1000 s so that 10^4 data points are recorded for peak values. The recorded data values may be then averaged over, say, 1 s (10 points averaged), generating a set of 10^3 averaged values, $A_1, A_2, A_3 \ldots$, etc. The single-value Allan variance of two adjacent averaged values in this set is defined by [97, 100]:

$$\sigma_A^2 = \frac{1}{2}\left(A_{s+1} - A_s\right)^2 \tag{A5.2.10}$$

This may be calculated for every adjacent pair in the set and then averaged to give the Allan variance as:

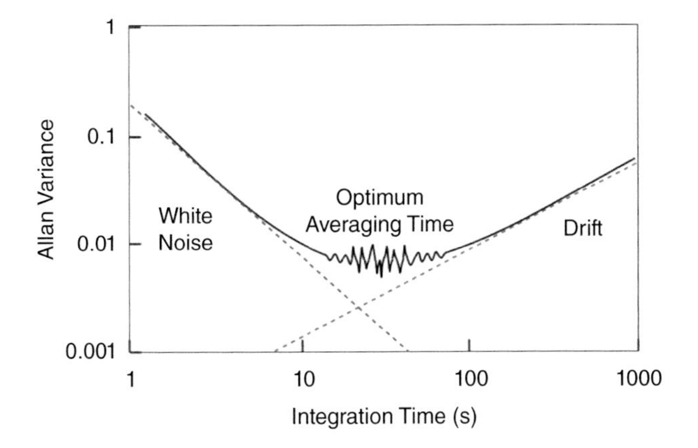

Figure A5.2.4 Typical form of the Allan–Werle plot for a spectroscopic measurement system.

$$\sigma_A^2 = \frac{1}{m}\sum\nolimits_{s=1}^{m} \frac{1}{2}(A_{s+1} - A_s)^2 \qquad \text{(A5.2.11)}$$

where $m = 999$ for the above choice of a 1 s averaging time.

Of course, the analysis may be repeated for a range of averaging (or integration) times, τ, so we may write generally:

$$\sigma_A^2(\tau) = \frac{1}{2m}\sum\nolimits_{s=1}^{m} [A_{s+1}(\tau) - A_s(\tau)]^2 \qquad \text{(A5.2.12)}$$

The Allan variance may then be plotted versus the integration time to give the characteristic Allan–Werle plot for a spectroscopic system, as illustrated in Figure A5.2.4.

Note from Figure A5.2.4 that at short averaging times, the sensitivity is dominated by white noise limitations, while at long averaging times, system drift becomes the limiting factor. Between these extremes, best sensitivity is attained at an optimum averaging time, which is dependent on the actual system and its noise characteristics. The Allan–Werle plot is a very useful tool for determining this optimum averaging or integration time.

References

1. C. S. Goldenstein, R. M. Spearrin, J. B. Jeffries and R. K. Hanson, Infrared laser-absorption sensing for combustion gases, *Prog. Energy Combust. Sci.*, 60, 132–176, 2016.
2. C. Liu and L. Xu, Laser absorption spectroscopy for combustion diagnosis in reactive flows: a review, *Appl. Spectr. Rev.*, 54, (1), 1–44, 2018.
3. F. K. Tittel, R. Lewicki, R. Lascola and S. McWhorter, Emerging infrared laser absorption spectroscopic techniques for gas analysis, in *Trace Analysis of Specialty and Electronic Gases*, W. M. Geiger and M. W. Raynor, Eds., Hoboken, New Jersey, John Wiley & Sons, Inc., ch. 4, 71–109, 2013.
4. J. Hodgkinson and R. P. Tatam, Optical gas sensing: a review, *Meas. Sci. Technol.*, 24, (1), 1–59, 2013.

5. P. T. Mosely, *Solid State Gas Sensors, (Adam Hilger Series on Sensors)*, B. C. Tofield, Ed., Bristol, UK, CRC Press, 1987.
6. J. U. White, Long optical paths of large aperture, *J. Opt. Soc. Am.*, 32, 285–288, 1942.
7. D. R. Herriott, H. Kogelnik and R. Kompfner, Off-axis paths in spherical mirror interferometers, *Appl. Opt.*, 3, (4), 523–526, 1964.
8. D. R. Herriott and H. J. Schulte, Folded optical delay lines, *Appl. Opt.,* 4, (8), 883–889, 1965.
9. J. Altmann, R. Baumgart and C. Weitkamp, Two-mirror multipass absorption cell, *Appl. Opt.*, 20, (6), 995–999, 1981.
10. J. B. McManus, P. L. Kebabian and M. S. Zahniser, Astigmatic mirror multipass absorption cells for long-path-length spectroscopy, *Appl. Opt.,* 34, (18), 3336–3348, 1995.
11. Aerodyne Research Inc. Astigmatic multipass absorption cells. 2019. [Online]. Available: www.aerodyne.com/products/astigmatic-multipass-absorption-cells (accessed April 2020)
12. Photonics Technologies Ltd. Herriott-type cell. 2018. [Online]. Available: www.photonicstechnologies.com/gas-cells/herriott-type-cell/cmp-30-st.html (accessed April 2020)
13. Wavelength References. Fiber-coupled reference cells. 2019. [Online]. Available: www.wavelengthreferences.com/product/gas-cells/ (accessed April 2020)
14. G. Stewart, A. Mencaglia, W. Philp and W. Jin, Interferometric signals in fibre optic methane sensors with wavelength modulation of the DFB laser, *IEEE J. Lightwave Technol.*, 16, (1), 43–53, 1998.
15. K. A. Busch and M. A. Busch, Eds., *Cavity-Ringdown Spectroscopy: An Ultratrace-Absorption Measurement Technique*, Oxford, UK, Oxford University Press, 1998.
16. D. Romanini, I. Ventrillard, G. Méjean, J. Morville and E. Kerstel, Introduction to cavity enhanced absorption spectroscopy, in *Cavity-Enhanced Spectroscopy and Sensing* (Springer Series in Optical Sciences, vol. 179), G. Gagliardi and H.-P. Loock, Eds., Berlin, Springer, 2014, ch. 1, 1–60.
17. M. Mazurenka, A. J. Orr-Ewing, R. Peverall and G. A. D. Ritchie, Cavity ring-down and cavity enhanced spectroscopy using diode lasers, *Annu. Rep. Prog. Chem., Sect. C*, 101, 100–142, 2005.
18. R. D. Van Zee, J. T. Hodges and J. P. Looney, Pulsed, single-mode cavity ringdown spectroscopy, *Appl. Opt.*, 38, (18), 3951–3960, 1999.
19. A.W. Liu, S. Kassi and A. Campargue, High sensitivity CW-cavity ring down spectroscopy of CH_4 in the 1.55μm transparency window, *Chem. Phys. Lett.*, 447, 16–20, 2007.
20. Y. Ding, P. Macko, D. Romanini, et al., High sensitivity CW-cavity ring down and Fourier transform absorption spectroscopies of $^{13}CO_2$, *J. Mol. Spectrosc.*, 226, 146–160, 2004.
21. R. Engeln, G. von Helden, G. Berden and G. Meijer, Phase shift cavity ring down spectroscopy, *Chem. Phys. Lett.*, 262, 105–109, 1996.
22. E. Hecht and A. Zajac, *Optics*, Reading, MA, Addison-Wesley, 306–309, 1974.
23. R. W. P. Drever, J. L. Hall, F. V. Kowalski, et al., Laser phase and frequency stabilization using an optical resonator, *Appl. Phys. B,* 31, 97–105, 1983.
24. J. B. Paul, L. Lapson and J. G. Anderson, Ultrasensitive absorption spectroscopy with a high-finesse optical cavity and off-axis alignment, *Appl. Opt.*, 40, (27), 4904–1020, 2001.
25. G. S. Engel, W. S. Drisdell, F. N. Keutsch, E. J. Moyer and J. G. Anderson, Ultrasensitive near-infrared integrated cavity output spectroscopy technique for detection of CO at 1.57μm: new sensitivity limits for absorption measurements in passive optical cavities, *Appl. Opt.*, 45, (36), 9221–9229, 2006.

26. J. Morville, D. Romanini, A. A. Kachanov and M. Chenevier, Two schemes for trace detection using cavity ringdown spectroscopy, *Appl. Phys. B*, 78, 465–476, 2004.
27. J. Morville, S. Kassi, M. Chenevier and D. Romanini, Fast, low-noise, mode-by-mode, cavity-enhanced absorption spectroscopy by diode-laser self-locking, *Appl. Phys. B*, 80, 1027–1038, 2005.
28. H. Li and N. B. Abraham, Analysis of the noise spectra of a laser diode with optical feedback from a high-finesse resonator, *IEEE J. Quant. Electron.*, 25, (8), 1782–1793, 1989.
29. D. Uttam and B. Culshaw, Precision time domain reflectometry in optical fiber systems using a frequency modulated continuous wave ranging technique, *IEEE J. Lightwave Technol.*, 3, (5), 971–977, 1985.
30. J. Morville, D. Romanini and E. Kerstel, Cavity enhanced absorption spectroscopy with optical feedback in *Cavity-Enhanced Spectroscopy and Sensing* (Springer Series in Optical Sciences, vol. 179), G. Gagliardi and H.-P. Loock, Eds., Berlin, Springer, ch. 5, 163–209, 2014.
31. Ap2e. Enhanced IR laser technology. 2019. [Online]. Available: www.ap2e.com/en/enhanced-ir-laser-technology/ (accessed April 2020)
32. J. Ye, L.-S. Ma and J. L. Hall, Ultrasensitive detections in atomic and molecular physics: demonstration in molecular overtone spectroscopy, *J. Opt. Soc. Am.*, 15, (1), 6–15, 1998.
33. L.-S. Ma, J. Ye, P. Dubé and J. L. Hall, Ultrasensitive frequency modulation spectroscopy enhanced by a high-finesse cavity: theory and application to overtone transitions of C_2H_2 and C_2HD, *J. Opt. Soc. Am.*, 16, (12), 2255–2268, 1999.
34. O. Axner, P. Ehlers, A. Foltynowicz, I. Silander and J. Wang, NICE-OHMS – frequency modulation cavity-enhanced spectroscopy – principles and performance in *Cavity-Enhanced Spectroscopy and Sensing* (Springer Series in Optical Sciences, vol. 179), G. Gagliardi and H.-P. Loock, Eds., Berlin, Springer, ch. 6, 210–251, 2014.
35. B. M. Siller and B. J. McCall, Applications of NICE-OHMS to molecular spectroscopy in *Cavity-Enhanced Spectroscopy and Sensing* (Springer Series in Optical Sciences, vol. 179), G. Gagliardi and H.-P. Loock, Eds., Berlin, Springer, ch. 7, 253–270, 2014.
36. C. L. Bell, G. Hancock, R. Peverall, et al., Characterization of an external cavity diode laser based ring cavity NICE-OHMS system, *Opt. Express*, 17, (12), 9834–9839, 2009.
37. A. Foltynowicz, F. M. Schmidt, W. Ma and O. Axner, Noise-immune cavity-enhanced optical heterodyne molecular spectroscopy: current status and future potential, *Appl. Phys. B*, 92, 313–316, 2008.
38. G. Stewart, F. A. Muhammad and B. Culshaw, Sensitivity improvement for evanescent wave gas sensors, *Sens. Actuators B: Chem*, 11, 521–524, 1993.
39. G. Stewart and B. Culshaw, Optical waveguide modelling and design for evanescent field chemical sensors, *Opt. Quant. Electron.*, 26, 249–259, 1994.
40. W. Jin, G. Stewart, M. Wilkinson, et al., Compensation for surface contamination in a D-fibre evanescent wave methane sensor, *IEEE J. Lightwave Technol.,* 13, (6), 1177–1183, 1995.
41. W. Jin, G. Stewart and B. Culshaw, A liquid contamination detector for D-fibre sensors using white light interferometry, *Meas. Sci. Technol.*, 6, 1471–1475, 1995.
42. G. Stewart, W. Jin and B. Culshaw, Prospects for fibre optic evanescent field gas sensors using absorption in the near-infrared, *Sens. Actuators B: Chem.*, 38, 42–7, 1997.
43. C. R. Lavers, K Itoh, S. C. Wu, et al., Planar optical waveguides for sensing applications, *Sens. Actuators B: Chem.*, 69, 85–95, 2000.
44. Phoenix Photonics. Side polished optical fibers. 2019. [Online]. Available: www.phoenixphotonics.com/website/technology/side-polished-fibers.html (accessed April 2020)

45. S. McCulloch, G. Stewart, R. M. Guppy and J. O. W. Norris, Characterisation of TiO_2-SiO_2 sol-gel films for optical chemical sensor applications, *Int. J. Optoelectron.*, 9, (3), 235–241, 1994.
46. L. Tombez, E. J. Zhang, J. S. Orcutt, S. Kamlapurkar and W. M. J. Green, Methane absorption spectroscopy on a silicon photonic chip, *Optica*, 4, (11), 1322–1325, 2017.
47. W. M. J. Green, E. J. Zhang, C. Xiong, et al., Silicon photonic gas sensing. *ICAn Optical Fiber Communication Conference (OFC) 2019*, San Diego, CA, USA, OSA Technical Digest 2019, paper M2J.5.
48. P. St. J. Russell, Photonic-crystal fibers, *IEEE J. Lightwave Technol.,* 24, (12), 4729–4749, 2006.
49. J. Knight, D. Hand and F. Yu, Hollow-core optical fibers offer advantages at any wavelength, *Photonics Spectra*, 53, (4), 53–57, 2019.
50. T. Ritari, J. Tuominen, H. Ludvigsen, et al., Gas sensing using air-guiding photonic bandgap fibers, *Opt. Express,* 12, 4080–4087, 2004.
51. Y. L. Hoo, S. Liu, H. L. Ho and W. Jin, Fast response microstructured optical fiber methane sensor with multiple side-openings, *IEEE Photon. Technol. Lett.*, 22, (5), 296–298, 2010.
52. W. Jin, H. L. Ho, Y. C. Cao, J. Ju and L.F. Qi, Gas detection with micro- and nano-engineered optical fibers, *Opt. Fiber Technol.*, 19, 741–759, 2013.
53. F. Yang, W. Jin, Y. Cao, H. L. Ho and Y. Wang, Towards high sensitivity gas detection with hollow-core photonic bandgap fibers, *Opt. Express*, 22, (20), 2014.
54. G. Stewart, C. Tandy, D. Moodie, M. A. Morante and F. Dong, Design of a fibre optic multi-point sensor for gas detection, *Sens. Actuators B: Chem.*, 51, (1–3), 227–232, 1998.
55. B. Culshaw, G. Stewart, F. Dong, C. Tandy and D. Moodie, Fibre optic techniques for remote spectroscopic methane detection: from concept to system realisation, *Sens. Actuators B: Chem.*, 51, 1–3, 25–37, 1998.
56. OptoSci Ltd. Innovative fibre optic gas detection systems. 2019. [Online]. Available: www.optosniff.com/ (accessed April 2020)
57. J. Shemshad, Design of a fibre optic sequential multipoint sensor for methane detection using a single tunable diode laser near 1666nm, *Sens. Actuators B: Chem.*, 186, 466–477, 2013.
58. H. L. Ho, W. Jin and M. S. Demokan, Sensitive, multipoint gas detection using TDM and wavelength modulation spectroscopy, *Electron. Lett.*, 36, (14), 1191–1193, 2000.
59. A. J. Hymans and J. Lait, Analysis of a frequency-modulated continuous-wave ranging system, *Proc. IEEE*, 107, 365–372, 1960.
60. J. Zheng, Analysis of optical frequency-modulated continuous-wave interference, *Appl. Opt.*, 43, (21), 4189–4198, 2004.
61. M. Završnik and G. Stewart, Coherent addressing of quasi-distributed absorption sensors by the FMCW method, *IEEE J. Lightwave Technol.*, 18, (1), 57–65, 2000.
62. M. Završnik and G. Stewart, Theoretical analysis of a quasi-distributed optical sensor system using FMCW for application to trace gas measurement, *Sens. Actuators B: Chem.*, 71, 31–35, 2000.
63. M. Završnik and G. Stewart, Analysis of quasi-distributed optical sensors combining rf modulation with the FMCW method, *Opt. Eng.*, 39, (11), 3053–3059, 2000.
64. H. L. Ho, W. Jin, H. B. Yu, et al., Experimental demonstration of a fiber-optic gas sensor network addressed by FMCW, *IEEE Photon. Technol. Lett.*, 12, (11), 1546–1548, 2000.
65. H. B. Yu, W. Jin, H. L. Ho, et al., Multiplexing of optical fiber gas sensors with a frequency-modulated continuous-wave technique, *Appl. Opt.*, 40, (7), 1011–1020, 2001

66. SPIE Optics.org. Laser methane sensor installed at Californian gas storage facility. 2016. [Online]. Available: https://optics.org/news/7/12/17 (accessed April 2020)
67. Environmental Defense Fund. Acutect: continuous open path methane monitors. 2019. [Online]. Available: http://business.edf.org/acutect-continuous-open-path-methane-monitors (accessed April 2020)
68. Physical Sciences Inc. Laser based sensors. 2019. [Online]. Available: www.psicorp.com/products/laser-based-sensors (accessed April 2020)
69. Heath Consultants Inc. Remote methane leak detector, 2019. [Online]. Available: https://heathus.com/products/remote-methane-leak-detector-rmld/ (accessed April 2020)
70. M. B. Frish, PSI – Emerging mobile and airborne TDLAS methods for detecting, locating and quantifying methane leakage, in *International Conference on Field Laser Applications in Industry and Research, (FLAIR 2016)*, Aix-les-Bains, France, 20, 2016.
71. G. M. Gibson, B. Sun, M. P. Edgar, et al., Real-time imaging of methane gas leaks using a single-pixel camera, *Opt. Express*, 25, (4) 2998–3005, 2017.
72. Physical Sciences Inc. Tunable diode laser gas sensors. 2019. [Online]. Available: www.psicorp.com/products/laser-based-sensors/tunable-diode-laser-tdl-gas-sensors (accessed April 2020)
73. Zolo Technologies. Combustion optimization. 2019. [Online]. Available: www.johnzinkhamworthy.com/products-applications/zolo-technologies/ (accessed April 2020)
74. W. Cai and C. F. Kaminski, Tomographic absorption spectroscopy for the study of gas dynamics and reactive flows, *Prog. Energy Combust. Sci.*, 59, 1–31, 2016.
75. T. Benoy, D. Wilson, M. Lengden, et al., Measurement of CO_2 concentration and temperature in an aero engine exhaust plume using wavelength modulation spectroscopy, *IEEE Sens. J.*, 17, (19), 6409–6417, 2017.
76. M. Lengden, D. Wilson, I. Armstrong, et al., Fibre laser imaging of gas turbine exhaust species – a review of CO_2 aero engine imaging in *OSA Advanced Photonics Congress, Boston, Massachusetts, USA, 27 June–1 July, 2015*, paper JM 3A.37.
77. D. Wilson, G. S. Humphries, T. Benoy, et al., Working towards cleaner air travel: the technology behind the FLITES project, in *International Conference on Field Laser Applications in Industry and Research, (FLAIR 2016), Aix-les-Bains, France, 12–16 Sept., 2016*, 124.
78. Y. Feng, J. Nilsson, S. Jain, et al., LD-seeded thulium-doped fibre amplifier for CO_2 measurements at 2μm in *6th EPS QEOD Europhoton Conference (Europhoton 2014), Neuchatel, Switzerland, 24–29 August 2014*, Poster TuP-T1-P-12.
79. C. Liu, Z. Cao, Y. Lin, L. Xu and H. McCann, Online cross-sectional monitoring of a swirling flame using TDLAS tomography, *IEEE Trans. Instrum. Meas.*, 67, (6), 1338–1348, 2018.
80. Q. Qu, Z. Cao, L. Xu, et al., Reconstruction of two-dimensional velocity distribution in scramjet by laser absorption spectroscopy tomography, *Appl. Opt.*, 58, (1), 201–212, 2019.
81. S.-A. Tsekenis, K. G. Ramaswamy, N. Tait, et al., Chemical species tomographic imaging of the vapour fuel distribution in a compression-ignition engine, *Int. J. Engine Res.*, 19, (7), 718–731, 2018.
82. P. Werle, *Laser Optical Sensors for In-Situ Gas Analysis*, Recent Research Developments in Optical Engineering, vol. 2, *Research Signpost*, 1–20, 1999.
83. D. M. Bailey, E. M. Adkins and J. H. Miller, An open-path tunable diode laser absorption spectrometer for detection of carbon dioxide at the Bonanza Creek Long−Term Ecological Research Site near Fairbanks, Alaska, *Appl. Phys. B.*, 123, (245), 2017.
84. S. B. Schoonbaert, D. R. Tyner and M. R. Johnson, Remote ambient methane monitoring using fiber−optically coupled optical sensors, *Appl. Phys. B.*, 119, 133–142, 2015.

85. B. L. Ford, Atmospheric sensing: TDLAS atmospheric water vapor sensing improves weather forecasting, *Laser Focus World*, 54, (8), 2018.
86. B. L. Ford. TDLAS atmospheric water vapor sensing improves weather forecasting. 2019. [Online]. Available: www.laserfocusworld.com/test-measurement/spectroscopy/article/16555201/photonics-applied-atmospheric-sensing-tdlas-atmospheric-water-vapor-sensing-improves-weather-forecasting (accessed April 2020)
87. SpectraSensors Inc. WVSS-II: Atmospheric water vapor sensing system. 2019. [Online]. Available: www.spectrasensors.com/wvss/ (accessed April 2020)
88. Knestel Technologie & Elektronik GmbH. Gasanalytics. 2019. [Online]. Available: www.knestel.de/en/technology-fields/gasanalytics.html (accessed April 2020)
89. Axetris A. G. Laser gas detection. 2019. [Online]. Available: www.axetris.com/en/lgd/products (accessed April 2020)
90. Los Gatos Research Inc. Trace gas analysers. 2019. [Online]. Available: www.lgrinc.com/ (accessed April 2020)
91. Nanosystems & Technologies GmbH. DFB lasers & applications. 2019. [Online]. Available: https://nanoplus.com/ (accessed April 2020)
92. Hamamatsu Photonics K. K. Infrared detectors. 2019. [Online]. Available: www.hamamatsu.com/eu/en/product/optical-sensors/infrared-detector/index.html (accessed April 2020)
93. Thorlabs Inc. Detectors. 2019. [Online]. Available: www.thorlabs.com/navigation.cfm?guide_id=36 (accessed April 2020)
94. J. Gowar, The receiver amplifier in *Optical Communication Systems*, London, England, Prentice Hall, ch. 14, 411–424, 1984.
95. G. P. Agrawal, Optical receivers, in *Fiber-Optic Communication Systems*, 3rd edn., New York, USA, John Wiley & Sons, Inc., ch. 4, 133–182, 2002.
96. Zurich Instruments. Principles of lock-in detection. [Online]. Available: www.zhinst.com/applications/principles-of-lock-in-detection (accessed April 2020)
97. P. Werle, R. Miicke and F. Slemr, The limits of signal averaging in atmospheric trace-gas monitoring by tunable diode laser absorption spectroscopy (TDLAS), *Appl. Phys. B*, 57, 131–139, 1993.
98. P. W. Werle, P. Mazzinghi, F. D'Amato, et al. Signal processing and calibration procedures for in situ diode laser absorption spectroscopy, *Spectrochim. Acta A*, 60, 1685–1705, 2004.
99. P. Werle, Accuracy and precision of laser spectrometers for trace gas sensing in the presence of optical fringes and atmospheric turbulence, *Appl. Phys. B*, 102, 313–329, 2011.
100. D. W. Allan, Statistics of atomic frequency standards, *Proc. IEEE*, 54, (2), 221–230, 1966.

6 Principles of Fibre Amplifiers and Lasers for Near-IR Spectroscopy

6.1 Introduction

Fibre amplifiers, particularly erbium-doped fibre amplifiers (EDFAs), are currently widely used in fibre optic data and communication systems, but there are a number of emerging applications for fibre amplifiers and lasers in near-IR gas spectroscopy. This includes basic applications, such as fibre amplifiers for optical power boosting in large-area and multiplexed fibre sensor networks or use of amplified spontaneous emission (ASE) and fibre lasers as sources in spectroscopic gas sensors. Of particular note has been the intense interest in recent years in the application of frequency combs to spectroscopy, and these combs are commonly generated by mode-locked and highly stabilised erbium-doped fibre lasers. Other applications for fibre amplifier and laser systems include intra-cavity laser absorption spectroscopy (ICLAS) and ring-down cavity spectroscopy. We shall discuss in Chapter 7 current and potential future applications, but here we present the fundamental principles involved in the operation of fibre amplifiers and lasers so that an in-depth understanding can be gained of the practical systems discussed in Chapter 7, as well as the potential for future research and development. However, since it is the objective of this book to provide a fundamental understanding of spectroscopy, it is beyond the scope of this chapter to give an exhaustive account of the theory, properties and characteristics of all the different types of rare-earth-doped fibre systems. Rather we focus on the most developed and widely used system, namely erbium-doped fibre amplifiers and lasers, with the knowledge that similar principles govern the operation of all types. Also in the presentation of the theory we do not consider in detail all the complex effects that may affect their operation, such as non-radiative processes and ion-to-ion interactions, but the reader is referred to the references [1–5] for further information if required. We conclude the chapter by reviewing the principles of fibre lasers based on Raman scattering, but which nevertheless make use of erbium fibre amplifier or laser systems as a pump source.

6.2 Rare Earth Elements for Fibre Amplifiers and Lasers

The most common rare earth elements used in the doping of fibre for amplifiers and lasers are found in the lanthanide series, which is normally appended as a horizontal strip at the foot of the periodic table, with atomic numbers ranging from 57 to 71. They are referred to as 'rare earths' because they are found in less common mineral rocks such as monazite, bastnasite and xenotime. The outer electronic structure of the atom is the same for all the rare earths in the lanthanide series and consists of the filled shells: $5s^2 5p^6 6s^2$. Across the elements of the lanthanide series, the *4f* shell is progressively filled, starting with lanthanum with no electrons in the *4f* shell and ending with lutetium with 14 electrons in the *4f* shell. Normally when doped in fibre, the rare earth element is in trivalent ionised form, such as Er^{3+}, and this corresponds to the removal of the two electrons from the $6s^2$ shell and one from the *f* shell (with the exception of gadolinium and lutetium, where an electron is removed from the *5d* shell) . However, it is important to note that the *4f* shell is buried within, or shielded by, the filled $5s^2$ $5p^6$ shells in the ion and it is the properties and energy levels of the *4f* shell that give rise to the rich optical spectra and gain characteristics of rare earth-doped fibre amplifiers and lasers.

As we shall explain in detail for erbium ions, the energy levels of the *4f* electrons are split into sub-levels by electron–electron interactions and spin–orbit coupling. In addition, the host material, in this case the silica glass fibre, has an important effect on the energy levels and the absorption and emission spectra of the rare-earth-doped fibre. The electric field surrounding the ion from the host material causes a Stark splitting of the energy levels and phonon interactions with these Stark levels produce a homogeneous broadening of the absorption and emission spectra. There is also a small degree of inhomogeneous broadening due to a 'site-to-site' variation in the environment of each ion. Clearly, differences in glass composition, as may be used for special types of optical fibre, will also have a bearing on the fine details of the energy levels and the on degree of broadening. The most common rare earth elements used for fibre amplifiers and lasers in the near-IR, along with their approximate range of wavelength operation, are listed in Table 6.1.

As noted, it is the energy levels and properties of the *4f* electrons that give rise to the spectral characteristics of rare-earth-doped fibre amplifiers and lasers. We shall examine in detail the *4f* electron energy levels and the consequent absorption and emission cross-sections for the particular case of erbium-doped fibre, but similar principles and notation [1–7] apply for all the rare earth elements, as given in Table 6.1.

Table 6.1 Some rare earth elements used for near-IR fibre amplifiers and lasers

Rare earth ion	Pump wavelengths (nm)	Emission bands (nm)
Ytterbium, Yb^{3+}	915–976	1010–1150
Neodymium, Nd^{3+}	810, 880	860–940, 1040–1080, 1280–1420
Erbium, Er^{3+}	980, 1480	1500–1620
Thulium, Tm^{3+}	790, 1660	1700–2100
Holmium, Ho^{3+}	1100	2000–2200

6.3 Spectral Characteristics of Erbium-Doped Fibre

6.3.1 Energy Levels of Erbium Ions in Erbium-Doped Fibre

In this section we provide a brief overview of the background and notation used for describing the most important energy levels associated with erbium-doped fibre. For erbium atoms with 68 electrons, the electron configuration is: [Xe] $4f^{12}$ $6s^2$, where the Xenon [Xe] configuration with 54 electrons is: $1s^2$ $2s^2$ $2p^6$ $3s^2$ $3p^6$ $3d^{10}$ $4s^2$ $4p^6$ $4d^{10}$ $5s^2$ $5p^6$. Here, *s, p, d, f* are the labels for the various electron orbitals which represent the probability functions, derived from quantum mechanics, of finding an electron in the space around an atom. According to the Pauli exclusion principle each electron has a unique set of quantum numbers (n, l, m, s), where n is the principal quantum number, l is the angular momentum quantum number, m is the magnetic quantum number and s is the spin quantum number. Table 6.2 shows the quantum numbers for each orbital up to *4f* and the number of combinations and hence electrons allowed for each orbital.

The removal of three electrons to form the erbium ion, Er^{3+}, results in the configuration [Xe] $4f^{11}$, but as noted earlier, it is important to realise that the filled $5s^2$ and $5p^6$ shells are actually outermost, shielding the *4f* electrons [2], which has an important influence on the absorption and emission properties of erbium-doped fibres.

In general, the principal quantum number, n, is the most important in determining the energy levels of electrons, but for multi-electron atoms the energy levels are influenced by other factors, particularly electron–electron interactions and spin–orbit coupling (an orbiting electron produces a small magnetic field in which the spin magnetic moment of an electron is immersed). For rare earth ions, such as Er^{3+}, the effect of electron–electron interactions is dominant over spin–orbit coupling and the $4f^{11}$ electron orbital in Er^{3+} splits into a number of energy levels, as shown in Figure 6.1, which are described according to the Russell–Saunders *LSJ* scheme, denoted as $^{2S+1}L_J$.

Table 6.2 Electron orbitals and quantum numbers

Orbital	n	l	m	s	No. of combinations
1s	1	0	0	+½, −½	2
2s	2	0	0	+½, −½	2
2p	2	1	+1, 0, −1	+½, −½	6
3s	3	0	0	+½, −½	2
3p	3	1	+1, 0, −1	+½, −½	6
3d	3	2	+2, +1, 0, −1, −2	+½, −½	10
4s	4	0	0	+½, −½	2
4p	4	1	+1, 0, −1	+½, −½	6
4d	4	2	+2, +1, 0, −1, −2	+½, −½	10
4f	4	3	+3, +2, +1, 0, −1, −2, −3	+½, −½	14

Table 6.3 Quantum number, L, of resultant orbital angular momentum and symbol

L	0	1	2	3	4	5	6	7
Symbol	S	P	D	F	G	H	I	J

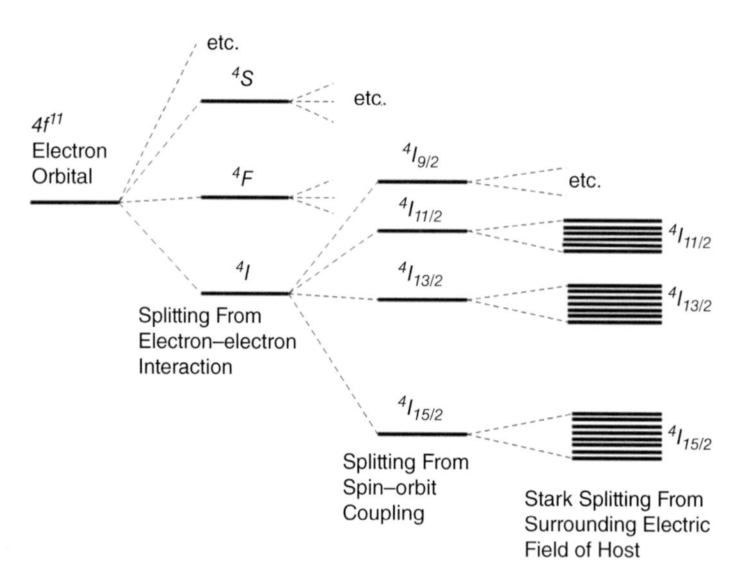

Figure 6.1 Energy levels for the $4f^{11}$ electrons of erbium ions in glass.

Here L is the quantum number of the resultant orbital angular momenta, $\tilde{L}$, of the $4f$ electron (only the $4f$ electrons are considered since the resultant angular momentum of filled shells vanishes due to cancellation from paired electrons), S is the quantum number of the resultant spin angular momenta, $\tilde{S}$, and J is the quantum number of the total angular momentum, $\tilde{J} = \tilde{L} + \tilde{S}$. As indicated in Figure 6.1, in an approximate description we can view the electron–electron interaction as causing the splitting into the ^{2S+1}L levels, which are then further split into the $^{2S+1}L_J$ levels through the spin–orbit interaction. Similar to the notation of Table 6.2, where the various values of the angular momentum quantum number, l, are assigned the small letters s, p, d, f, etc., the various values of the quantum number, L, are assigned a capital letter according to the scheme shown in Table 6.3.

For example, the energy level of the $4f^{11}$ electron of Er^{3+} with quantum numbers $L = 6$, $S = 3/2$ and $J = 15/2$ is denoted as $^4I_{15/2}$. The most important energy levels for erbium fibre amplifiers and lasers are the $^4I_{15/2}$, $^4I_{13/2}$, and $^4I_{11/2}$ levels, where the energy differences between the first and the second, and the first and the third levels correspond to photon energies with wavelengths of ~1550 nm and ~980 nm respectively, as shown in Figure 6.2.

There is a further important effect that we need to consider, which is also illustrated in Figures 6.1 and 6.2. So far, we have considered the Er^{3+} ion in isolation, but when the

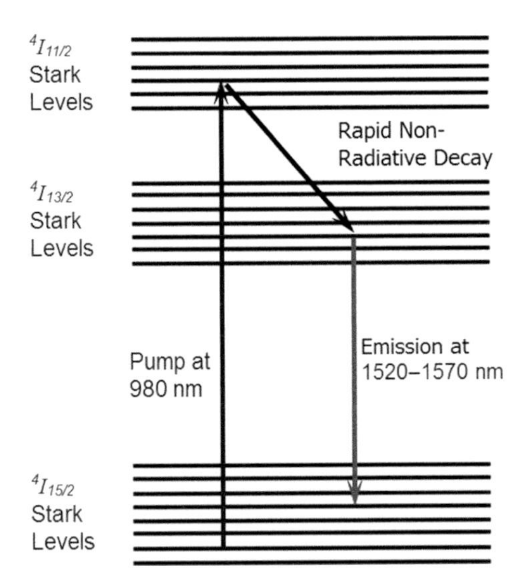

Figure 6.2 The most important energy levels for erbium-doped fibre amplifiers and lasers.

ions are embedded in a glass or crystalline host, such as in glass fibres, the electric field surrounding the ions causes a Stark splitting of the $^{2S+1}L_J$ levels. The splitting arises because the charge surrounding the Er^{3+} ions is non-spherically symmetric and the non-symmetric electric field lifts the degeneracy of the $^{2S+1}L_J$ level and splits it into a maximum of (J + ½) components. As noted earlier, however, since the *4f* electron orbitals are electrostatically shielded by the filled $5s^2$ and $5p^6$ shells, this host field perturbation is about one to two orders of magnitude weaker than the effects of electron–electron interaction and spin–orbit coupling. As a result, the manifold widths from the Stark splitting are relatively small, typically ~400 cm^{-1} in wavenumber units (or 0.8×10^{-20} J in energy units), which is similar in magnitude to $kT \approx 200$ cm^{-1} at room temperature, where k is the Boltzmann constant. We shall explain shortly the significance of this similarity on the absorption and emission characteristics of erbium-doped fibres, but one other feature is worth restating at this point when considering the properties of rare earth ions doped into a glassy network. Due to the random overall structure of glass, each ion experiences a slightly different field and this 'site-to-site' variation of the field gives rise to a small degree of inhomogeneous broadening in the gain characteristics of rare-earth-doped fibre amplifiers or lasers.

6.3.2 Absorption and Emission Properties of Rare-Earth-Doped Fibre

In this section we discuss the principles which give rise to the absorption and emission characteristics of rare-earth-doped fibre and the relationship between the absorption and emission cross-sections, known as the McCumber relationship [8, 9]. This is essential for understanding the properties of fibre amplifiers and lasers.

Figure 6.2 shows the most important energy levels for an erbium-doped fibre amplifier or laser pumped at a wavelength of 980 nm with gain over the typical wavelength range of 1520–1570 nm. (Note that pumping may alternatively be performed at ~1480 nm, between the $^4I_{15/2}$ and $^4I_{13/2}$ levels). An electron excited to the $^4I_{11/2}$ level by 980 nm pumping decays very rapidly to the $^4I_{13/2}$ level through non-radiative decay with a lifetime of ~1 μs or so, whereas the lifetime of the $^4I_{13/2}$ level is very much longer, ~10 ms. Hence, to a good approximation, we need only focus our attention on transitions between and within the $^4I_{13/2}$ and $^4I_{15/2}$ levels in determining the basic gain characteristics of erbium-doped fibres. A very important phenomenon affecting the gain characteristics is the role of phonons and their interaction with the Stark energy levels of each manifold, as depicted in Figure 6.3.

Phonons are acoustic waves or thermal vibrations [10] within the host material in which the erbium ions are embedded. In crystalline materials these thermal acoustic waves, or phonons, can propagate in different directions and over a range of frequencies. Although in glass the propagation characteristics of these acoustic waves are not so regular or clearly defined due to the random nature of the glassy network, they nevertheless have an important influence on the erbium ions. In the same way as an electromagnetic (optical) wave at a particular frequency can interact with a pair of appropriately spaced energy levels and either be absorbed or amplified (by stimulated emission), acoustic phonons can likewise interact with appropriately spaced energy levels resulting in either absorption or emission of phonons with the consequent promotion or demotion of an electron to a higher or lower level. If the separation between the levels involved can be spanned by one or two phonons, then the phonon absorption/emission process occurs very rapidly, on the order of picosecond or sub-picosecond timescales. Since, as noted earlier, the widths of the manifolds in Figure 6.3 are around 400 cm^{-1} (spacing between individual Stark levels of ~60–80 cm^{-1}), which is of the same order as the thermal energy factor, $kT \approx 200$ cm^{-1} at room temperature, then phonon interactions occur rapidly, producing a thermal

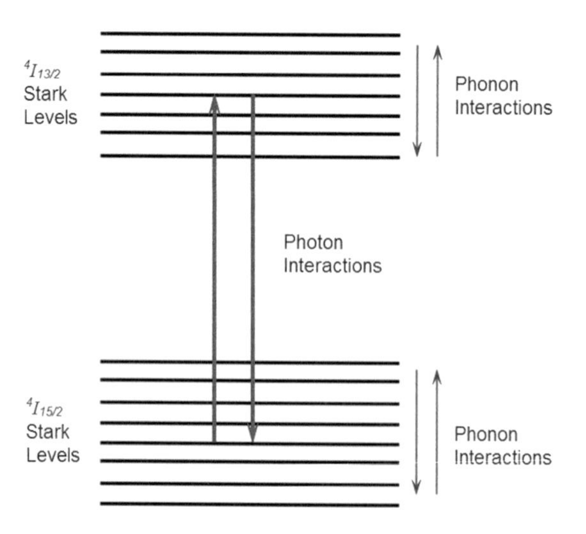

Figure 6.3 Phonon and photon interactions with Stark energy levels.

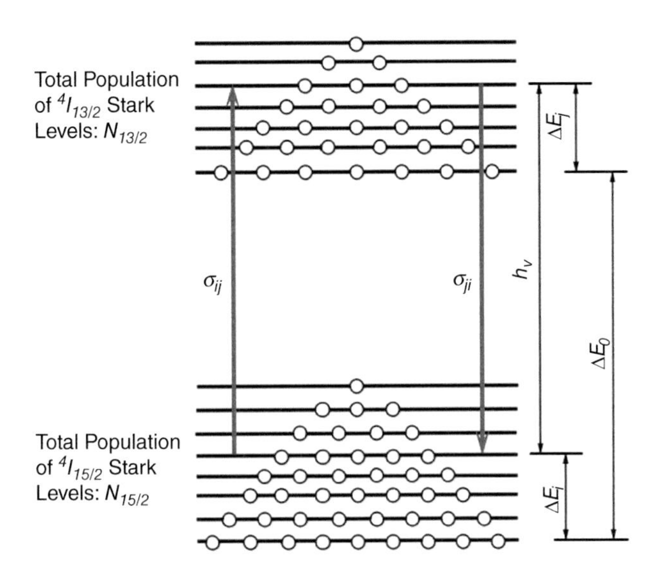

Figure 6.4 Illustration of the thermalisation of the population of the Stark levels for each manifold.

distribution ('thermalisation') of the population of the Stark levels *within* each of the $^4I_{13/2}$ and $^4I_{15/2}$ manifolds. Compare this with interactions *between* the $^4I_{13/2}$ and $^4I_{15/2}$ levels, where the spacing is very much greater, corresponding to photon energies, as shown in Figure 6.3. Thus if we have a collection of erbium ions in glass and a number of their $4f^{11}$ electrons are either raised to the $^4I_{13/2}$ level through photon absorption or lowered to the $^4I_{15/2}$ level through photon emission, the relative occupation of the Stark levels within the $^4I_{13/2}$ and $^4I_{15/2}$ levels will be rapidly thermalised through phonon interactions on a timescale much faster than the optical processes. This is illustrated in Figure 6.4.

As depicted in Figure 6.4 and explained in Appendices 6.1 and 6.2, photons will interact between the individual Stark levels of the $^4I_{15/2}$ and $^4I_{13/2}$ manifolds with individual emission and absorption cross-sections. However, the overall process can be described by an 'effective' emission cross-section, $\sigma_e(\nu)$, and absorption cross-section, $\sigma_a(\nu)$, which are related through the McCumber equation:

$$\sigma_e(\nu) = \sigma_a(\nu)\exp\left\{\frac{\varepsilon - h\nu}{kT}\right\} \tag{6.1}$$

To emphasise, this McCumber relationship arises as a result of the rapid thermalisation of the Stark energy levels within each manifold through phonon interactions on a timescale much faster than the optical processes and is the appropriate description for the absorption and emission properties of rare-earth-doped glasses. In contrast, the simpler Einstein relation (see Appendices 6.1 and 6.2) is based on equal occupation of all the sub-levels within a manifold, but is not valid for typical rare-earth-doped fibres. Also, as shown in Appendix 6.2, the McCumber relationship for erbium-doped fibre may be approximated [11] as:

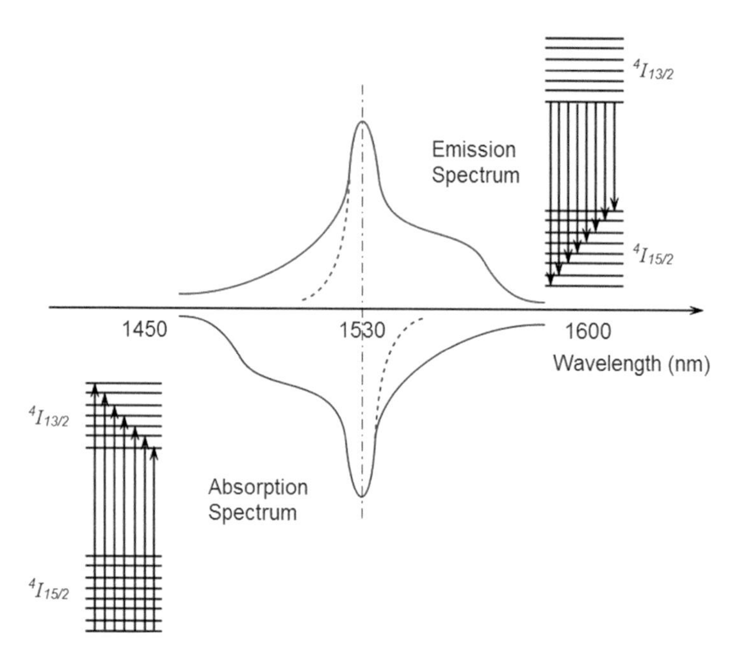

Figure 6.5 Typical form of the absorption and emission spectra for erbium-doped fibre at room temperature and narrowing at low temperature (dashed line), with the corresponding low-temperature transitions between the Stark levels.

$$\sigma_e(\nu) = \sigma_a(\nu)\exp\left\{\frac{6550 - \lambda^{-1}}{kT}\right\} \tag{6.2}$$

with $kT \sim 200\ \mathrm{cm}^{-1}$ and the inverse wavelength in units of cm^{-1}.

Figure 6.5 illustrates the typical shape of the absorption and emission spectra at room temperature for erbium-doped fibre. At a low temperature, the absorption spectrum is narrowed on the low-energy (long-wavelength) side and the emission spectrum is narrowed on the high-energy (short-wavelength side) as indicated by the dashed lines in Figure 6.5. At a very low temperature, only the lowest Stark level of each manifold is occupied and the corresponding transitions between the Stark levels are shown in Figure 6.5. The spectra broaden at room temperature since the occupation of upper Stark levels increases and transitions can occur between many more levels. Transitions between the individual Stark levels are broadened from natural and inhomogeneous broadening effects so the overall absorption and emission spectra are smeared into a continuous distribution.

6.4 Principles of Operation of Fibre Amplifiers and Lasers

For applications in spectroscopy, as discussed in Chapter 7, fibre amplifiers and lasers may be used in various modes of operation. These include basic CW operation for use as a source in spectroscopy, but also multi-wavelength, mode-locked, Q-switched or transient operation depending on the application. In this section we present the fundamental principles of operation based on the atomic rate and cavity rate equations, which enable modelling of the various operating regimes.

6.4.1 Atomic Rate Equation for Fibre Amplifiers

As noted earlier, the most important energy levels for erbium-doped fibre amplifiers or lasers are the three levels shown in Figure 6.2. An electron excited to the $^4I_{11/2}$ level by 980 nm pumping decays very rapidly to the $^4I_{13/2}$ level and so, to a good approximation, we need only focus our attention on the $^4I_{13/2}$ and $^4I_{15/2}$ levels, which we shall now refer to simply as level 2 and level 1, respectively. In Appendix 6.3 it is shown that for an optical fibre of length l with the core doped with erbium with a density of $\rho(r,\phi) = \rho_e \cdot f_e(r,\phi)$, the rate of change of the level 2 population, without pumping, is described by:

$$\rho_e S_e l\left(\frac{\mathrm{d}\overline{N_2}}{\mathrm{d}t}\right) = P_{\mathrm{in}}\left\{\mathrm{e}^{\overline{g(\nu)}\cdot l} - 1\right\} - \rho_e S_e l\frac{\overline{N_2}}{\tau_{21}} \tag{6.3}$$

where P_{in} is the input power to the fibre in photons per second, τ_{21} is the upper state lifetime, S_e is the effective area of the erbium-doping region:

$$S_e = \iint f_e(r,\phi) r \mathrm{d}r \mathrm{d}\varphi \tag{6.4}$$

and is equal to the area of the core, for uniform doping of the core only, if $f_e(r,\phi) = 1$ and zero elsewhere.

The length-averaged fractional inversion level, $\overline{N_2}$, is defined by:

$$\overline{N_2} = \frac{1}{l}\int_0^l N_2^f \mathrm{d}z \tag{6.5}$$

The length-averaged gain $\overline{g(\nu)}$ is:

$$\overline{g(\nu)} = \{\gamma_e(\nu) + \alpha_a(\nu)\}\overline{N_2} - \alpha_a(\nu) \tag{6.6}$$

where $\alpha_a(\nu) = \Gamma\rho_e\sigma_a(\nu)$ is the absorption coefficient, $\gamma_e(\nu) = \Gamma\rho_e\sigma_e(\nu)$ is the emission coefficient and Γ is the overlap between the erbium density distribution and the optical intensity distribution:

$$\Gamma = \frac{\iint f_e(r,\phi) f_i(r,\phi) r \mathrm{d}r \mathrm{d}\phi}{\iint f_i(r,\phi) r \mathrm{d}r \mathrm{d}\phi} \tag{6.7}$$

For a fibre amplifier, as shown in Figure 6.6, we need to include the pump source in (6.3). If the absorption coefficient for the pump light at 980 nm is α_p then the pump power absorbed over the length l of the fibre is $\Delta P_p = P_p\{1 - \mathrm{e}^{-\overline{g_p}\cdot l}\}$ so (6.3) becomes (see also (A6.3.13) in Appendix 6.3):

$$\rho_e S_e l\left(\frac{\mathrm{d}\overline{N_2}}{\mathrm{d}t}\right) = P_p\{1 - \mathrm{e}^{-\overline{g_p}\cdot l}\} - P_{\mathrm{in}}\left\{\mathrm{e}^{\overline{g(\nu)}\cdot l} - 1\right\} - \rho_e S_e l\frac{\overline{N_2}}{\tau_{21}} \tag{6.8}$$

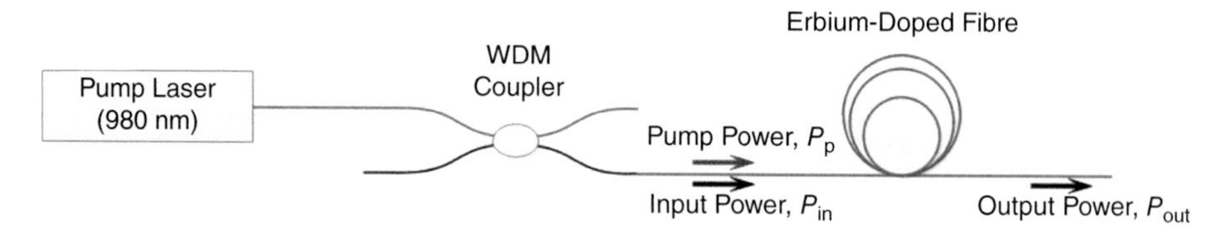

Figure 6.6 Erbium-doped fibre amplifier with a wavelength division multiplexer (WDM) to couple the output from the pump laser to the erbium-doped fibre.

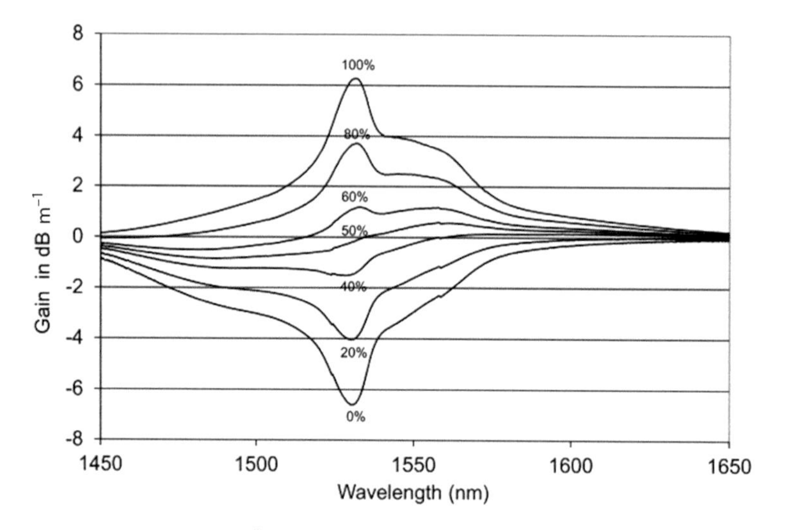

Figure 6.7 Gain in dB m^{-1} for different values of length-averaged inversion level ranging from 0 to 100% inversion.

where $\overline{g_p} = \alpha_p\overline{N_1} = \alpha_p\{1 - \overline{N_2}\}$ and the input pump power, P_p, is also in photons per second.

Equation (6.8) is the basic atomic rate equation describing a fibre amplifier. The output power from the fibre is: $P_{\text{out}} = P_{\text{in}}e^{\overline{g(\nu)}\cdot l}$ and hence gain occurs at inversion levels and wavelengths for which $\overline{g(\nu)} > 0$. Figure 6.7 shows a plot (for typical erbium absorption and emission coefficients) of the gain, $G = e^{\overline{g(\nu)}\cdot l}$, expressed in dB m^{-1} as $10\log\left(e^{\overline{g(\nu)}\cdot l}\right)/l = 4.34\cdot\overline{g(\nu)}$, as a function of wavelength for different values of the length-averaged inversion level, $\overline{N_2}$, using (6.6).

As shown in Figure 6.7, the erbium fibre can potentially provide gain over a large range of near-IR wavelengths between about 1500 to 1600 nm, with greatest gain around 1530 nm. Note that at average inversion levels of <50% there is still gain at longer wavelengths. This can be understood with reference to Figure 6.4, where it can be seen that even if the overall population inversion between the $^4I_{13/2}$ and $^4I_{15/2}$ bands is less than 50%, there still can be population inversion between the *individual* lower Stark levels of the $^4I_{13/2}$ band and the individual upper Stark levels of the $^4I_{15/2}$ band, due to the rapid thermalisation of the Stark level populations. These lower-energy transitions give rise to gain at longer wavelengths. At 50% inversion level, (6.6) gives $\overline{g(\nu)} = 0.5\{\gamma_e(\nu) - \alpha_a(\nu)\}$ and from the McCumber relation (6.2) we see that the gain is: (i) zero at a wavelength of 1527 nm (wavenumber 6550 cm^{-1}), (ii) less than zero (i.e. net absorption) at wavelengths less than 1527 nm and (iii) greater than zero for wavelengths greater than 1527 nm. As discussed later, these principles apply to the operation of L-band erbium-fibre lasers, where the longer-wavelength outputs are achieved by a long length of erbium fibre operating at a low average inversion level.

In applications of fibre amplifiers for optical power boosting in spectroscopy, (6.8) or the more comprehensive version (6.17) given later may be used to calculate the signal gain for given input pump and signal power levels by iteration to find the associated length-averaged inversion level. If more detailed knowledge is required of the power distribution and inversion levels, etc., as a function of position along the length of the erbium-doped fibre, then finite element analysis may be used, where the fibre is subdivided into short lengths of Δl and (6.8) or (6.17) then applies for each length element with the exponential terms replaced by the linear approximation, $e^x \cong 1 + x$. The output of one element then becomes the input of the next. This procedure also allows the modelling of forward and backward pumping or multiple pump sources positioned at various points along the fibre amplifier.

6.4.2 Cavity and Atomic Rate Equations for Fibre Lasers

A simple ring-cavity fibre laser, as shown in Figure 6.8, may be formed from a fibre amplifier by connecting the output of the amplifier to its input and using various optical components, such as isolators, to ensure unidirectional laser operation, a coupler to tap light out of the ring and a tuneable filter if required to select the wavelength of operation.

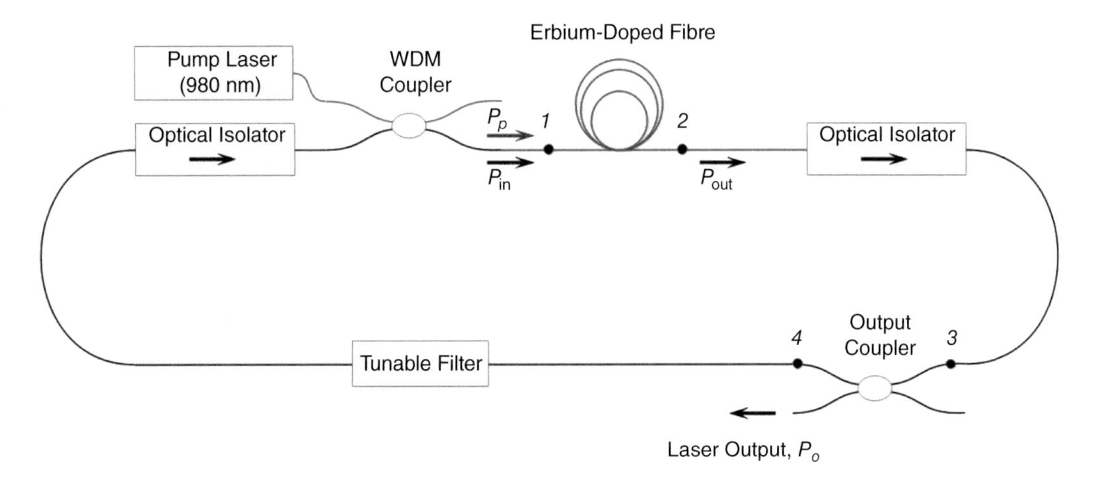

Figure 6.8 Typical fibre ring cavity laser including a tuneable filter and optical isolators.

We now derive a cavity rate equation to describe the laser operation. For simplicity let us initially assume that only a single longitudinal mode of the ring cavity is allowed to operate. Consider one circulation of the light around the fibre ring-cavity. The output power from the erbium-doped fibre amplifier section, P_{out}, at point (2) in Figure 6.8 circulates around the loop, back to the amplifier input at point (1) and in doing so experiences a total optical attenuation of $e^{-\alpha_c(\nu)}$ (or $4.34\alpha_c$ dB) from the various optical components, filters, connectors, couplers, etc., in the loop. The input light to the amplifier at point (1) is therefore $\overline{P_{\text{out}}}\mathrm{e}^{-\alpha_c(\nu)}$, which is then amplified by the erbium-doped fibre to give $\overline{P_{\text{out}}}\mathrm{e}^{-\alpha_c(\nu)}\mathrm{e}^{\overline{g(\nu)}\cdot l} = \overline{P_{\text{out}}}\mathrm{e}^{\overline{g(\nu)}\cdot l-\alpha_c(\nu)}$ at the output, point (2). Hence the change in power over one circulation is $\overline{P_{\text{out}}}\left\{\mathrm{e}^{\overline{g(\nu)}\cdot l-\alpha_c(\nu)}-1\right\}$, with a round-trip time of $\tau = nL_c/c$, where L_c is the total length of the fibre ring cavity, including the erbium fibre length l. Hence, we can write the rate of change of power at the amplifier output as:

$$\frac{\mathrm{d}P_{\text{out}}}{\mathrm{d}t} = \frac{P_{\text{out}}}{\tau}\left\{\mathrm{e}^{\overline{g(\nu)}\cdot l-\alpha_c(\nu)}-1\right\} \tag{6.9}$$

However, there are two further important factors that we need to take into account in the operation of fibre lasers. First, the fibre loop cavity can support a number of longitudinal modes, which may affect the laser operation depending on the degree of optical filtering. Second, there is coupling of the spontaneous emission (SE) to the cavity modes and this SE will experience gain in the fibre amplifier section, resulting in amplified spontaneous emission (ASE). In fact without the coupling of SE to cavity modes, lasing operation could not start spontaneously.

The longitudinal modes of the loop cavity are simply defined by the resonant condition that the optical loop length is equal to an integer number of wavelengths, $nL_c = q\lambda$, where q is an integer, or equivalently, in terms of frequency, $\nu_m = q(c/nL_c)$, with longitudinal mode spacing of $\Delta\nu_m = (c/nL_c)$. For a fibre cavity of typically 20 m in length, the mode spacing is ~10 MHz, so the cavity can support a large number of closely spaced modes.

As explained in Appendix 6.1, spontaneous emission is in fact stimulated emission by the fluctuating vacuum field [12] and the spontaneous emission rate, from (A6.1.13) in Appendix 6.1, over the bandwidth ν to $\nu + \mathrm{d}\nu$ is:

$$\mathrm{d}P_{\text{SE}} = -\left(\frac{\mathrm{d}N_2}{\mathrm{d}t}\right)_{\text{SE}} = \left(\frac{c}{n}\right)N_2\sigma_e(\nu)f_m(\nu)\mathrm{d}\nu \tag{6.10}$$

where N_2 is the population of level 2 and, for the fibre volume element $\mathrm{d}V = S_e\mathrm{d}z$, is given by $N_2 = \rho_e S_e N_2^f \mathrm{d}z$. For emission into free space, $f_m(\nu)\mathrm{d}\nu$ is the number of free-space modes per unit volume over the bandwidth ν to $\nu + \mathrm{d}\nu$. However, in the presence of an optical resonator, spontaneous emission also couples to the cavity modes [13]. Using the expression $\nu_m = q(c/nL_c)$ for the longitudinal modes of the fibre loop cavity, the number of modes $\mathrm{d}q$ in the frequency range ν to $\nu + \mathrm{d}\nu$ is $\mathrm{d}q = 2(nL_c/c)\mathrm{d}\nu = 2\tau\mathrm{d}\nu$, where the factor of two arises from the two possible polarisation states of each mode. The volume of the cavity is $S_m L_c$, where S_m is the cross-sectional area of the mode. Hence the cavity mode density is $f_m^c(\nu) = 2n/(cS_m)$ and, using this expression in

(6.10), the spontaneous emission power, $\mathrm{d}P_{\mathrm{SE}}$, coupled to cavity modes from a length element, $\mathrm{d}z$, of the fibre is:

$$\mathrm{d}P_{\mathrm{SE}} = 2\rho_e\left(\frac{S_e}{S_m}\right)\{\sigma_e(\nu)\mathrm{d}\nu\}N_2^f\mathrm{d}z = 2\{\gamma_e(\nu)\mathrm{d}\nu\}N_2^f\mathrm{d}z = \gamma_e(\nu)\left\{\frac{\mathrm{d}q}{\tau}\right\}N_2^f\mathrm{d}z \quad (6.11)$$

where, as defined above, $\mathrm{d}q$ is the number of modes over the bandwidth ν to $\nu + \mathrm{d}\nu$.

The total SE coupled to the cavity modes in one direction over the erbium fibre length and over the full bandwidth will therefore be:

$$P_{\mathrm{SE}} = \frac{1}{\tau}\left\{\int \gamma_e(\nu)\mathrm{d}q\right\}\left\{\int_0^l N_2^f\mathrm{d}z\right\} = 2\overline{N_2}l\int \gamma_e(\nu)\mathrm{d}\nu \quad (6.12)$$

where we have used $\mathrm{d}q = 2\tau\mathrm{d}\nu$.

However, the spontaneous emission into cavity modes at a position, z, along the fibre will experience gain as it traverses the remaining length, $(l - z)$, of the doped fibre according to $G = \mathrm{e}^{g(\nu)\cdot(l-z)}$. Hence the amplified spontaneous emission (ASE) in the cavity modes, $\mathrm{d}q$, emerging from one end of the fibre is:

$$\mathrm{d}P_{\mathrm{ASE}} = \gamma_e(\nu)\left\{\frac{\mathrm{d}q}{\tau}\right\}\int_0^l N_2^f(z)\mathrm{e}^{g(\nu)\cdot(l-z)}\mathrm{d}z \quad (6.13)$$

To simplify the integration, we replace $N_2^f(z)$by the length-averaged value, $\overline{N_2}$, from (6.5) and obtain the result [14]:

$$\mathrm{d}P_{\mathrm{ASE}} = \left(\frac{\overline{N_2}l}{\tau}\right)\gamma_e(\nu)A_{\mathrm{SE}}(\nu)\mathrm{d}q \quad (6.14)$$

where the amplification factor is given by:

$$A_{\mathrm{SE}}(\nu) = \frac{\mathrm{e}^{\overline{g(\nu)}\cdot l} - 1}{\overline{g(\nu)}\cdot l} \quad (6.15)$$

The total ASE power coupled to all the cavity modes in one direction over the full emission bandwidth is then:

$$P_{\mathrm{ASE}} = 2\overline{N_2}l\int \gamma_e(\nu)A_{SE}(\nu)\mathrm{d}\nu \quad (6.16)$$

We can now modify the atomic rate equation (6.8) and the cavity rate equation (6.9) to take into account the above considerations. The SE coupled to the cavity modes and its subsequent amplification act to reduce the upper level population, so the atomic rate equation becomes:

$$\rho_e S_e l\left(\frac{\mathrm{d}\overline{N_2}}{\mathrm{d}t}\right) = P_p\{1 - \mathrm{e}^{-\overline{g_p}\cdot l}\} - \sum\nolimits_m P_{\mathrm{out}}\left\{1 - \mathrm{e}^{-\overline{g(\nu)}\cdot l}\right\} - R_{\mathrm{ASE}} - \rho_e S_e l\frac{\overline{N_2}}{\tau_{21}} \quad (6.17)$$

The ASE rate, $R_{\mathrm{ASE}} = 2\{P_{\mathrm{ASE}} - P_{\mathrm{SE}}\}$, in (6.17) includes a factor of two to take into account that the ASE occurs in both directions in the fibre (although only one direction affects laser action). Note that the spontaneous emission power, P_{SE}, in the modes as given by (6.12) is subtracted in formulating the expression for R_{ASE} since all the spontaneous emission is already taken into account through the lifetime in the last term, and only the effect of the amplification of the SE on the upper level population needs to be added. Also, in (6.17) there is now a summation over the modes that are excited in the fibre ring cavity and, for convenience, we have replaced P_{in} in (6.8) using $P_{out} = P_{in}e^{\overline{g(\nu)\cdot l}}$.

For each mode that contributes to lasing action there is a cavity rate equation. If there are a very large number of closely spaced longitudinal modes involved in the laser operation, in order to reduce the total number of equations, we can sub-divide these into mode groups and assume that the absorption and emission parameters are reasonably uniform within the group, and hence write a cavity rate equation for each mode group. Considering the spontaneous emission coupled to the modes, we have an additional contribution of dP_{ASE} as given by (6.14) per round-trip cycle, giving a rate of power increase of dP_{ASE}/τ, so the cavity rate equation for each mode or mode group becomes:

$$\frac{dP_{out}}{dt} = \frac{P_{out}}{\tau}\left\{e^{\overline{g(\nu)\cdot l} - \alpha_c(\nu)} - 1\right\} + \left(\frac{\overline{N_2 l}}{\tau^2}\right)\gamma_e(\nu)A_{SE}(\nu)dq \tag{6.18}$$

where dq is the number of modes in the group (equals 2 for single-mode operation to take into account the two polarisation states) and P_{out} is the output power of a single mode, or the combined output power of a mode group, in units of photons per second at the output side of the rare-earth-doped fibre amplifier, that is point (2) in Figure 6.8.

6.5 Regimes of Operation of Fibre Lasers

In this section we show how the atomic rate and cavity rate equations may be used to model the various ways in which fibre lasers may be operated, including CW, transient and mode-locked regimes.

6.5.1 CW Operation

For applications of fibre lasers as a source in spectroscopy, the properties of CW operation are obtained from the steady-state solutions of the rate equations, that is, by setting the LHS of (6.17) and (6.18) to zero. We will assume at this stage that the rare-earth- or erbium-doped fibre is predominantly homogeneously broadened, which means that under steady-state conditions all longitudinal modes are extinguished, except for the most favourable mode(s) at the peak of the gain curve or at the centre of the filter bandwidth. Above threshold, to sustain CW laser operation, the difference between the erbium fibre gain and the cavity loss is very small, so the exponential term in (6.18) can be approximated as, $e^x \simeq 1 + x$, and the steady-state situation becomes:

$$0 = P_{out}\left\{\overline{g_{ss}(\nu_0)\cdot l} - \alpha_c(\nu_0)\right\} + 2\left(\frac{\overline{N_{2ss} l}}{\tau}\right)\gamma_e(\nu_0)A_{SE}(\nu_0) \tag{6.19}$$

where the subscripts denote steady-state values and we assume a single mode of operation at optical frequency ν_0.

Rearranging (6.19), we obtain an expression relating the fibre amplifier gain to the round-trip cavity loss:

$$\overline{g_{ss}(\nu_0)\cdot l} = \alpha_c(\nu_0) - \delta_{\text{ASE}} \tag{6.20}$$

where

$$\delta_{\text{ASE}} = \frac{2\gamma_e(\nu_0)A_{\text{ASE}}(\nu_0)\overline{N_{2ss}}l}{\tau P_{\text{out}}} \tag{6.21}$$

From these expressions we note that the gain at, and above, threshold is slightly less than the cavity loss due to the ASE contribution to the cavity mode. As we will show later, this ASE 'noise' also determines the ultimate limit in linewidth of the laser output [15].

However, when the laser output power is sufficiently high, we can reasonably neglect δ_{ASE} in (6.20) to obtain an expression using (6.6) for the steady-state inversion level as:

$$\overline{N_{2ss}} = \frac{\alpha_a(\nu_0) + \dfrac{\alpha_c(\nu_0)}{l}}{\alpha_a(\nu_0) + \gamma_e(\nu_0)} \tag{6.22}$$

With this value in the atomic rate equation (6.17) under state-state conditions, we obtain the laser power at the coupler output (see Figure 6.8) as:

$$P_o = \eta(P_p - P_{th}) \tag{6.23}$$

with the threshold power as:

$$P_{th} = \frac{hc}{\lambda_p}\left(\frac{1}{a_p}\right)\left[\frac{(\rho_e S_e l)\overline{N_{2ss}}}{\tau_{21}} + R_{\text{ASE}}\right] \tag{6.24}$$

and the slope efficiency:

$$\eta = R\mathrm{e}^{-\alpha_{23}}\left(\frac{\lambda_p}{\lambda_l}\right)\frac{a_p}{1 - \mathrm{e}^{-\alpha_c(\nu_0)}} \tag{6.25}$$

The quantity $a_p = \left\{1 - \mathrm{e}^{-\overline{g_p \cdot l}}\right\}$ in (6.24) and (6.25) represents the degree of pump power absorption and, for most well-designed cases, is approximately unity. In (6.25), the attenuation in the loop between the fibre amplifier output and the coupler, that is, between points (2) and (3) in Figure 6.8, is represented by α_{23} and R is the out-coupled power fraction of the coupler. Note that in the rate equations (6.17) and (6.18), the power is in units of photons per second, but for (6.23) and (6.24) the pump and laser powers have been converted to units of watts using the pump and laser wavelengths of λ_p and λ_l respectively.

Figure 6.9 shows the output power characteristics for a typical low-power erbium ring fibre laser using the parameters given in Table 6.4 for various values of the cavity attenuation, α_c.

Table 6.4 Typical parameters for the erbium fibre ring laser shown in Figure 6.8

Lasing wavelength	**1560 nm**
Absorption coefficient, α_a	1.85 dB m^{-1}
Emission coefficient, γ_e	3.23 dB m^{-1}
Pump wavelength	980 nm
Pump absorption coefficient, α_p	4.5 dB m^{-1}
Upper state lifetime, τ_{21}	10 ms
Erbium fibre length, l	10 m
Erbium ion density, ρ_e	5×10^{24} ions m^{-3}
Effective area of erbium doping, S_e	10^{-11} m^2
Total length of fibre ring cavity, L_c	20 m
Longitudinal mode spacing, $\Delta\nu_m$	10 MHz
Round-trip time of cavity, τ	100 ns
Values for cavity loss, α_c	6, 8 or 13 dB
Out-coupled ratio of coupler, R	10%
Erbium fibre output-to-coupler loss, α_{23}	3 dB

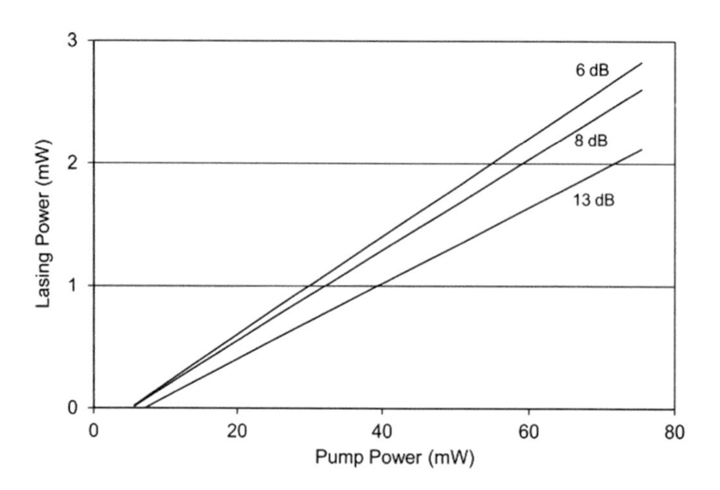

Figure 6.9 Typical output power characteristics from an erbium fibre ring laser operating at 1560 nm as a function of pump power at 980 nm for various values of the cavity attenuation (other parameters given in Table 6.4).

For a free-running fibre ring laser with no optical filter in the cavity, the operation wavelength will correspond to the most favourable gain position of the erbium fibre amplifier, that is, on the peak of the appropriate gain curve shown in Figure 6.7, depending on the average inversion level. Note from (6.22) that the average inversion level is dependent on the cavity loss factor, α_c, so a variable optical attenuator may be inserted in the loop to adjust the inversion level and hence the wavelength position of the most favourable gain. If the loop has a relatively high attenuation resulting in an inversion level greater than ~60%, then, as shown in Figure 6.7, the gain peak is at

~1530 nm and the laser will naturally operate at this wavelength. On the other hand, if the loop attenuation is reduced so that the inversion level is less than 60%, there is a broad flat peak in the gain curve centred at ~1560 nm and the laser wavelength will switch to this value. With the parameters given in Table 6.4, this switchover occurs at a loop attenuation of ~11 dB. The operation wavelength of erbium-doped fibre lasers may be further extended [16] into the so-called 'L-band' region (1560–1620 nm) by working at low inversion levels of 30–40%, where, as indicated in Figure 6.7, this is the region of most favourable gain, albeit quite small. Note from (6.22) that the inversion level depends on the erbium fibre length, so low (length-averaged) inversion levels can be achieved by using a long length of erbium-doped fibre (typically ~50 m) combined with multiple pump sources, including forward and backward pumping. As noted earlier, the fact that gain is possible at long wavelengths at low inversion levels can be understood from Figure 6.4, which illustrates the thermalisation of the Stark levels by phonon interactions. Even if the total upper band population is less than the lower band (i.e. inversion levels of <50%), there can still be population inversion between the lower Stark levels of the upper band and the upper Stark levels of the lower band, giving gain at the lower-energy (longer-wavelength) side of the emission spectrum.

However, in most practical cases of CW operation, some form of wavelength filtering is included in the fibre loop to set or tune the operation wavelength and this may be modelled in the above description of laser operation by including the filter transmission function as part of the cavity loss function, $\alpha_c(\nu)$. However, even with a filter, unless special measures are taken, it is difficult to ensure single longitudinal mode operation, since for a typical fibre loop with length of ~20 m the longitudinal mode spacing is only ~10 MHz and small thermal and mechanical disturbances of the loop result in mode-hopping. Single-mode operation has been demonstrated using feedback control loops [17] or by use of a saturable absorber in the fibre cavity [18]. A short cavity, such as employed in DFB fibre lasers [19], increases the longitudinal mode spacing and hence facilitates single-mode operation. A range of tunable, single-mode, narrow-linewidth fibre lasers and ASE sources have been developed and marketed by companies such as NP Photonics [20].

For use of fibre lasers with WMS in gas sensing, wavelength scanning and modulation is more difficult as compared with DFB diode lasers, where this is simply performed by diode current modulation. Pump current modulation gives amplitude modulation with fibre lasers, but wavelength scanning and modulation requires a wavelength scanning and modulation element within the cavity. This may be, for example, a fibre Bragg grating mounted on a PZT which is used to stretch the grating and hence modulate or scan the Bragg wavelength position, as illustrated in Figure 6.10. Even so, because of the close spacing of the longitudinal modes in a standard loop cavity fibre laser, scanning and modulation causes transiting across modes, resulting in fluctuations of the output power.

For single-mode lasing, the limit in the theoretical linewidth is determined by the ASE noise, the Schawlow–Townes limit [15]. We can derive this limit for the fibre ring cavity laser from (6.20) and (6.21) on the basis that the gain is slightly less than the cavity loss and hence the coherent laser radiation is circulating in a cavity with a small net round-trip loss factor of $e^{-\delta}$. The electric field in the fibre loop resonator may be

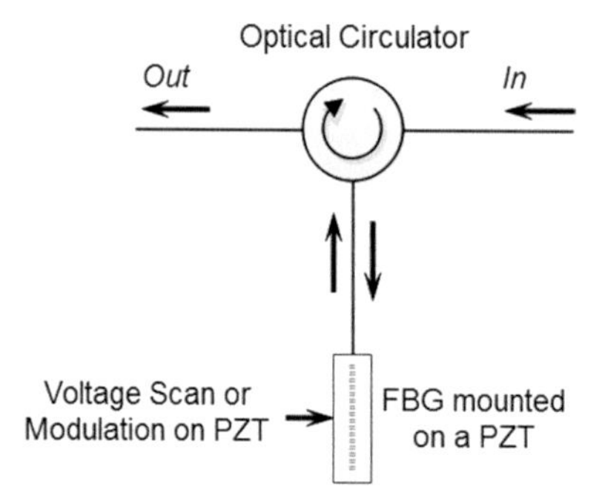

Figure 6.10 Fibre Bragg grating (FBG) and PZT for tuning, modulation or scanning of the output wavelength of a fibre laser.

expressed as the sum of waves delayed in phase by βL_c and attenuated by the factor $e^{-\delta/2}$ (for the electric field) for each circulation. (See also (5.14) to (5.18) in Chapter 5, where a similar analysis is performed for a high-finesse cell for cavity-enhanced absorption spectroscopy)

Hence we have:

$$E_t = E_0 e^{j2\pi\nu t}\left\{1 + e^{\frac{-\delta}{2}}e^{-j\beta L_c} + e^{-\delta}e^{-2j\beta L_c} + \cdots\right\} \tag{6.26}$$

which is a geometric series with sum of:

$$\frac{E_t}{E_0} = \frac{e^{j2\pi\nu t}}{1 - e^{\frac{-\delta}{2}}e^{-j\beta L_c}} \tag{6.27}$$

The intensity is proportional to $\langle E.\ E^*\rangle$, so after some algebraic manipulation we obtain an expression for the intensity in the general form [21]:

$$I = \frac{I_{\max}}{1 + F\sin^2\left(\frac{\beta L_c}{2}\right)} \tag{6.28}$$

where F is the coefficient of finesse given by:

$$F = \frac{4e^{\frac{-\delta}{2}}}{\left(1 - e^{\frac{-\delta}{2}}\right)^2} \simeq \frac{16}{\delta^2} \tag{6.29}$$

The longitudinal modes or resonance peaks correspond to $\beta L_c = 2m\pi$ in (6.28). The half width at half maximum is determined from the condition that $F\sin^2(\beta L_c/2) = 1$ in (6.28), which gives the full width at half maximum (FWHM) frequency linewidth, $\Delta\nu_l$, as:

$$\Delta\nu_l = \frac{2c}{\pi n L_c\sqrt{F}} = \Delta\nu_m\left(\frac{\delta}{2\pi}\right) \tag{6.30}$$

where $\Delta\nu_m = c/nL_c = 1/\tau$ is the longitudinal mode spacing of the loop resonator and we have used the approximation for F from (6.29). Note that this is the same as the linewidth of the Lorentzian function obtained from the Fourier transform of an exponentially decaying sinewave with decay constant of $\delta/2\tau$.

Using the expression for δ_{ASE} from (6.21) in (6.30) we obtain the Schawlow–Townes limit in the laser linewidth as:

$$\Delta\nu_l = \frac{h\nu_0(\Delta\nu_m)^2 R e^{-\alpha_{23}}}{\pi P_l} \cdot \left(e^{\alpha_c(\nu_0)} - 1\right)\left(\frac{\gamma_e(\nu_0)}{\alpha_c(\nu_0)}\right)\left(\frac{\alpha_a(\nu_0)l + \alpha_c(\nu_0)}{\alpha_a(\nu_0) + \gamma_e(\nu_0)}\right) \tag{6.31}$$

where P_l is the power coupled out from the laser in watts and we have used (6.22) for the inversion level and $A_{SE}(\nu_0) \simeq (e^{\alpha_c(\nu_0)} - 1)/\alpha_c(\nu_0)$ from (6.15) with $\overline{g_{ss}(\nu_0)} \cdot l \cong \alpha_c(\nu_0)$ from (6.20).

This limit is very small, ~5×10^{-3} Hz at 1 mW output power, using the parameters given in Table 6.4, but in practice the linewidth is dominated by other factors.

6.5.2 Transient Operation

As will be discussed in Chapter 7, fibre lasers may be used for intra-cavity laser absorption spectroscopy (ICLAS), where a gas cell is placed within the fibre cavity [22–27]. During the transient period of laser start-up, when light is building up within the fibre cavity, the multiple circulations of light through the gas cell vastly enhance the effective absorption length of the cell, thus enabling high-sensitivity detection of gas species in the near-IR.

The properties of the fibre laser during build-up and the spectral evolution of the modes during the transient period can be modelled from the rate equations (6.17) and (6.18) by now retaining the time-dependent terms for the inversion level and the power on the LHS of the equations. For transient operation, the pump power is abruptly stepped from zero to a value above the threshold level, with the assumption that there is no initial power in the fibre cavity. Under these conditions, there is an initial delay before any lasing output is obtained, which is the time required for the inversion level to build up to the threshold value. This delay time can be estimated from the rate equation (6.17) by setting the laser power to zero, neglecting the ASE and assuming good pump power absorption (a_p set to unity), to give:

$$\rho_e S_e l\left(\frac{d\overline{N_2}}{dt}\right) = P_p - \rho_e S_e l\frac{\overline{N_2}}{\tau_{21}} \tag{6.32}$$

which has the solution:

$$\frac{\rho_e S_e l}{\tau_{21}}\overline{N_2} = P_p\left\{1 - \exp\left(-\frac{t}{\tau_{21}}\right)\right\} \tag{6.33}$$

Hence, using (6.24) for the threshold power, with a_p unity and neglecting the ASE, the delay time to reach threshold is:

$$t_{th} \cong \tau_{21}\ln\left\{\frac{\xi}{\xi - 1}\right\} \tag{6.34}$$

where $\xi = P_p/P_{th}$.

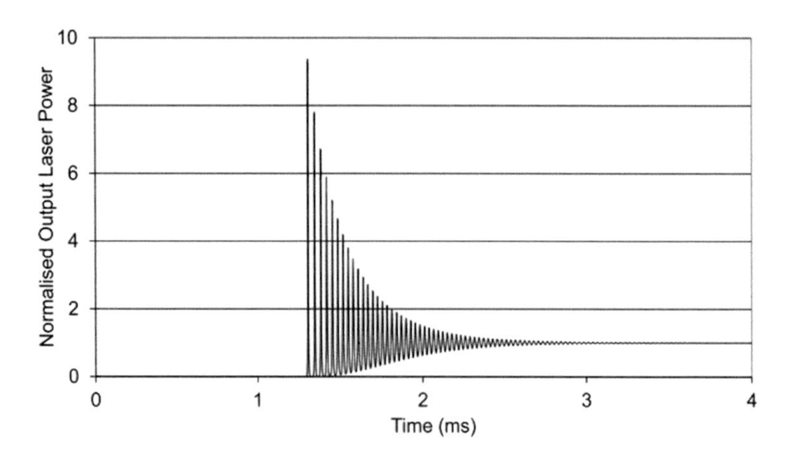

Figure 6.11 Theoretical laser output power as a function of time after application (at $t = 0$) of a step input of ~$8P_{th}$ to the pump power (cavity loss of 13 dB and other parameters as given in Table 6.4).

As can be seen for a given upper-state lifetime, the delay time is solely dependent on the magnitude of the step in the pump power input and typically has a value of a few milliseconds. After this initial delay, the power output from the laser goes through a period of spiking and relaxation oscillations before settling to its steady-state value. This is illustrated in Figure 6.11 where the magnitude of the pump step is ~$8P_{th}$ and the output is normalised to the steady-state power. Note from Figure 6.11 that the delay time to reach threshold is t_{th} ~ 1.3 ms, as predicted by (6.34), and that it takes ~$3t_{th}$ for the relaxation oscillations to die out and the output to settle to its steady-state value.

The characteristics of the relaxation oscillations can also be derived analytically from the rate equations and can be understood on the basis of the following description. If the laser power is slightly perturbed above its steady-state value, the inversion level starts to fall below its steady-state value due to the increased level of stimulated emission. This has the effect of reducing the laser power, which undershoots and falls to a value below its steady-state level. The inversion level now increases and the cycle continues, so an oscillation occurs between the inversion level and the laser power until they both settle to the steady-state values. The characteristics of this oscillation are derived by perturbation analysis, where small perturbations around the steady-state values are applied to the inversion level and laser power, namely, $\overline{N_2}(t) = \overline{N_{2ss}} + \delta\bar{N}(t)$ and $P(t) = P_{ss} + \delta P(t)$, and these quantities are substituted into the rate equations. After considerable algebraic manipulation [25] the following second-order differential equation is obtained for the perturbation in power:

$$\frac{d^2(\delta P)}{dt^2} + 2\alpha\frac{d(\delta P)}{dt} + \omega_0^2(\delta P) = 0 \tag{6.35}$$

The solution to (6.35) is a decaying sine wave function representing the relaxation oscillations. The parameters of the relaxation oscillations, particularly the decay constant, are affected when a number of modes participate in the oscillations and the full analysis, including the effects of ASE, is given by Stewart [25]. For many situations the simplified case, where ASE is neglected and only a single mode is considered, provides an adequate description and in this situation the decay constant and the relaxation oscillation frequency, $\omega = \sqrt{\omega_0^2 - \alpha^2} \approx \omega_0$ are given by:

$$\alpha \cong \frac{1}{2\tau_{21}} \left\{ 1 + \frac{\tau}{t_c(1 - e^{-\alpha_c})} \left(\frac{P_p}{P_{th}} - 1 \right) \right\} \tag{6.36}$$

$$\omega_0^2 = \frac{1}{t_c \tau_{21}} \left(\frac{P_p}{P_{th}} - 1 \right) \tag{6.37}$$

where an effective cavity lifetime is defined by $t_c = \tau/(\alpha_c + \alpha_a l)$ and has typical values of ~14–18 ns, using the parameters in Table 6.4. The relaxation oscillation frequency is thus typically ~20 kHz with a decay time of ~1 ms at a pump power of ~$4P_{th}$.

The transient process during which the laser power builds up in the cavity and undergoes spiking and relaxation oscillations, as illustrated in Figure 6.11, is accompanied by a spectral evolution of the cavity modes. Since the inversion level initially spikes above the steady-state value, a large number of modes are initially excited across the spectral width of the gain, but as the laser output approaches the steady-state condition, spectral narrowing occurs until the final state of a single or a few modes is reached, as determined by the optical filtering characteristics present in the cavity. This spectral narrowing occurs over a typical period of a few milliseconds, during which there are many thousands of circulations of light within the cavity. When a gas absorption feature is present within the fibre laser cavity, the spectral evolution is dramatically altered and a small dip from the absorption line, superimposed on the modal intensity, is greatly amplified and grows in depth during this period until spectral saturation is reached, effectively enhancing the length of the gas cell by several orders of magnitude. This technique and its limitations for gas sensing are discussed more fully in Chapter 7 on the applications of fibre laser systems.

The whole process of power build-up and the spectral evolution of the modes during the transient period can be modelled by numerical analysis of the rate equations (6.17) and (6.18). This can be done using, for example, the internal differential equation solver in MATLAB (function reference: ode15s) which is capable of solving more than 100 coupled differential equations. Typically, around 2000 modes can be included in the simulation by considering 80 mode groups each containing 25 modes with, as noted earlier, the fibre laser parameters assumed constant within a group, but different for each group. Hence there is a cavity rate equation (6.18) for each mode group, giving 80 differential equations plus the atomic rate equation (6.17), with its summation term taken over the mode groups. The cavity loss function $\alpha_c(\nu)$ is used to model the optical filtering, as well as any gas absorption features present in the fibre cavity [25]. Using this model, Figure 6.12a shows the theoretical spectral distribution of the laser output when a step input is applied to the pump power at $t = 0$, with a filter of bandwidth ~1 nm included in the fibre cavity. As can be seen, spectral narrowing, as determined by the filter characteristics, occurs over typical timescales of a few milliseconds during the transient period.

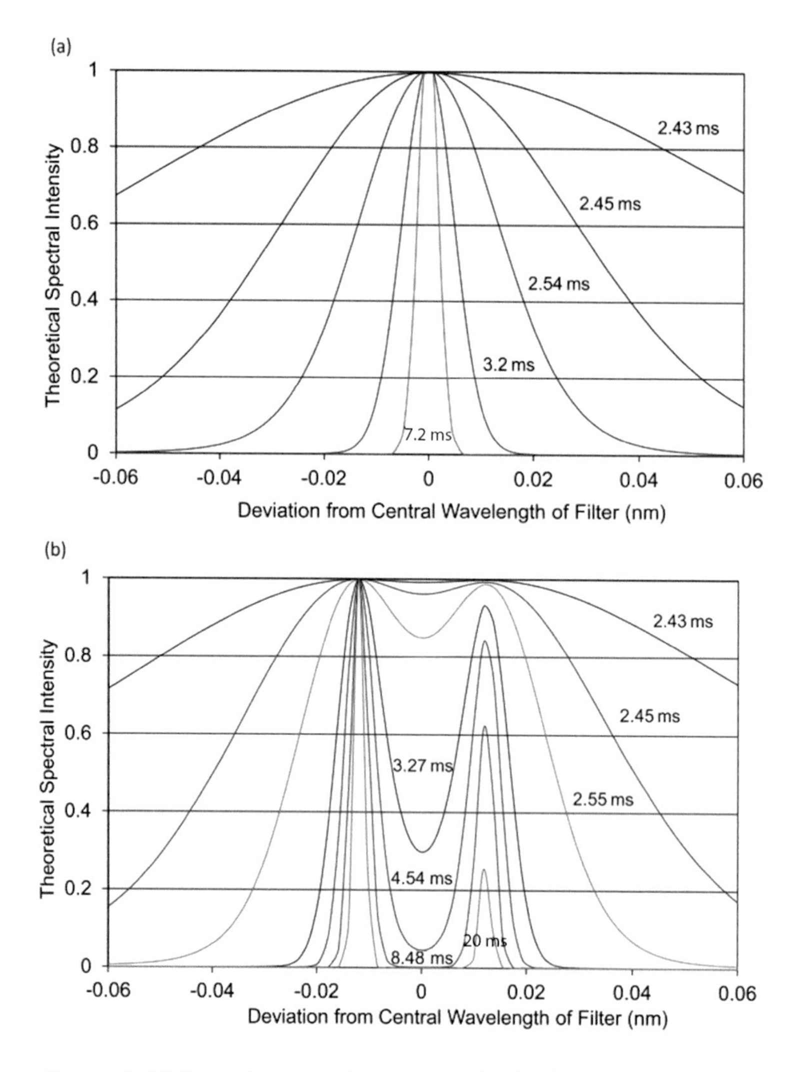

Figure 6.12 (a) Theoretical normalised spectral distribution of the laser output during the transient period after the application of a 25 mW step input to the pump power, showing the spectral narrowing during the transient period (cavity loss of 8 dB, filter bandwidth ~1 nm, other parameters as in Table 6.4) and (b) the effect of including a 5 GHz linewidth Lorentzian absorption line with line centre absorbance of 0.01 dB. © [2007] IEEE. Reprinted, with permission, from [25].

Figure 6.12b further illustrates the effect of including a Lorentzian absorption line, aligned with the filter centre wavelength, with a full linewidth of 5 GHz and line centre absorbance of 0.01 dB. As can be seen, the intensity in the central region is totally extinguished as the small attenuation dip from the absorption line is amplified from the multiple circulations of light in the cavity and grows in depth over a period of ~2 ms. Since the round-trip time for a 20 m fibre cavity is ~100 ns, there are some ~20 000 circulations over this 2 ms period, which indicates the theoretical degree of path length enhancement possible with an intra-cavity cell. Of course in practice, other factors limit the degree of sensitivity enhancement that can be achieved, notably the effects of Rayleigh scattering and the practical difficulties of capturing the time-evolving modal spectrum in the presence of relaxation oscillations, but the practical implementation of ICLAS with fibre lasers will be further discussed in Chapter 7.

6.5.3 Multi-Wavelength Operation

Although erbium-doped and other rare-earth-doped fibre lasers are predominantly homogeneously broadened, they can be induced to operate as a multi-wavelength source by the inclusion of two elements in the fibre cavity, namely, a wavelength-periodic filtering element to select the desired wavelengths and a suitable 'active' element to distribute the gain to the selected wavelength channels. Multi-wavelength sources are of interest in correlation spectroscopy, where the multi-wavelength outputs are closely matched to the multiple rotational lines in the absorption spectrum of a particular gas. This makes for better selectivity and measurement of that gas in the presence of other interfering species which have absorption lines in a similar wavelength region but with different line spacings.

A simple, low-cost design for multi-wavelength operation of fibre lasers employs a piezo-electric transducer (PZT) to generate multi-wavelength outputs at the peaks of a Sagnac loop periodic filter in the cavity [28, 29], as illustrated in Figure 6.13.

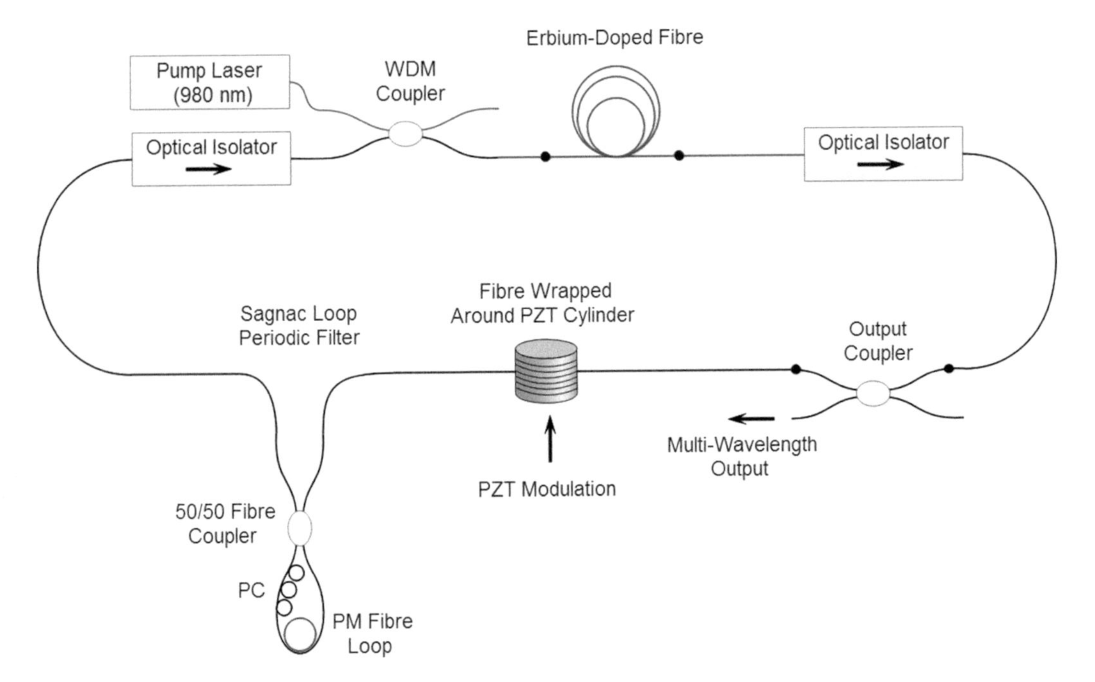

Figure 6.13 Multi-wavelength fibre laser with PZT and Sagnac loop filter formed from a 50/50 fibre coupler, a polarisation-maintaining (PM) fibre loop and polarisation controllers (PC). Adapted from [28].

With a sinusoidal modulation applied to the PZT phase modulator, the cavity length and hence the wavelength (or frequency) position, of the longitudinal modes of the cavity become a periodic function of time. The ratio of the amplitude of the frequency shift, $\delta\nu_{\mathrm{PZT}}$, of the modes induced by the PZT to the axial mode spacing defines a modulation index of $\delta\nu_{\mathrm{PZT}}/\Delta\nu_m$. The modulation is performed at a rate much slower than the round-trip time of the cavity, τ, and at less than the relaxation oscillation frequency. The disturbance of the mode positions by the PZT modulation in relation to the fixed transmission function of the periodic wavelength filter results in a periodic modulation of the attenuation experienced by the modes. This in turn creates a periodic perturbation of the inversion level around its steady-state value (somewhat akin to the spiking and relaxation oscillation processes described earlier), enabling multi-wavelength operation at the peaks of the periodic filter. Stable operation is achieved for particular combinations of pump power, modulation frequency and modulation index with typical values [28] of 100 mW, 20 kHz and modulation indices in the range of 1–2, respectively. When a stable multi-wavelength output is obtained, a stable periodic pulse train appears in the time domain with a repetition rate at the PZT modulation frequency, akin to sustained spiking in laser operation. Similar to the methods described earlier for transient analysis, multi-wavelength operation may be modelled by numerical solution of the rate equations with mode groups selected around the peaks of the filter and a cavity rate equation for each group. The filter is modelled in the form of a wavelength periodic transmission function in the fibre cavity and the PZT modelled as a modulation on the attenuation experienced by a mode group as a result of the variation of the position of the mode group in relation to the fixed transmission function of the periodic filter.

6.5.4 Mode-Locked Operation

As will be discussed in Chapter 7, mode-locked operation of fibre lasers is extensively used in spectroscopy for the generation of frequency combs. The fundamental principles of mode locking have been extensively studied for a variety of laser types [15, 30–34]. Fibre lasers may be either actively mode-locked by placing a modulator within the cavity, or passively mode-locked by a saturable absorber in the cavity. For the active case, mode-locking occurs when the modulation frequency, f_{ml}, is chosen to correspond to the frequency spacing of the longitudinal modes of the fibre cavity (or an integer multiple of the spacing), that is, when $f_{ml} = m_h\Delta\nu_{\mathrm{m}} = m_h/\tau$ where m_h is the harmonic order of the mode-locking. The basic principle of mode-locking can be simply understood in terms of the effect of amplitude modulation (AM) on a carrier wave. The AM transfers power to sidebands on either side of the carrier frequency, separated from the carrier by the modulation frequency. Hence in the mode-locked fibre laser, the intra-cavity modulation results in power being transferred from the favoured mode(s) in CW operation to the adjacent longitudinal modes, which in turn experience the same modulation and transfer power to their neighbours. This causes a large number of longitudinal modes (with a limit determined by the effects of dispersion on mode spacing) to be sustained in laser operation and locked in phase, creating a narrow pulse in the time domain with a repetition rate of $1/\tau$ for the fundamental mode-lock frequency. For higher-order mode-locking frequencies, there are m_h pulses circulating in the cavity, each corresponding to a set of interlacing modes and the total pulse repetition rate is m_h/τ at the laser output. From the standpoint of the time domain, the modulator acts as a 'shutter', periodically opening to allow the passage of the circulating pulse(s). For passive mode-locking with a saturable absorber, the 'shutter' is opened by the initial part of the pulse, which saturates or bleaches the absorption so that the rest of the pulse may pass. After passage of the pulse, the saturable absorber returns to its high absorption state, effectively closing the shutter. For applications to gas spectroscopy, the phase-locked modes of the laser provide a uniform comb, consisting of a large number of closely spaced multi-wavelength outputs across a wide spectral range for the simultaneous and high-precision interrogation of multiple absorption lines. Of course the requirements on the mode-locked fibre laser design and operation are very stringent in order to achieve stability in wavelength and time, and this aspect will be discussed further in Chapter 7 in relation to applications.

6.6 Raman Fibre Amplifiers and Lasers

The process of stimulated Raman scattering (SRS) may be employed to extend the available wavelength range of fibre lasers and amplifiers to access near-IR absorption lines at longer wavelengths [35–37]. Raman fibre lasers and amplifiers frequently use high-power erbium- or other rare-earth-doped fibre lasers as a pump source and hence SRS allows the wavelength range of rare-earth-doped fibre lasers to be shifted or extended by the Stokes shift (the peak gain is at a Stokes shift of ~100 nm for optical fibres) or by multiples of the Stokes shift with cascaded systems [37].

The basic principles of Raman scattering can be understood with reference to Figure 6.14, where three vibrational states of a molecule are shown and a necessary condition is that the vibrational states cause a change in the polarisability of the molecule. As shown in Figure 6.14, some incident photons from the pump source are momentarily absorbed and excite the molecule to a 'virtual state'. Thereafter there are three possibilities, as follows:

(i) The molecule may return to the initial state, releasing a photon of the same energy. This is Rayleigh scattering and is the most probable outcome.
(ii) The molecule may return to an upper vibrational state, releasing a photon of less energy. This is Raman scattering, corresponding to a Stokes wave at a longer wavelength.
(iii) A molecule may return to a lower state, releasing a photon of higher energy, corresponding to an anti-Stokes wave at a shorter wavelength. This is the least likely outcome due to the thermal population distribution of energy levels.

We can also understand the process and the condition for a Raman-active molecule that its polarisability must change under its vibration by considering a simple classic electromagnetic model. The oscillating electric field of the pump wave at frequency ν_o

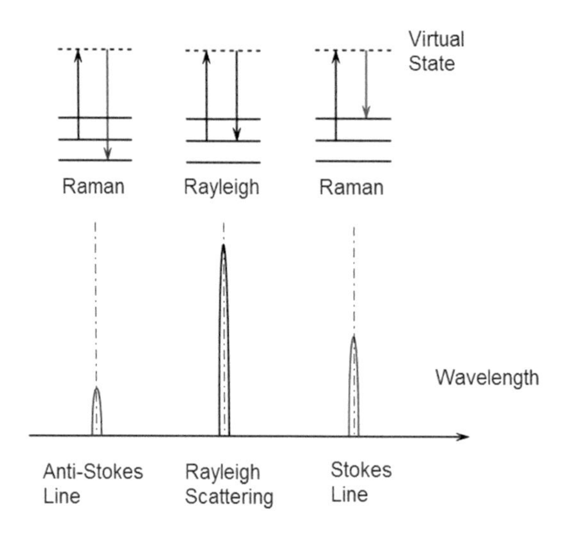

Figure 6.14 Schematic illustration of Rayleigh scattering, and Stokes and anti-Stokes Raman scattering.

induces a dipole moment in the molecule by distorting the electron clouds with respect to the nuclei, described by $\mu = pEe^{j2\pi\nu_o t}$, where p is the polarisability. If the molecular vibration at frequency, ν_m, modulates the polarisability then p will be a function of time of the form:

$$p(t) = p_0 + p_m \cos(2\pi\nu_m t) = p_0 + \frac{p_m}{2}\left(e^{j2\pi\nu_m t} + e^{-j2\pi\nu_m t}\right) \tag{6.38}$$

Hence the induced dipole moment will have the form:

$$\mu(t) = \left\{p_0 + \frac{p_m}{2}\left(e^{j2\pi\nu_m t} + e^{-j2\pi\nu_m t}\right)\right\}Ee^{j2\pi\nu_o t} \tag{6.39}$$

and radiation from the dipole will generate frequency components of ν_o and $\nu_o \pm \nu_m$ corresponding to Rayleigh scattering, and Raman Stokes and anti-Stokes waves. Clearly, if p_m is zero and the polarisability does not change under molecular vibration, then no Raman shifted waves are generated. Although we have illustrated only three levels in Figure 6.14, there may be a series of vibrational states giving rise to a series of Stokes and anti-Stokes lines of varying strengths. In addition to the pump source, if an additional (seed) source is added at a Stokes-shifted wavelength the process of SRS occurs.

Of particular interest is SRS from silica in optical fibres, where Raman scattering occurs over a broad frequency band [35–37], shifted in frequency up to 40 THz with a broad peak in the Raman gain occurring at a shift of ~13 THz (or ~100 nm in terms of wavelength) and width of ~5 THz. The Raman gain can be expressed as $G_R = \exp(g_R I_p \cdot l)$ where I_p is the pump intensity in W m^{-2}, l is the interaction length and g_R is the gain coefficient with a peak value of ~6×10^{-14} m W^{-1} at a pump wavelength of ~1550 nm. Although the gain coefficient is very small, the long interaction lengths of several kilometres possible with optical fibres, along with the high pump intensities that can be achieved in the fibre core, have led to the development and application of Raman fibre amplifiers in fibre optic communications systems [35, 36].

There are several noise sources that must be dealt with in Raman fibre amplifiers [35, 36] and, for high-power outputs, it is necessary to suppress stimulated Brillouin scattering (SBS), which removes power from both the pump and the Raman-shifted output. In SBS, the pump wave creates a periodic variation in the density and hence the refractive index through the process of electrostriction and this pump-induced grating, moving at the acoustic velocity, creates a backward scattered wave shifted in frequency from the pump [38, 39]. The frequency shift from Brillouin scattering is $\Delta\nu_B = 2nV_a/\lambda_p$ where V_a is the acoustic velocity ($V_a \sim 5945$ m s^{-1} for silica glass fibre) giving a shift of ~11 GHz at a pump wavelength of ~1550 nm. Compared with Raman scattering, which occurs in both directions, Brillouin scattering only occurs in the direction opposite to that of the pump wave, with a much larger peak gain coefficient of $g_B \sim 4\times10^{-11}$ m W^{-1}, but the SBS linewidth is very narrow, ~20 MHz. Effective SBS suppression in Raman amplifiers can thus be achieved by ensuring that both the pump and the Raman output have a broad linewidth compared with the SBS linewidth. An example of the use of Raman amplifiers in near-IR gas sensing is given by Bauer [40], where a high-power Raman output of ~1 W at 1651 nm is generated for photoacoustic spectroscopy of methane. The 4.5 km long Raman fibre amplifier is counter-pumped by a relatively wideband, ~2.5 THz, filtered output of an ASE source, amplified by a high-power erbium-doped fibre amplifier. The Raman-active fibre is also seeded from the output of a 1651 nm laser diode, modulated as for standard photoacoustic spectroscopy, but with an additional higher-frequency modulation at 500 kHz for SBS suppression.

6.7 Conclusion

This chapter has reviewed the essential principles underlying the operation of rare-earth-doped fibre amplifiers and lasers. Of fundamental importance is the special nature of the *4f* electron shell in rare earths and the rapid thermalisation of the Stark energy levels through phonon interactions, which gives rise to the homogeneously broadened absorption and emission spectra of rare-earth-doped fibres, leading to the McCumber relationship between the absorption and emission cross-sections. The cavity and atomic rate equations derived in this chapter for fibre amplifiers and lasers provide the basic tools to model their behaviour under various operating conditions, including steady-state, multi-wavelength, transient and mode-locked regimes. We shall draw on the principles and understanding developed in this chapter when we consider the applications of fibre lasers to gas spectroscopy in Chapter 7, such as for tuneable sources, frequency combs and intra-cavity laser absorption spectroscopy.

Appendix 6.1 Einstein Relations and the Absorption and Emission Cross-Sections

For the two energy levels as shown in Figure A6.1.1 with populations of N_1 and N_2, the basic Einstein relations with the A and B coefficients are defined by:

$$\text{Absorption:} \left(\frac{\mathrm{d}N_2}{\mathrm{d}t}\right)_{\text{abs}} = -\frac{\mathrm{d}N_1}{\mathrm{d}t} = B_{12}N_1\rho(\nu) \tag{A6.1.1}$$

$$\text{Stimulated emission:} \left(\frac{\mathrm{d}N_2}{\mathrm{d}t}\right)_{\text{stim}} = -B_{21}N_2\rho(\nu) \tag{A6.1.2}$$

$$\text{Spontaneous emission:} \left(\frac{\mathrm{d}N_2}{\mathrm{d}t}\right)_{\text{spon}} = -A_{21}N_2 = -\frac{N_2}{\tau_{21}} \tag{A6.1.3}$$

where $\rho(\nu) = h\nu \cdot D(\nu)$ is the photon energy density, $D(\nu)$ is the number of photons per unit volume per unit frequency bandwidth and τ_{21} is the spontaneous emission lifetime.

Combining (A6.1.1) to (A6.1.3) gives the total rate of change of the upper level population as:

$$\frac{\mathrm{d}N_2}{\mathrm{d}t} = (B_{12}N_1 - B_{21}N_2)\rho(\nu) - A_{21}N_2 \tag{A6.1.4}$$

The B coefficients between non-degenerate discrete atomic levels can be determined theoretically from the atomic wave functions ψ_1 and ψ_2 of the two states using Fermi's golden rule [12]. Specifically, for non-degenerate levels:

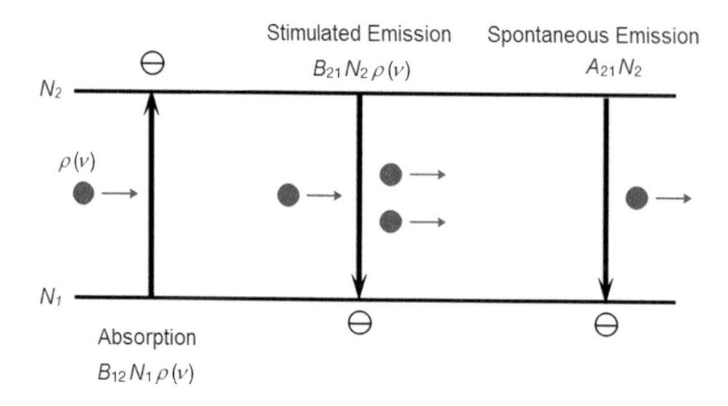

Figure A6.1.1 Photon absorption and emission processes for two electron energy levels.

$$B_{12} = B_{21} = B = \frac{\mu_{12}^2}{6\varepsilon_0 \hbar^2} \tag{A6.1.5}$$

where the electric dipole of the transition is given by: $\mu_{12} = -e \int \psi_2^* x \psi_1 \mathrm{d}\tilde{r}^3$.

This relationship links the classical approach for the emission and absorption of radiation, where the atom is viewed as an oscillating electric dipole, with the quantum mechanical description where the wave functions ψ_1 and ψ_2 describe the two states.

The A coefficient may also be calculated theoretically from Fermi's golden rule based on the concept of the vacuum field. In this approach, spontaneous emission is in fact emission that is stimulated by the randomly fluctuating vacuum field. The vacuum field arises from the quantisation of the electromagnetic field as a harmonic oscillator with energy levels of $(n + ½)h\nu$ so that even in the lowest $n = 0$ state, the energy is non-zero and is given by $h\nu/2$. Since the time-averaged energy density (per unit volume) of an electromagnetic field is $\varepsilon_0 E^2$, the vacuum field for volume V may be computed from $\int \varepsilon_0 E_{\mathrm{vac}}^2 \mathrm{d}V = h\nu/2$ giving $E_{\mathrm{vac}}^2 = h\nu/(2\varepsilon_0 V)$. The spontaneous emission coefficient from Fermi's golden rule is then given by [12]:

$$A_{21} = \frac{|M_{12}|^2}{\hbar^2} f_m(\nu) V \tag{A6.1.6}$$

Here $f_m(\nu)$ is a density of states (or modes) function which we will discuss below. M_{12} is the matrix element of the transition, calculated from the time-averaged scalar product of the electric dipole vector of the atomic transition with the vacuum electric field vector, $M_{12} = \langle \tilde{p} \cdot \tilde{E}_{vac} \rangle$. This gives $|M_{12}|^2 = \mu_{12}^2 E_{vac}^2/3$, where the division by the factor of three arises from averaging over all possible directions of the dipole with respect to the electric field direction.

Combining the above expressions and using (A6.1.5) gives the spontaneous emission coefficient as:

$$A_{21} = f_m(\nu) \cdot h\nu \cdot B \tag{A6.1.7}$$

The density of states (modes) function $f_m(\nu)$ refers to the density of final states or modes for the photon per unit frequency so that $f_m(\nu)\mathrm{d}\nu$ is the number of states per unit volume in the frequency range ν to $\nu + \mathrm{d}\nu$. For spontaneous emission into free space or a medium of index n, the photons are emitted into a continuum of states with density of $f_m(\nu) = 8\pi(n^3\nu^2/c^3)$. (This is computed [12] by deriving the number of modes per unit frequency bandwidth in a cubic region of space of side L and then dividing by the volume of the cube.) In other situations, the density of photon states may be different, for example, if emission occurs into an optical cavity or resonator. Since the direction of spontaneous emission is random, emission may occur both to cavity and to free space modes.

Comparison of the fundamental definitions of stimulated and spontaneous emission given by (A6.1.2) and (A6.1.3) with (A6.1.7) shows that spontaneous emission can be described as stimulated emission with an energy density of $f_m(\nu) \cdot h\nu$ per unit volume per unit frequency bandwidth, i.e. one photon per mode (giving rise to the so-called 'extra photon' in laser rate equations [13]).

The above analysis was based on transitions between two non-degenerate levels, but if level 1 and level 2 consist of g_1 and g_2 degenerate sub-levels then we need to take into account the individual transition rates between each of the sub-levels, with $B_{ij} = B_{ji}$ for each sub-level, and hence the above rate equations will become a summation of terms. If the simplifying assumption is made that the sub-levels have equal populations of $N_{1s} = N_1/g_1$ and $N_{2s} = N_2/g_2$ then it follows that $g_1B_{12} = g_2B_{21}$, where the B coefficients now represent averaged transition strengths over the sub-levels.

We note in passing that for the particular case of black-body radiation at thermal equilibrium where $dN_2/dt = 0$ in (A6.1.4) and the relative populations follow the Boltzmann distribution, $N_2/N_1 = (g_2/g_1) \exp(-h\nu/kT)$, then, with the above relationships for the A and B coefficients, the standard black-body energy density distribution is obtained:

$$\rho(\nu) = \frac{8\pi n^3 h\nu^3}{c^3\left\{\exp\left(\frac{h\nu}{kT}\right) - 1\right\}} \tag{A6.1.8}$$

In the above description using the Einstein coefficients, a single frequency of absorption or emission is tacitly assumed, but of course, rather than absorption and emission occurring at a single frequency, it will occur over a narrow range of frequencies, defining a lineshape function. Heisenberg's uncertainty principle gives rise to natural (or lifetime) broadening, which is homogeneous in the sense that all atoms involved in the same transition are indistinguishable. Considering erbium-doped or other rare-earth-doped glasses, a small degree of inhomogeneous broadening occurs because of variations in the environment of the erbium ion from one lattice site to the next, so that each erbium ion is unique. However, the main cause of broadening of the absorption and emission characteristics for rare-earth-doped glasses is the homogeneous broadening arising from Stark splitting of the energy levels and phonon interactions. Rapid thermalisation of the Stark sub-levels through phonon interactions means that the earlier assumption of equal population of the sub-levels is not valid and a better description of the relationship between the absorption and emission characteristics is provided by the McCumber analysis given in Appendix 6.2.

When using the rate equations (A6.1.1) to (A6.1.4) to describe amplifier and laser operation with rare-earth-doped fibres, it is convenient to use light intensity rather than the photon energy density, and absorption and emission cross-sections rather than the Einstein A and B coefficients. These quantities may be related as follows.

The number of photons per unit time per unit frequency bandwidth traversing a cross-sectional element of area dA in a medium of index n is given by: $(c/n) \cdot \{\rho(\nu)/h\nu\} \cdot dA$. Hence the photon intensity, expressed as the number of photons per unit cross-sectional area, per unit frequency bandwidth is:

$$I_\nu(\nu) = \left(\frac{c}{n}\right) \cdot \frac{\rho(\nu)}{h\nu} \tag{A6.1.9}$$

Also, to take into account the finite absorption linewidth, the Einstein coefficient is written as $B_{12}g_a(\nu)$, where $g_a(\nu)$ is the absorption lineshape function, normalised by $\int g_a(\nu) d\nu = 1$.

Hence with the above definitions and considering the frequency range, ν to $\nu + \mathrm{d}\nu$, with photon density, $\rho(\nu)\mathrm{d}\nu$, and photon intensity, $I_\nu(\nu)\mathrm{d}\nu$, the contribution to the rate of absorption from (A6.1.1) over this frequency range is:

$$\left.\frac{\mathrm{d}N_1}{\mathrm{d}t}\right|_{\mathrm{d}\nu} = -N_1 \cdot \{B_{12}g_a(\nu)\} \cdot \rho(\nu)\mathrm{d}\nu = -\sigma_a(\nu)N_1 I_\nu(\nu)\mathrm{d}\nu \tag{A6.1.10}$$

where the absorption cross-section is given by:

$$\sigma_a(\nu) = h\nu\left(\frac{n}{c}\right)B_{12}g_a(\nu) \tag{A6.1.11}$$

Note that the description of the absorption by a cross-section in (A6.1.10) is the same as that given in Chapter 1, where the probability of capture of a photon by an atom in the lower state over a beam area of A is σ/A.

A similar definition may be made for the stimulated emission cross-section:

$$\sigma_e(\nu) = h\nu\left(\frac{n}{c}\right)B_{21}g_e(\nu) \tag{A6.1.12}$$

Also, using (A6.1.7) with the modified B coefficient, the contribution to the spontaneous emission rate over the frequency range, ν to $\nu + \mathrm{d}\nu$, is, from (A6.1.3):

$$\left.\frac{\mathrm{d}N_2}{\mathrm{d}t}\right|_{\mathrm{d}\nu} = -h\nu N_2\{B_{21}g_e(\nu)\}f_m(\nu)\mathrm{d}\nu = -\left(\frac{c}{n}\right)N_2\sigma_e(\nu)f_m(\nu)\mathrm{d}\nu \tag{A6.1.13}$$

The total spontaneous emission rate is given by integrating (A6.1.13) over the full bandwidth which, using the expression given earlier for the density of free space modes, gives the free space lifetime as:

$$A_{21} = \frac{1}{\tau_{21}} = \frac{8\pi n^2}{c^2}\int \nu^2\sigma_e(\nu)\mathrm{d}\nu \tag{A6.1.14}$$

However, as noted earlier, the free-space lifetime may be modified by the presence of cavity modes.

With the above definitions the rate equation (A6.1.4) may be written in terms of the cross-sections as:

$$\frac{\mathrm{d}N_2}{\mathrm{d}t} = -\int g_\sigma(\nu) \cdot I_\nu(\nu)\mathrm{d}\nu - \frac{N_2}{\tau_{21}} \tag{A6.1.15}$$

where $g_\sigma(\nu) = \{\sigma_e(\nu)N_2 - \sigma_a(\nu)N_1\}$.

Appendix 6.2 McCumber Relationship for the Absorption and Emission Cross-Sections

In Appendix 6.1 it was noted that if level 1 and level 2 consist of g_1 and g_2 degenerate sub-levels and if the assumption is made that the sub-levels have equal populations then it follows that $g_1B_{12} = g_2B_{21}$. However, this assumption is not valid for the Stark sub-levels of rare-earth-doped fibre due to the rapid thermalisation of the population through phonon interactions. Hence, we first derive an expression for the relative population of the Stark levels (or equivalently the probability of occupation for an individual Stark level) of each manifold shown in Figure 6.4. This leads to the McCumber relationship [8, 9] between the absorption and emission cross-sections of rare-earth-doped fibres.

Consider a collection of N ions in glass, with $N_{13/2}$ ions having their $4f^{11}$ electrons in the $^4I_{13/2}$ manifold and $N_{15/2}$ ions in the $^4I_{15/2}$ manifold, where $N = N_{13/2} + N_{15/2}$. As illustrated in Figure 6.4, the population of the Stark levels within each manifold will follow a thermal distribution due to phonon interactions.

For the $^4I_{13/2}$ manifold in Figure 6.4 with seven Stark levels for erbium, if ΔE_j represents the increased energy of the jth Stark level over the lowest Stark level of this manifold, then the population, N_j^S, of the jth Stark level relative to the population, $N_{j=1}^S$, of the lowest level of this manifold follows the classical thermal distribution function:

$$\frac{N_j^S}{N_{j=1}^S} = \exp\left\{-\frac{\Delta E_j}{kT}\right\} \tag{A6.2.1}$$

The total population of the $^4I_{13/2}$ manifold is: $N_{13/2} = N_{j=1}^S + N_{j=2}^S + N_{j=3}^S + \cdots + N_{j=7}^S$ and using (A6.2.1) this can be written as:

$$N_{13/2} = N_{j=1}^S + N_{j=1}^S \cdot \exp\left\{-\frac{\Delta E_{j=2}}{kT}\right\} + N_{j=1}^S \cdot \exp\left\{-\frac{\Delta E_{j=3}}{kT}\right\} + \cdots = N_{j=1}^S \cdot S_j \tag{A6.2.2}$$

where $S_j = 1 + \sum_{j=2}^{7} \exp\{-\Delta E_j/kT\}$.

Similarly the total population of the $^4I_{15/2}$ manifold with eight Stark levels is $N_{15/2} = N_{i=1}^S \cdot S_i$, where $S_i = 1 + \sum_{i=2}^{8} \exp\{-\Delta E_i/kT\}$.

Using (A6.2.1) and (A6.2.2), we can write the probability, p_j, of occupation for the jth Stark level, i.e. the population of an individual Stark level relative to the total population of the $^4I_{13/2}$ manifold, as:

$$p_j = \frac{N_j^S}{N_{13/2}} = \frac{1}{S_j} \cdot \exp\left\{-\frac{\Delta E_j}{kT}\right\} \tag{A6.2.3}$$

Similarly, the probability, p_i, of occupation for the ith Stark level of the lower $^4I_{15/2}$ manifold is:

$$p_i = \frac{N_i^S}{N_{15/2}} = \frac{1}{S_i} \cdot \exp\left\{-\frac{\Delta E_i}{kT}\right\} \tag{A6.2.4}$$

A 'mean energy' difference, ε, between the $^4I_{13/2}$ and $^4I_{15/2}$ manifolds may also be defined through the expression:

$$\frac{N_{13/2}}{N_{15/2}} = \exp\left\{-\frac{\varepsilon}{kT}\right\} \tag{A6.2.5}$$

Since, as derived earlier, $N_{13/2} = N_{j=1}^S \cdot S_j$ and $N_{15/2} = N_{i=1}^S \cdot S_i$ and if, as shown in Figure 6.4, ΔE_0 is the actual energy separation between the lowest Stark level of each of the manifolds, then $N_{j=1}^S / N_{i=1}^S = \exp\{-\Delta E_0/kT\}$ so we can write:

$$\frac{N_{13/2}}{N_{15/2}} = \exp\left\{-\frac{\varepsilon}{kT}\right\} = \frac{S_j}{S_i} \cdot \exp\left\{-\frac{\Delta E_0}{kT}\right\} \tag{A6.2.6}$$

Now let us consider the interaction of photons of energy, $E = h\nu$, with the $^4I_{15/2}$ and $^4I_{13/2}$ manifolds, as depicted in Figure 6.4. We assume that the photons interact between the individual Stark levels i and j of each manifold, with emission and absorption cross-sections of σ_{ji} and σ_{ij} respectively, and, by the Einstein relations (see Appendix 6.1), these cross-sections will be equal, $\sigma_{ji} = \sigma_{ij}$, so there appears at first sight to be equal probability of an upward or downward transition. However, the probability of occupation for an individual Stark level will have a bearing on whether a photon is absorbed (raising an electron from the lower to the upper level) or a photon is emitted (downward transition from the upper level). The key factor is the probability of occupation of the Stark level where the transition *originates*, i.e. if the photon is to be absorbed, it is the probability of occupation of the lower Stark level that is important and vice versa if stimulated emission is to occur. Hence an 'effective' emission cross-section, σ_e, and absorption cross-section, σ_a, may be defined as: $\sigma_e = p_j \cdot \sigma_{ji}$ and $\sigma_a = p_i \cdot \sigma_{ij}$. Since $\sigma_{ji} = \sigma_{ij}$ we can write the relation:

$$\sigma_e = \frac{p_j}{p_i} \cdot \sigma_a \tag{A6.2.7}$$

An expression for the probability ratio can be obtained from (A6.2.3) and (A6.2.4) as follows:

$$\frac{p_j}{p_i} = \frac{S_i}{S_j} \cdot \exp\left\{-\frac{(\Delta E_j - \Delta E_i)}{kT}\right\} = \frac{S_i}{S_j} \cdot \exp\left\{\frac{(\Delta E_0 - h\nu)}{kT}\right\} \tag{A6.2.8}$$

where we have used $\Delta E_i + h\nu = \Delta E_0 + \Delta E_j$, as can be seen from Figure 6.4, to derive the right hand term in (A6.2.8).

Substituting (A6.2.8) into (A6.2.7) and making use of (A6.2.6) we obtain the McCumber equation [8, 9] relating the emission and absorption cross-sections at the optical frequency of ν:

$$\sigma_e(\nu) = \sigma_a(\nu)\exp\left\{\frac{\varepsilon - h\nu}{kT}\right\} \tag{A6.2.9}$$

This expression may be compared with the simpler Einstein relations (see Appendix 6.1), based on equal population of all the sub-levels within a manifold where $\sigma_e = (g_1/g_2)\sigma_a$. This simple expression only applies in the limit of $kT \rightarrow \infty$ in which case $S_i = 8$ and $S_j = 7$ with probabilities $p_i = 1/8$ and $p_j = 1/7$ in the above equations, giving $\sigma_e(\nu) = \{8/7\}\sigma_a(\nu)$.

In practice, the absorption cross-section of the doped glass may be measured and (A6.2.9) allows the determination of the theoretical emission coefficient, knowing the value of ε. Note that the emission and absorption cross-sections are equal at a wavelength corresponding to $h\nu = \varepsilon$. The value of ε may be determined using (A6.2.6) along with the definitions of S_i and S_j given earlier as follows:

$$\exp\left\{-\frac{\varepsilon}{kT}\right\} = \frac{1 + \sum_{j=2}^{7}\exp\left\{-\frac{\Delta E_j}{kT}\right\}}{1 + \sum_{i=2}^{8}\exp\left\{-\frac{\Delta E_i}{kT}\right\}} \cdot \exp\left\{-\frac{\Delta E_0}{kT}\right\} \tag{A6.2.10}$$

An approximation [11] for evaluating (A6.2.10) may be made by assuming that the Stark levels in each manifold are equally spaced so that $\Delta E_j = (j - 1)\Delta E_u$ and $\Delta E_i = (i - 1)\Delta E_l$ where ΔE_u and ΔE_l are an averaged value of the Stark spacing for the upper and lower manifolds, respectively. The summations in (A6.2.10) may then be evaluated as a geometric series, giving:

$$\exp\left\{-\frac{\varepsilon}{kT}\right\} = \left[\frac{1 - \exp\left\{-7\frac{\Delta E_u}{kT}\right\}}{1 - \exp\left\{-\frac{\Delta E_u}{kT}\right\}}\right]\left[\frac{1 - \exp\left\{-\frac{\Delta E_l}{kT}\right\}}{1 - \exp\left\{-8\frac{\Delta E_l}{kT}\right\}}\right] \cdot \exp\left\{-\frac{\Delta E_0}{kT}\right\} \tag{A6.2.11}$$

A value for ΔE_0 (see Figure 6.5) may be approximated from the wavelength of the absorption and emission peaks, which are typically around 1530 nm for erbium-doped fibre and within ~1 nm (5 cm^{-1}) of each other at room temperature. Approximate values for the averaged Stark spacings ΔE_u and ΔE_l may be obtained from low-temperature measurements of the absorption and emission spectra [11]. The width of the lower manifold, $7\Delta E_l$, is taken as the width of the emission spectrum on the low-energy side of the peak and the width of the upper manifold $6\Delta E_u$ is taken as the width of the absorption spectrum on the high-energy side of the peak, as illustrated in Figure 6.5. Typically [3], the manifold widths for erbium-doped fibre are 300–400 cm^{-1} with averaged Stark spacings of ~50 cm^{-1}. With these typical values, $\varepsilon \approx 6550$ cm^{-1} and the relationship between the emission and absorption cross-sections for erbium-doped fibre may be approximated as:

$$\sigma_e(\nu) = \sigma_a(\nu)\exp\left\{\frac{6550 - \lambda^{-1}}{kT}\right\} \tag{A6.2.12}$$

with $kT \sim 200$ cm^{-1} and the inverse wavelength in units of cm^{-1}.

Appendix 6.3 Atomic Rate Equation for Rare-Earth-Doped Fibre

In Appendix 6.1, see (A6.1.15), it is shown that the rate equation governing absorption, stimulated and spontaneous emission between two energy levels with population levels of N_1 and N_2 is given by:

$$\frac{\mathrm{d}N_2}{\mathrm{d}t} = -\int g_\sigma(\nu)\cdot I_\nu(\nu)\mathrm{d}\nu - \frac{N_2}{\tau_{21}} \tag{A6.3.1}$$

where τ_{21} is the spontaneous emission lifetime, $I_\nu(\nu)$ is the photon intensity expressed as the number of photons per unit cross-sectional area per unit time per unit frequency bandwidth and $g_\sigma(\nu) = \{\sigma_e(\nu)N_2 - \sigma_a(\nu)N_1\}$.

We wish to apply this equation to rare-earth-doped fibres to describe the properties of fibre amplifiers and lasers. For most cases of interest, the spectral extent of the photon intensity, $I_\nu(\nu)$, which may be the laser beam or optical signal(s) for amplification, is narrow compared to the linewidth of the cross-sections, so we can write:

$$\int g_\sigma(\nu)\cdot I_\nu(\nu)\mathrm{d}\nu \cong g_\sigma(\nu)\int I_\nu(\nu)\mathrm{d}\nu = g_\sigma(\nu_1)I_1 + g_\sigma(\nu_2)I_2 + \cdots \tag{A6.3.2}$$

where I_1, I_2, etc. represent the intensities of narrow linewidth optical signals at optical frequencies of ν_1, ν_2, etc., in units of number of photons per unit cross-sectional area per unit time. For simplicity in the following derivation, we shall just consider the presence of one signal intensity, denoted by $I(r,\varphi)$ to represent its transverse intensity distribution, but this may be easily extended for the case of multiple optical signals.

Consider the optical fibre shown in Figure A6.3.1, doped with a rare earth element such as erbium, with a density distribution of $\rho(r,\phi) = \rho_e \cdot f_e(r,\phi)$. (If only the core of radius, a, is doped with a uniform distribution, then $f_e(r,\phi) = 1$ for $r < a$ and zero elsewhere). The photon intensity may be written as: $I(r,\varphi) = I(\nu)\cdot f_i(r,\phi)$, where $f_i(r,\phi)$ represents the transverse distribution of the mode intensity across the core and cladding of the fibre. We are assuming for simplicity that this distribution does not change over the range of wavelengths of interest.

Consider a volume element $\mathrm{d}V$ consisting of a fibre length element of $\mathrm{d}z$ and cross-sectional area element of $\mathrm{d}A = r\mathrm{d}r\mathrm{d}\phi$ as shown in Figure A6.3.1 so that $\mathrm{d}V = r\mathrm{d}r\mathrm{d}\phi\mathrm{d}z$. The number of states at level 1 within this volume element will be: $N_{f1}\cdot\rho(r,\phi)\cdot\mathrm{d}V$ where N_{f1} represents the fraction of the total states that are at level 1. Similarly the number of states at level 2 will be $N_{f2}\cdot\rho(r,\phi)\cdot\mathrm{d}V$. Note that $N_1^f + N_2^f = 1$.

From the definition of absorption cross-section, the probability of photon absorption or capture by an electron at level 1 for the element of area $\mathrm{d}A$ is $\sigma_a/\mathrm{d}A$, so the rate of

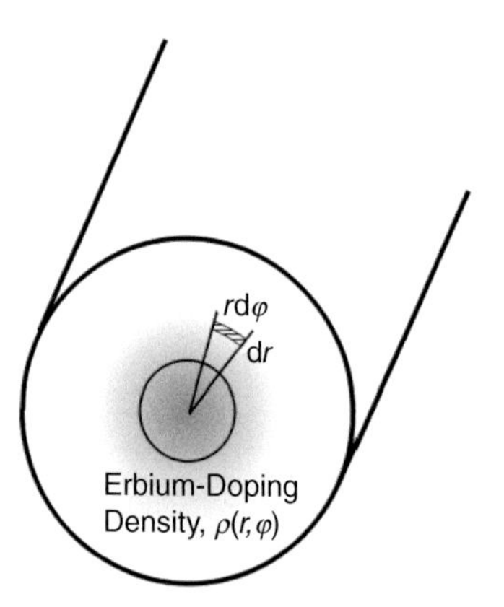

Figure A6.3.1 Optical fibre doped with erbium with a density distribution of $\rho(r, \phi)$ and element, dA, of cross-sectional area.

photon absorption over the length dz (or equivalently the rate of change in level 1 population in the volume element dV) will be this probability multiplied by the number of states at level 1 and multiplied by the number of photons per second crossing area dA, giving:

$$\rho(r,\phi)\mathrm{d}V\frac{\mathrm{d}N_1^f}{\mathrm{d}t} = -\left(\frac{\sigma_a}{\mathrm{d}A}\right)\cdot\left[N_1^f\rho(r,\phi)\mathrm{d}V\right]\cdot[I(r,\phi)dA] = -\sigma_a N_1^f[\rho(r,\phi)I(r,\phi)r\mathrm{d}r\mathrm{d}\phi]\mathrm{d}z \tag{A6.3.3}$$

Integrating (A6.3.3) over the full cross-sectional area of the fibre, the power, dP_a, absorbed over the length element dz will be:

$$\rho_e S_e\left(\frac{\mathrm{d}N_1^f}{\mathrm{d}t}\right)\mathrm{d}z = -\alpha_a N_1^f P\cdot\mathrm{d}z = \mathrm{d}P_a \tag{A6.3.4}$$

where $\alpha_a(\nu) = \Gamma\rho_e\sigma_a(\nu)$ is the absorption coefficient, $P = \iint I(r,\phi)r\mathrm{d}r\mathrm{d}\phi$ is the total power in the mode and S_e is the effective area of the erbium doping:

$$S_e = \iint f_e(r,\phi)r\mathrm{d}r\mathrm{d}\varphi \tag{A6.3.5}$$

Note that S_e is simply the area of the core for uniform doping of the core only.

Γ is the overlap between the erbium density distribution and the optical intensity distribution:

$$\Gamma = \frac{\iint f_e(r,\phi)f_i(r,\phi)r\mathrm{d}r\mathrm{d}\phi}{\iint f_i(r,\phi)r\mathrm{d}r\mathrm{d}\phi} \tag{A6.3.6}$$

A similar result may be obtained for the emission coefficient defined by, $\gamma_e(\nu) = \Gamma\rho_e\sigma_e(\nu)$ so that:

$$\rho_e S_e\left(\frac{\mathrm{d}N_2^f}{\mathrm{d}t}\right)\mathrm{d}z = -\gamma_e N_2^f P\cdot\mathrm{d}z = -\mathrm{d}P_e \tag{A6.3.7}$$

Hence combining (A6.3.4) and (A6.3.7), the change in power from both absorption and emission, $dP = dP_a + dP_e$, over the length element dz is:

$$\frac{dP}{P} = g(\nu)dz \tag{A6.3.8}$$

where $g(\nu) = \left\{\gamma_e(\nu)N_2^f - \alpha_a(\nu)N_1^f\right\} = \{\gamma_e(\nu) + \alpha_a(\nu)\}N_2^f - \alpha_a(\nu)$.

For an erbium-doped fibre of length l, the inversion level and power will vary as a function of position, z, along the fibre. However, a length-averaged fractional inversion level may be defined by:

$$\overline{N_2} = \frac{1}{l}\int_0^l N_2^f dz \tag{A6.3.9}$$

Integrating (A6.3.8) and using (A6.3.9), the output power from length l is:

$$P_{out} = P_{in}e^{\overline{g(\nu)}\cdot l} \tag{A6.3.10}$$

where the length-averaged gain is: $\overline{g(\nu)} = \{\gamma_e(\nu) + \alpha_a(\nu)\}\overline{N_2} - \alpha_a(\nu)$ and P_{in} is the input power.

Also from (A6.3.4), (A6.3.7) and the atomic rate equation (A6.3.1), we may write for the length element dz:

$$\left(\frac{dN_2^f}{dt}\right)\rho_e S_e dz = -dP - \frac{N_2^f}{\tau_{21}}\rho_e S_e dz \tag{A6.3.11}$$

where we have used the fact that $N_1^f = 1 - N_2^f$ in (A6.3.4).

If we integrate (A6.3.11) over the erbium-doped fibre length and use (A6.3.10) to calculate the total change in power between the input and output of the fibre we obtain:

$$\rho_e S_e l\left(\frac{d\overline{N_2}}{dt}\right) = -P_{in}\left\{e^{\overline{g(\nu)}\cdot l} - 1\right\} - \rho_e S_e l\frac{\overline{N_2}}{\tau_{21}} \tag{A6.3.12}$$

As noted earlier, this may be extended for multiple narrow-linewidth optical signals as:

$$\rho_e S_e l\left(\frac{d\overline{N_2}}{dt}\right) = -\left[P_{in}(\nu_1)\left\{e^{\overline{g(\nu_1)}\cdot l} - 1\right\} + P_{in}(\nu_2)\left\{e^{\overline{g(\nu_2)}\cdot l} - 1\right\} + \cdots\right] - \rho_e S_e l\frac{\overline{N_2}}{\tau_{21}} \tag{A6.3.13}$$

As explained in the main text, for fibre amplifiers one of these input signals will be the pump source, see (6.8).

References

1. A. Bjarklev, *Optical Fiber Amplifiers: Design and System Applications*, London, Artech House, 81–107, 1993.
2. P. C. Becker, N. A. Olsson and J. R. Simpson, Rare earth ions – introductory survey, in *Erbium Doped Fibre Amplifiers*, San Diego, Academic Press, ch. 4, 87–129, 1999.

3. M. J. F. Digonnet, *Rare Earth Doped Fibre Lasers and Amplifiers*, New York, Marcel Dekker, 2001.
4. E. Desurvire, *Erbium Doped Fiber Amplifiers: Principles and Applications*, Hoboken, New Jersey, Wiley, 2002.
5. L. Dong and B. Samson, *Fiber Lasers: Basics, Technology and Applications*, Boca Raton, FL, CRC Press, Taylor & Francis Group, 2017.
6. S. D. Agger and J. H. Povlsen, Emission and absorption cross section of thulium doped silica fibers, *Opt. Express*, 14, (1), 50–57, 2006.
7. Z. Li, A. M. Heidt, J. M. O. Daniel, et al., Thulium-doped fiber amplifier for optical communications at 2 μm, *Opt. Express*, 21, (8), 9289–9297, 2013.
8. D. E. McCumber, Theory of phonon-terminated optical masers, *Phys. Rev.*, 134, (2A), A299–A303, 1964.
9. D. E. McCumber, Einstein relations connecting broadband emission and absorption spectra, *Phys. Rev.*, 136, (4A), A954–A957, 1964.
10. C. Kittel, *Introduction to Solid State Physics*, 7th edn., New York, Wiley, 97–140, 1996.
11. W. J. Miniscalo and R. S. Quimby, General procedure for the analysis of Er^{3+} cross sections, *Opt. Lett.*, 16, (4), 258–260, 1991.
12. M. Fox, *Quantum Optics. An Introduction*, Oxford, England, Oxford University Press, 51–55, 174–177, 200–204, 2006.
13. A. E. Siegman, *Lasers*, Sausalita, CA, University Science Books, 502–503, 1986.
14. R. E. Tench and M. Shimizu, Fluorescence-based measurement of $g^*(\lambda)$ for erbium-doped fluoride fiber amplifiers, *IEEE J. Lightwave Technol.,* 15, (8), 1559–1564, 1997.
15. J. T. Verdeyen, *Laser Electronics*, 3rd edn., New York, Prentice Hall, 145–159, 235–243, 296–311, 1995.
16. J. Marshall, G. Stewart and G. Whitenett, Design of a tuneable L-band multi-wavelength laser system for application to gas spectroscopy, *Meas. Sci. Technol.*, 17, 1023–1031, 2006.
17. S. K. Kim, G. Stewart, W. Johnstone and B. Culshaw, Mode-hop-free single-longitudinal-mode erbium-doped fibre laser frequency scanned with a fibre ring resonator, *Appl. Opt.*, 38, (24), 5154–5157, 1999.
18. Z. Meng, G. Stewart and G. Whitenett, Stable single-mode operation of a narrow-linewidth, linear-polarization erbium fibre ring laser using a saturable absorber, *IEEE J. Lightwave Technol.*, 24, (5), 2179–2183, 2006.
19. G. A. Cranch, G. M. H. Flockhart and C. K. Kirkendall, Distributed feedback fiber laser strain sensors, *IEEE Sens. J.*, 8, (7), 1161–1172, 2008.
20. NP Photonics. Single-frequency fibre lasers. 2019. [Online]. Available: www.npphotonics .com/ (accessed April 2020)
21. E. Hecht and A. Zajac, *Optics*, Reading, MA, Addison-Wesley, 306–309, 1974.
22. V. M. Baev, T. Latz and P. E. Toschek, Laser intra-cavity absorption spectroscopy, *Appl. Phys. B*, 69, 171–202, 1999.
23. Y. O. Barmenkov, A. Ortigosa-Blanch, A. Diez, J. L. Cruz and M. V. Andres, Time-domain fiber laser hydrogen sensor, *Opt. Lett.*, 29, (21), 2461–2463, 2004.
24. G. Stewart, P. Shields and B. Culshaw, Development of fibre laser systems for ring-down and intra-cavity gas spectroscopy in the near-IR, *Meas. Sci. Technol.*, 15, (8), 1621–1628, 2004.
25. G. Stewart, G. Whitenett, S. Sridaran and V. Karthik, Investigation of the dynamic response of erbium fibre lasers with potential application for sensors, *IEEE J. Lightwave Technol.*, 25, (7), 1786–1796, 2007.

26. B. Löhden, S. Kuznetsova, K. Sengstock, et al., Fiber laser intracavity absorption spectroscopy for in situ multicomponent gas analysis in the atmosphere and combustion environments, *Appl. Phys. B.*, 102, (2), 331–344, 2011.
27. P. Fjodorow, O. Hellmig, V. M. Baev, H. B. Levinsky and A. V. Mokhov, Intracavity absorption spectroscopy of formaldehyde from 6230 to 6420 cm^{-1}, *Appl. Phys. B.*, 123, 147, 2017.
28. M. A. Mirza and G. Stewart, Multi-wavelength operation of erbium-doped fibre lasers by periodic filtering and phase modulation, *IEEE J. Lightwave Technol.*, 27, (8), 1034–1044, 2009.
29. A. Mirza and G. Stewart, Theory and design of a simple tunable Sagnac loop filter for multi-wavelength fibre lasers, *Appl. Opt.*, 47, (29), 5242–5252, 2008.
30. D. J. Kuizenga and A. E. Siegman, FM and AM mode locking of the homogeneous laser – part 1: theory, *IEEE J. Quantum. Electron.,* QE-6, (11), 694–709, 1970.
31. J. S. Wey, J. Goldhar and G. L. Burdge, Active harmonic mode-locking of an erbium fibre laser with intra-cavity Fabry-Perot filter, *IEEE J. Lightwave Technol.*, 15, (7), 1171–1180, 1997.
32. H. A. Haus, Mode-locking of lasers, *IEEE J. Sel. Top. Quantum. Electron.*, 6, (6), 1173–1185, 2000.
33. M. Horowitz, C. R. Menyuk, T. F. Carruthers and I. N. Duling, Theoretical and experimental study of harmonically mode locked fibre lasers for optical communication systems, *IEEE J. Lightwave Technol.*, 18, (11), 1565–1574, 2000.
34. G. Whitenett, G. Stewart, H. Yu and B. Culshaw, Investigation of a tuneable mode-locked fibre laser for application to multi-point gas spectroscopy, *IEEE J Lightwave Technol.*, 22, (3), 813–819, 2004.
35. M. N. Islam, Raman amplifiers for telecommunications, *IEEE J. Sel. Top. Quantum. Electron.*, 8, (3), 548–559, 2002.
36. J. Bromage, Raman amplification for fibre communication systems, *IEEE J. Lightwave Technol.*, 22, (1), 79–93, 2004.
37. V. R. Supradeepa, Y. Feng and J. W. Nicholson, Raman fibre lasers, *J. Opt.*, 19, 1–26, 2017.
38. A. Yeniay, J.-M. Delavaux and J. Toulouse, Spontaneous and stimulated Brillouin scattering gain spectra in optical fibers, *IEEE J. Lightwave Technol.*, 20, (8) 1425–1432, 2002.
39. A. Kobyakov, M. Sauer and D. Chowdhury, Stimulated Brillouin scattering in optical fibers, *Adv. Opt. Photonics*, 2, 1–59, 2010.
40. R. Bauer, T. Legg, D. Mitchell, et al., Miniaturized photoacoustic trace gas sensing using a Raman fiber amplifier *IEEE J. Lightwave Technol.*, 33, (18), 3773–3780, 2015.

7 Applications of Fibre Amplifiers and Lasers in Spectroscopy

7.1 Introduction

There are potentially a number of different ways in which fibre amplifiers and fibre lasers may be deployed in near-IR spectroscopy, ranging from the basic use of fibre amplifiers to boost signal levels, to more complex systems such as the creation of high-finesse cavities for ring-down, cavity-enhanced and intra-cavity laser absorption spectroscopy. To date, however, applications of these techniques have been somewhat limited to laboratory demonstrations. Perhaps of greater practical importance has been the development in recent years of fibre laser combs, with the associated technique of dual-comb spectroscopy. Also, as discussed in Chapter 8, high-power CW fibre lasers and fibre laser combs in the near-IR provide an essential pump source for the generation of coherent mid-IR light and mid-IR combs through non-linear processes. In this chapter we shall review the various techniques for applying fibre lasers to spectroscopy and although most of the systems discussed are based on erbium-doped fibre lasers, the techniques are equally applicable to all the various types of rare-earth-doped fibre amplifiers and lasers.

7.2 Basic Applications as Amplifiers or Sources in Near-IR Spectroscopy

We first discuss the simplest applications where fibre amplifiers may be used to boost the input power in fibre optic multiplexed systems, or to improve the signal levels in photoacoustic spectroscopy. Also, in principle, fibre lasers may be directly used as an alternative to DFB laser sources for near-IR spectroscopy.

7.2.1 Applications of Fibre Amplifiers in Near-IR Absorption Spectroscopy

In fibre optic networks, such as those discussed in Chapter 5, for gas sensing at a large number of locations over a wide area, the output from a single laser source is split into a large number of paths and a fibre amplifier may be used to boost signal levels as required. A standard erbium-doped fibre amplifier (EDFA) may be used if the source wavelength is in the 1500–1600 nm region or a thulium-doped fibre amplifier for longer wavelengths in the 1700–2100 nm region [1, 2].

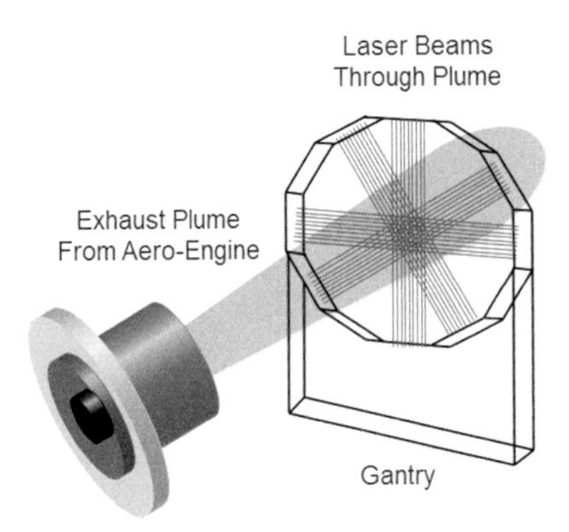

Figure 7.1 System for creating tomographic images of CO_2 emission from aero-engines.

Another example concerns the measurement of CO_2 emission from jet engines (see Chapter 5), which is important for engine diagnostics, determination of fuel efficiencies and CO_2 reduction. A 2-D map of the CO_2 concentration in the exhaust plume from an aero-engine is created by simultaneously passing ~100 beams through the plume, distributed uniformly around its circumference, as illustrated in Figure 7.1. Measurements of the integrated absorption along each path length of a metre or more through the plume are then processed to generate a 2-D tomographic image of the CO_2 concentration. To perform the absorption measurements, the methods of wavelength modulation spectroscopy (WMS) described in Chapter 3 are used, where the drive current of a DFB diode laser, operating around 2 μm, is sinusoidally modulated while the wavelength is scanned through the absorption line. To do this with ~100 separate lasers would be impractical, so the modulated output of a single DFB laser is amplified from a few milliwatts to a few watts by a thulium-doped fibre amplifier [3], which is then split and fibre-coupled to the multiple paths through the exhaust plume.

As was noted in Chapter 4, one important advantage of PAS over standard WMS is that large optical powers of several watts can be used to improve the signal magnitude and hence the sensitivity of gas detection, since it is the absorbed power that is being measured in PAS through the acoustic signal and not the transmitted power as in WMS (which saturates an optical detector at high levels). Hence in PAS systems, an EDFA or other type of fibre amplifier may be employed to boost the power and hence the sensitivity by increasing the power output from a modulated DFB laser to levels of several watts. This has been demonstrated by Ma et al. [4] for ultra-high sensitivity for acetylene on the 1530.37 nm absorption line, where the output power of a modulated DFB laser diode is boosted from ~7 mW to 1500 mW. The methods of quartz-enhanced photoacoustic spectroscopy (QEPAS), as discussed in Chapter 4, along with WMS are also employed in [4] to give a minimum detection limit of 33 ppb with a noise-equivalent absorption coefficient of 3.54×10^{-8} cm^{-1} $Hz^{-1/2}$. Bauer [5] makes use of a fibre Raman amplifier to generate an output of ~1 W at 1651 nm for high-sensitivity photoacoustic spectroscopy of methane. The system is illustrated in Figure 7.2.

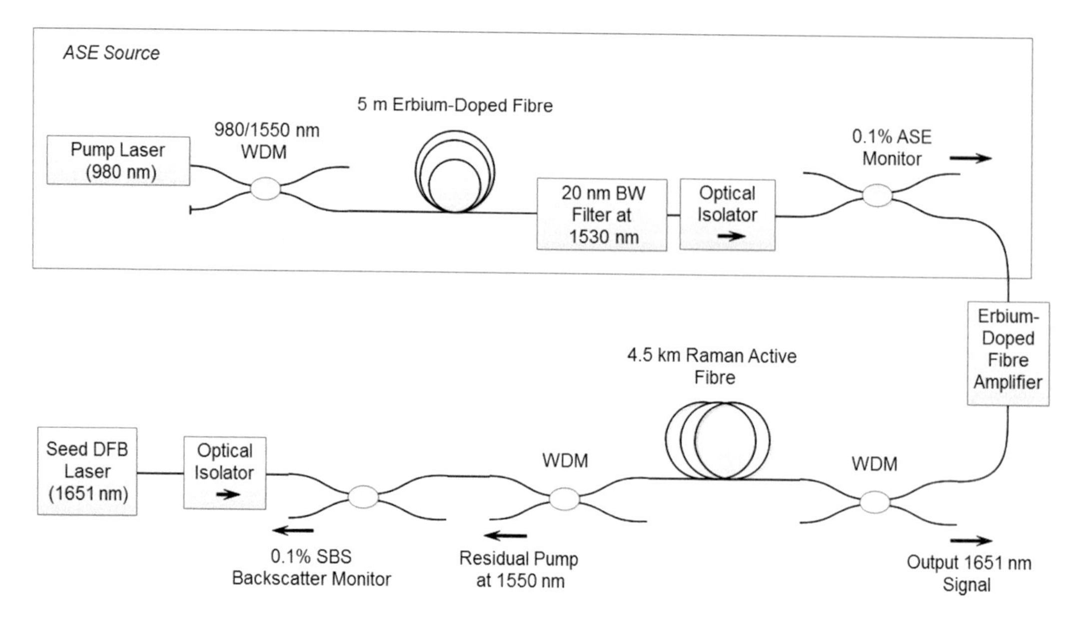

Figure 7.2 Raman fibre amplifier system to generate a high power output at 1651 nm for the photoacoustic detection of methane. Adapted from [5].

The Raman fibre amplifier is seeded with the output of a 1651 nm laser diode, modulated as for standard photoacoustic spectroscopy, as described in Chapter 4, but with an additional higher-frequency modulation at 500 kHz for suppression of stimulated Brillouin scattering (SBS). An ASE source, amplified by a high-power erbium-doped fibre amplifier, is used to pump the Raman active fibre. A minimum detection limit of a few tens of ppb for methane at 1651 nm with a normalised noise-equivalent absorption coefficient (NNEA) of 4.1×10^{-9} cm^{-1} W Hz$^{-1/2}$ was demonstrated with a miniaturised 3-D printed photoacoustic cell (resonator length of 10 mm, 0.9 mm radius with 5 mm long buffer regions).

7.2.2 Fibre Laser Sources for Near-IR Absorption Spectroscopy

At first sight, the use of fibre lasers as sources for spectroscopic systems appears to offer a number of benefits when compared with the distributed feedback (DFB) lasers discussed in Chapter 2. DFB lasers need to be custom designed at a specific wavelength for the particular gas of interest and their tuning range is limited to a scan over a few absorption lines of the gas, so a different DFB laser is normally required for each gas to be monitored. By contrast, a single fibre laser has a broad tuning range of 100 nm or more, so potentially may be used to monitor a number of species. However, as noted in Chapter 6, there are a number of problems associated with the direct application of fibre lasers as sources in spectroscopy. Fibre cavities may be several metres in length, giving rise to closely spaced longitudinal modes and hence instabilities from mode-hopping due to thermal and mechanical

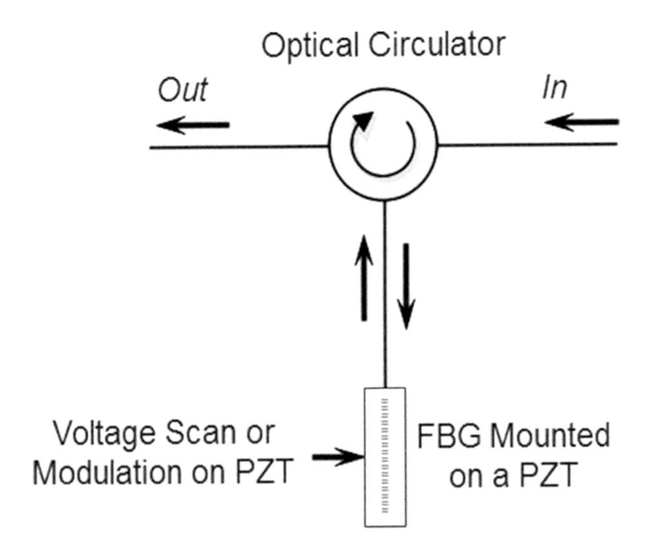

Figure 7.3 Fibre Bragg grating (FBG) and piezo-electric transducer (PZT) combined with an optical circulator for wavelength scanning and tuning of fibre lasers.

perturbations. Whereas wavelength scanning and modulation with DFB lasers can be simply performed by diode current modulation, it is more difficult and cumbersome with fibre lasers. Intensity modulation can be simply performed with fibre lasers by pump current modulation, but wavelength scanning and modulation requires, for example, a custom-designed fibre Bragg grating (FBG) and piezo-electric transducer (PZT) arrangement, as illustrated in Figure 7.3. In addition, scanning and modulation may result in transiting across the longitudinal modes, resulting in undesirable fluctuations of the output power.

Several authors have reported proof-of-principle demonstrations of spectroscopy systems based on CW fibre laser sources. In early work, Cousin et al. [6] described a near-infrared spectrometer based on a standard CW erbium-doped fiber laser. Continuous mode-hop-free temperature tuning in conjunction with mechanical tuning was achieved over a range of ~5–10 GHz. High-resolution absorption spectra of acetylene were measured around 1544 nm using a 100 m long Herriot multi-pass cell, giving a minimum detectable absorption coefficient of 3.5×10^{-7} cm^{-1} $Hz^{-1/2}$ in a direct absorption configuration or 6.6×10^{-11} cm^{-1} $Hz^{-1/2}$ in ring-down operation. More recently, thulium-doped fibre lasers [7, 8] operating around 2 μm have been investigated to target the CO_2 absorption lines in this wavelength region. Bremer [7] reported CO_2 detection with a tunable thulium-doped fiber laser consisting of a 15 cm length of thulium-doped alumino-silicate fibre, pumped by an erbium-doped fibre laser operating at 1.6 μm, as illustrated in Figure 7.4.

In Figure 7.4, a pair of FBGs at 1600 nm defines the cavity and lasing wavelength of the erbium fibre laser and similarly another pair of FBGs at 1995 nm defines the cavity and lasing wavelength of the thulium-doped laser. Wavelength tuning over a range of ~15 nm was achieved by compression of the FBG cavity of

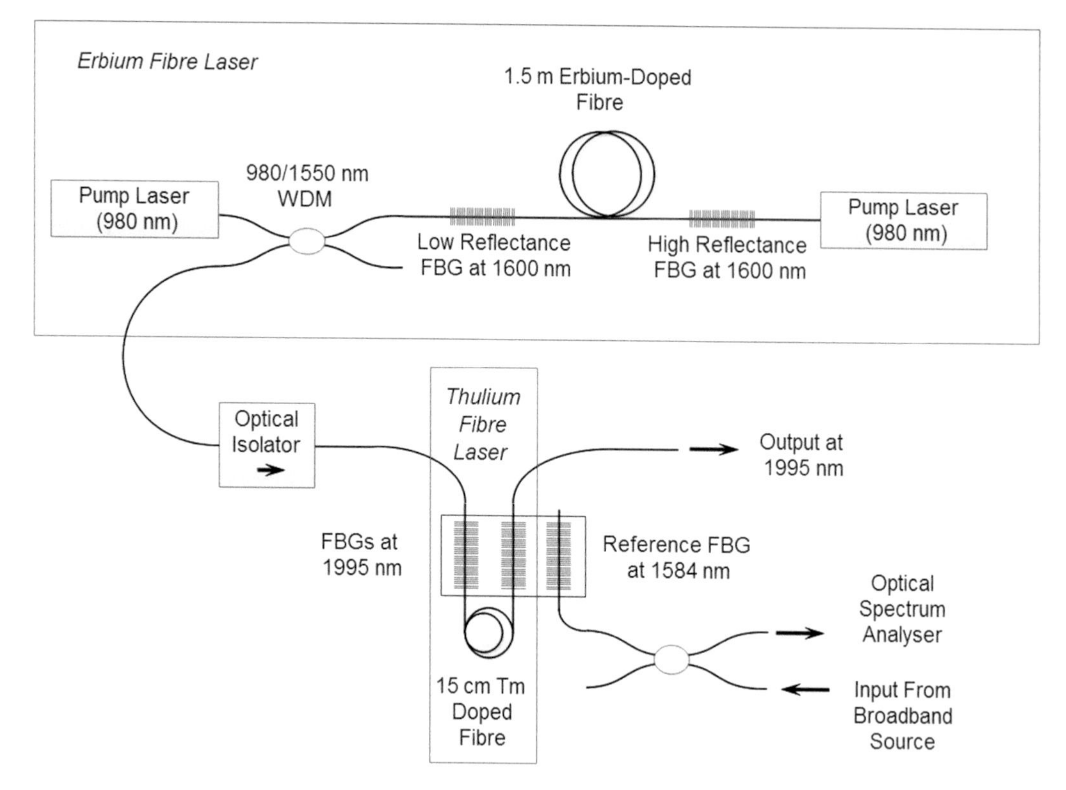

Figure 7.4 Thulium fibre laser for carbon dioxide detection at ~2 μm wavelength. Adapted from [7].

the thulium-doped fibre laser, with a reference FBG at 1584 nm on the same compression mount to monitor the wavelength shift. The output intensity of the thulium-doped fibre laser was modulated at 1 kHz by sinusoidal modulation of the current driving the 980 nm pump laser for the erbium-doped fibre. The change in the magnitude of the intensity modulation as a result of absorption from the CO_2 lines around 2 μm was captured by a fast Fourier transform (FFT) of the output with lock-in detection. With a 36 cm long gas cell, a minimum detectable concentration of ~1% volume CO_2 was achieved.

Improvements in the stability and tunability of fibre lasers have been made over the years and, as of 2018, a range of CW tunable, single-mode, narrow-linewidth fibre lasers have been commercialised for various applications, including use as sources for gas spectroscopy or environmental monitoring applications [9, 10]. For example, NP Photonics [10] market low-noise, narrow-linewidth, single-frequency, CW fibre lasers based on phosphate fibres doped with erbium (1.55 μm band), ytterbium (1 μm band), thulium or holmium (2 μm band). Acoustically damped packages are used to give improved stability, while short-length fibre cavities are employed to give mode-hop-free thermal tuning of ~60 GHz (0.5 nm) for erbium- or ytterbium-doped fibre lasers or

~10 GHz for holmium-doped lasers. PZT tuning over a range of 8 GHz or modulation at rates up to 40 kHz can also be performed on the erbium- or ytterbium-doped fibre lasers.

Instead of CW operation, fibre laser sources may be readily operated in multi-wavelength, pulsed or mode-locked regimes, which potentially offer a number of unique features for spectroscopy. We shall discuss in detail in Section 7.3 the important application of mode-locked fibre lasers to generate frequency combs for high-precision spectroscopy, but here we mention some other possibilities for these operational regimes. As discussed in Chapter 6, Section 6.5.3, multi-wavelength sources [11] have potential application in correlation spectroscopy where the multi-wavelength outputs are matched to multiple rotational lines in the absorption spectrum of the gas under observation to give better selectivity in the presence of other interfering species. Pulsed operation may be useful in photoacoustic spectroscopy, where the gas cell is placed within the fibre cavity to take advantage of the high peak powers within a laser cavity. Only a small amount of light needs to be tapped off from the cavity for optical power monitoring purposes since it is the acoustic signal that is measured. Mode-locked operation [12] may be used to select a particular wavelength of operation using the FBG arrangement shown in Figure 7.5a or a particular gas cell may be selected in a multiplexed system containing several gas cells, as illustrated in Figure 7.5(b).

In the fibre Bragg grating wavelength selector shown in Figure 7.5a each FBG is separated by a length of fibre so that each corresponds to a different round-trip time of the cavity and hence each FBG wavelength may be selected by choice of the appropriate mode lock frequency. Wavelength tuning may be performed by including a chirped FBG [13], where the grating adds a small time delay, dependent on wavelength through the relation, $\Delta\tau = D \cdot \Delta\lambda$ where D is the dispersion of the chirped FBG. Since the mode lock frequency is given by $f_{ml} = m_h/\tau$, the wavelength tuning is related to a shift in the mode-lock frequency, Δf_{ml}, by:

$$\Delta\lambda = -\Delta f_{ml} \frac{\tau^2}{m_h D} \tag{7.1}$$

where m_h is the harmonic order of the mode-locking.

Similarly, as shown in Figure 7.5b, multiple cells arranged in a ladder network may be individually addressed by selecting the mode-lock frequency corresponding to the cavity length of each path. However, this involves the use of long optical fibre cavities with the consequent instability problems noted earlier.

7.3 Frequency Comb Spectroscopy with Mode-Locked Fibre Lasers

The technique of direct frequency comb spectroscopy (DFCS) employs a frequency comb to probe absorption features over a very large bandwidth of typically 10–100 nm and at a very high resolution of ~200 MHz, determined by the comb line spacing and the width of a single comb line. The generation of frequency combs for spectroscopy (and other applications) using mode-locked fibre lasers is particularly attractive when

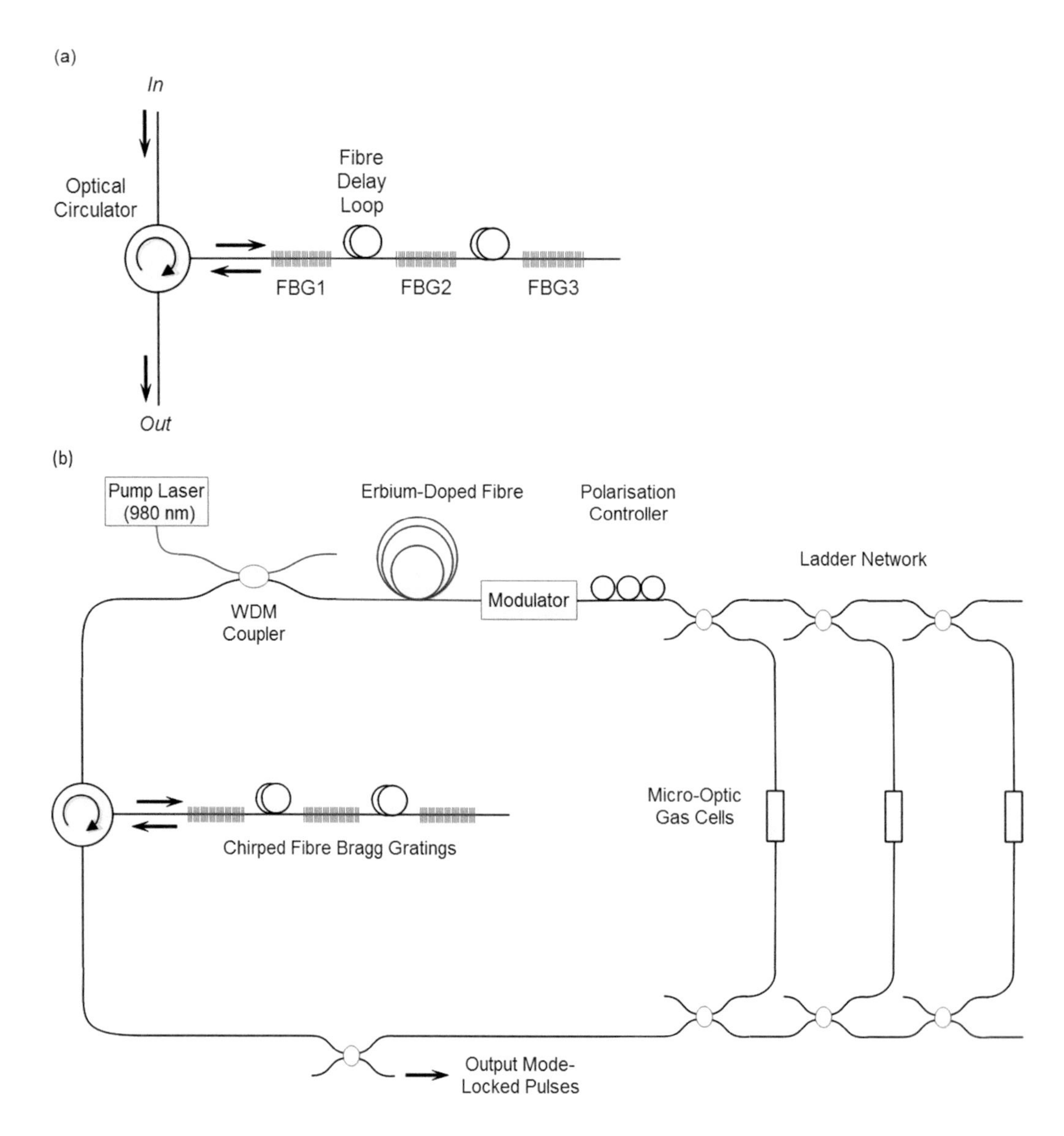

Figure 7.5 Mode-locked fibre laser system: (a) wavelength selection through fibre Bragg gratings, (b) addressing multiple cells arranged in a ladder network though the mode-lock frequency. After [12].

compared with solid-state laser combs and has received much attention in recent years [14–40]. With properly stabilised operation, the comb output of a fibre laser effectively acts as a parallel set of thousands of narrow-linewidth CW laser sources, simultaneously probing the spectral features over the full bandwidth of the fibre laser source. Fibre laser comb sources based on erbium- and ytterbium-doped fibre lasers are available commercially from several companies [41, 42] and, as discussed later in Chapter 8, Section 8.2.4, these near-IR combs also provide a basis for the generation of mid-IR combs by down-conversion in a non-linear crystal.

7.3.1 Generation of Frequency Combs

A detailed theoretical description of pulse formation and dynamics in mode-locked lasers with active or passive mode-locking (using either a fast or a slow saturable absorber) is given by Haus [14]. Passive mode-locking in fibre lasers may be achieved by use of an artificial (fast) saturable absorber, such as additive pulse mode-locking (APM) based on non-linear polarisation rotation (NPR). This makes use of the non-linear Kerr effect, where there is an intensity-dependent shift in phase and hence in the polarisation state. Alternatively a non-linear amplifying loop mirror (NALM) may be used, where the counter-propagating beams in the loop generate an intensity-dependent shift in the phase and hence in the transmittance of the loop. The advantage of using the NALM over the NPR method is that with NALM the fibre laser may be constructed with polarisation-maintaining fibre and components for greater long-term stability. Examples of real (slow) saturable absorbers for mode-locking fibre lasers include semiconductor saturable absorber mirrors (SESAM), and carbon nanotube or graphene saturable absorbers. Compared with the use of artificial saturable absorbers, real saturable absorbers give the advantage of easy self-starting, but generally exhibit larger timing jitter.

As was discussed in Chapter 6, Section 6.5.4, when viewed in the frequency domain, the mode-locked fibre laser generates a comb of optical frequencies corresponding to the longitudinal modes of the fibre cavity. The optical frequency, ν_{cq}, of each comb line may be written as:

$$\nu_{cq} = f_{ceo} + q \cdot f_{rep} \tag{7.2}$$

where q is an integer, $q \sim 10^7$, f_{rep} is the pulse repetition rate in the time domain, and $f_{rep} = \Delta\nu_m = (c/nL_c)$ is the longitudinal mode spacing in the frequency domain, assuming locking at the fundamental frequency.

Here f_{ceo} represents the carrier-envelope-offset frequency, that is, the rate of pulse-to-pulse phase slip between the carrier and the envelope. This arises as a result of unequal phase and group velocities with respect to the carrier and the pulse envelope. Hence the carrier phase slips with respect to the envelope by an amount of ψ_{ce} on each round trip, corresponding to a frequency of: $f_{ceo} = (\psi_{ce}/2\pi) \cdot f_{rep}$. These parameters are illustrated in Figure 7.6.

Note that the two frequencies f_{ceo} and f_{rep} are in the frequency range of 10^6–10^9 Hz, so a stable comb of well-defined optical frequencies may be realised through control of these two RF frequencies. Since f_{rep} is dependent on the cavity length, it is very important to isolate the cavity from mechanical and temperature perturbations. By phase-locking a single comb line to a stabilised external reference CW laser, the entire comb spectrum may be stabilised due to the phase-locked nature of the mode spectrum. Also, f_{ceo} may be stabilised by phase-locking to a stable RF frequency with feedback control to the pump laser current.

For a stable, high-performance frequency comb there are various noise sources in mode-locked fibre lasers that must be minimised [15–18]. Below relaxation oscillation frequencies, relative intensity noise (RIN), defined as the ratio of the mean-square

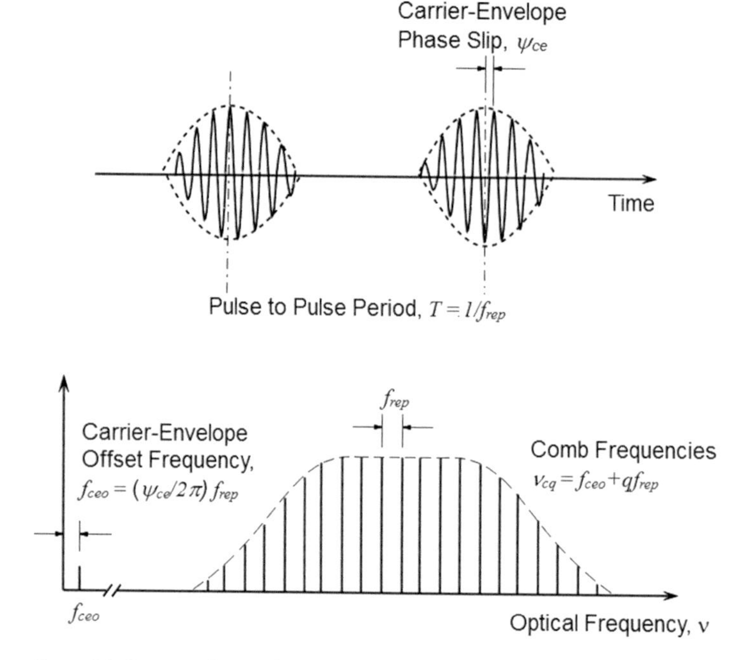

Figure 7.6 Output of a mode-locked fibre laser in the time and frequency domains.

optical power fluctuation of the pulses to the pulse train average power, arises from ASE noise and pump intensity noise, but is usually dominated by the pump noise and may be minimised by feedback control of the pump current. It is also important to minimise timing jitter (i.e. the deviation of the optical pulse train from a perfectly periodic pattern) since the optical comb-line frequency noise is highly correlated with timing jitter. The fundamental limit in timing jitter is set by ASE noise but other factors, particularly intensity and acoustic noise, also contribute to timing jitter and hence the importance of acoustic shielding. Due to the random-walk characteristics of timing in a free-running oscillator, it may be necessary to lock the pulse repetition rate to a stable RF oscillator. The fundamental limit in the comb-line frequency noise is set by the ASE noise or Schawlow–Townes limit (as discussed in Chapter 6, Section 6.5.1) with inverse dependence on the total power in all the comb lines. Pump intensity noise contributes to noise in f_{ceo} and f_{rep} through complex intra-cavity dynamics and hence to comb-line frequency noise. Fluctuations in ψ_{ce} induced by noise also contribute to the comb-line frequency noise. However, the primary cause of comb-line broadening is cavity length fluctuations from environmental mechanical and temperature disturbances, which must be minimised.

Several examples of high-performance erbium fibre laser combs are given in the literature [19–22]. Fehrenbacher [19] describes a polarisation-maintaining, femtosecond erbium fibre laser comb using a semiconductor saturable absorber for mode-locking. The carrier-envelope phase slip is eliminated through a difference frequency generation scheme in a periodically poled lithium niobate crystal (PPLN). Active linewidth narrowing is employed to achieve a sub-hertz optical linewidth by locking to an external 193 THz single-frequency laser. Kuse [20] demonstrates a self-starting, polarisation-maintaining, erbium fibre frequency comb with a NALM for mode-locking with very low timing jitter and ultra-low noise performance.

However, of primary importance for practical applications in spectroscopy is the development of compact, field-deployable, fibre laser combs. The detailed design and performance of a laptop-sized, self-referenced comb suitable for a range of applications outside the laboratory is given by Sinclair [21, 22]. A schematic of the system is illustrated in Figure 7.7.

The frequency comb is generated in a linear erbium fibre cavity with a SESAM for mode-locking. The output is amplified by an erbium-doped fibre amplifier and passed through a ~25 cm length of highly non-linear fibre to broaden the spectrum over the 1–2 μm region. In order to determine and control f_{ceo}, the broadened spectrum is coupled to a PPLN waveguide to frequency-double light at a wavelength of 2128 nm to 1064 nm. The RF frequency, f_{ceo}, is then obtained as the heterodyne signal from mixing this doubled light at 1064 nm with undoubled light at 1064 nm, i.e. mixing light from both ends of the spectrum, using an in-line interferometer. A phase-locked loop is used to stabilise f_{ceo} by locking to a stable RF frequency with feedback to the pump current of the erbium fibre laser comb. Stabilisation of the comb optical frequencies is achieved by mixing a single comb line with the output of a stabilised CW laser. The RF

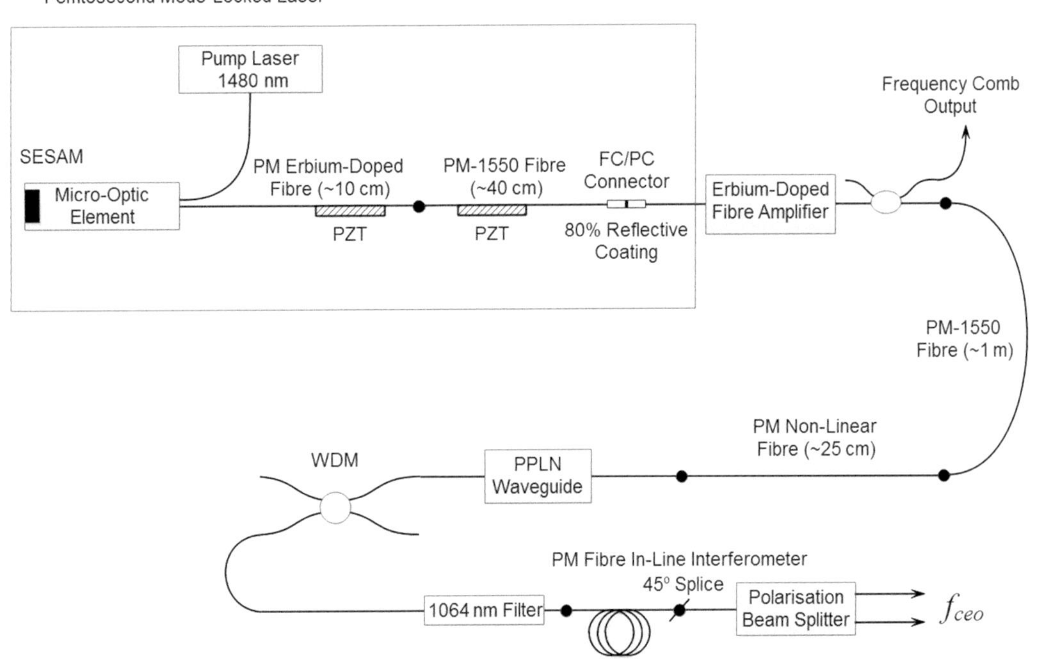

Figure 7.7 Schematic of a field-deployable fibre laser comb system. Adapted from [21, 22].

heterodyne signal generated is locked to a stable RF frequency via feedback control of the cavity length of the erbium fibre laser comb using the PZT elements. Both RF reference frequencies for locking purposes are derived from the pulse repetition rate, f_{rep}. The system provides a phase-coherent output spectrum over the 1–2 μm region, with a pulse repetition rate, or comb mode spacing, of ~200 MHz and a residual pulse-to-pulse timing jitter of less than 3 fs. The robust construction suitable for field deployment is ensured through a carefully designed optics package with polarisation-maintaining (PM) fibre throughout and high signal-to-noise ratio detection of the control signals, with digitally based phase detection and feedback for the control loops.

7.3.2 Interrogation of Absorption Lines by Frequency Combs

When the output of a frequency comb is passed through a gas cell, the absorption lines of the gases are directly imprinted on the comb spectrum, but the challenge is to measure this high-information-content spectrum at a sufficiently high resolution and in a short measurement time. A further challenge, especially for near-IR spectroscopy, may be to enhance the sensitivity, since direct comb frequency spectroscopy does not make use of the sensitivity advantages of WMS, as discussed in earlier chapters. We shall discuss in Section 7.3.4 the use of a high-finesse optical cavity for sensitivity enhancement in comb spectroscopy, but first we consider how to achieve broadband and rapid measurements with high resolution. A few techniques have been demonstrated, such as through the use of Vernier spectroscopy with a high-finesse optical cavity in combination with synchronous scanning of the detector [23] or by mapping the frequency comb onto a CCD array or camera by use of a virtually imaged phase array (VIPA) disperser. For example, Scholten et al. [24] have used a commercial laser comb from Menlo Systems [41] operating in the 1.5–1.6 μm range, along with a VIPA for accurate, real-time determination of the temperature and number density of CO_2, achieving an accuracy of better than 1% and a precision of 0.04% in less than one second. Another very important method is that of dual-comb spectroscopy [25–29], which we shall now discuss in detail.

7.3.3 Dual-Comb Frequency Spectroscopy

To fully exploit the resolution advantages of frequency combs in spectroscopy it is necessary to spectrally resolve every tooth of the comb, which cannot normally be done in conventional spectrometers due to their limited resolution. A Fourier transform (FT) spectrometer can accomplish this task, but is slow due to the need for scanning of the delay line. However, this problem is solved in dual-comb spectroscopy, which effectively performs the function of an FT spectrometer but without the moving parts. The basic idea behind dual comb spectroscopy is illustrated in Figure 7.8.

Here, two mode-locked lasers generate frequency combs with a small difference in their repetition rate of typically 100–1000 Hz and one (or both) combs are then passed through the gas cell. On the detector, interference between the comb lines produces an RF comb corresponding to the heterodyne or beat signal from the difference in

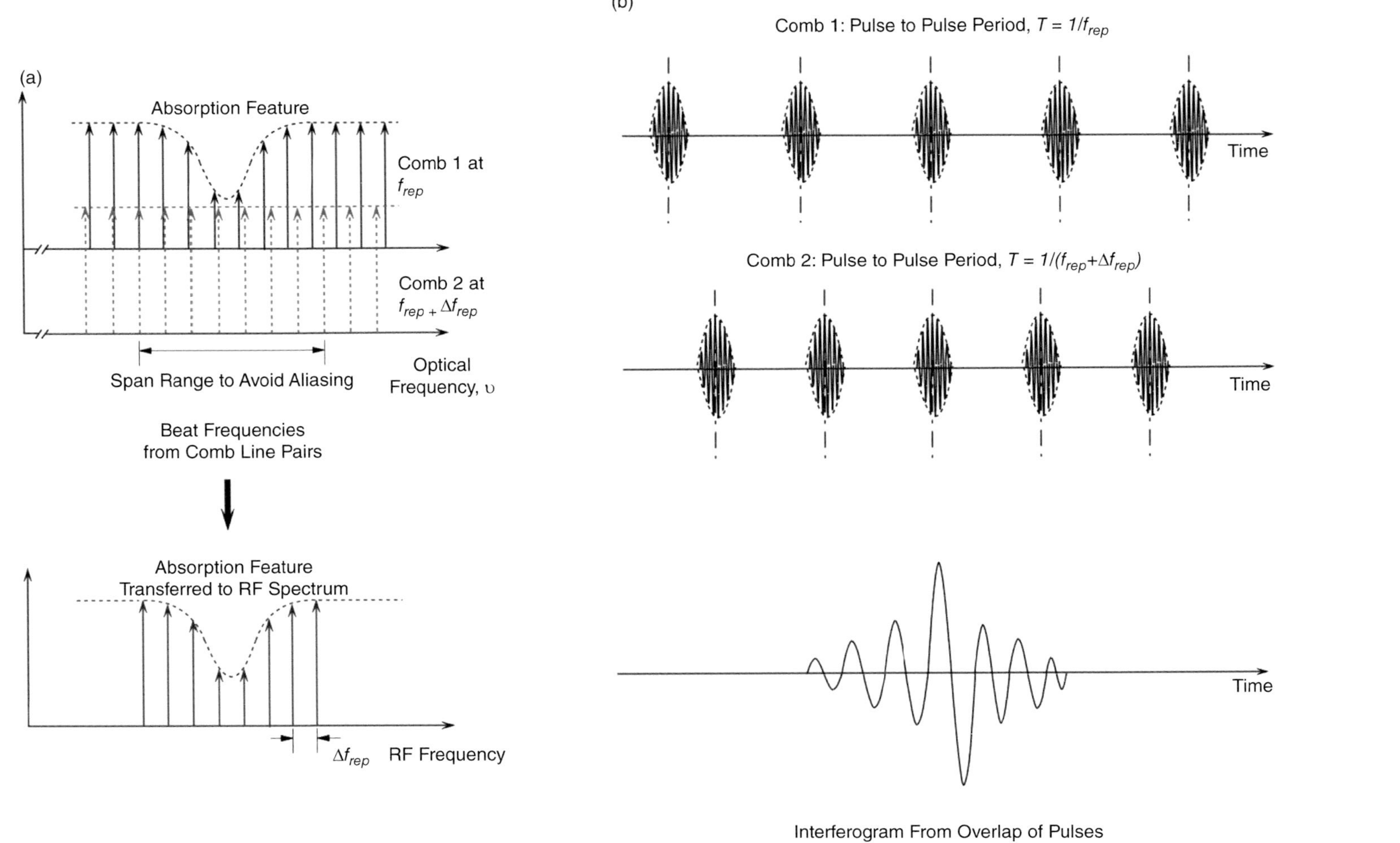

Figure 7.8 Principles of dual-comb spectroscopy: (a) frequency domain view showing the mapping of two optical combs to an a RF comb, (b) time domain view showing interferogram obtained from convolution of the pulses.

frequency of individual comb-line pairs, as shown in Figure 7.8a. The spectral absorption characteristics are transferred to the RF comb, which may be readily accessed through RF electronics and digitisers to analyse the down-converted signal. If only one comb is passed through the gas sample, then the other comb serves the function of a 'local oscillator', similar to dispersive FT spectroscopy, giving full phase and amplitude information for the absorption spectra. Passing both combs through the sample is more robust in turbulent situations, but yields only absorption data.

If the two optical combs, as shown in Figure 7.8a, have repetition rates of f_{rep} and $f_{rep} + \Delta f_{rep}$ then m comb lines will span an optical spectrum of width $\Delta\nu = mf_{rep}$, which is mapped to an RF spectrum of width: $\Delta f_{RF} = m\Delta f_{rep}$. Hence the optical spectrum is compressed to an RF spectrum of width:

$$\Delta f_{RF} = \Delta\nu\left(\frac{\Delta f_{rep}}{f_{rep}}\right) \tag{7.3}$$

With typical values for repetition rates of 100 MHz and 100 MHz + 1 kHz, an optical spectrum spanning 10 nm (~1200 GHz) is compressed and mapped to an RF spectrum of width 12 MHz. However, in order to avoid aliasing in mapping the optical comb to the RF comb, then, as shown in Figure 7.8a, we require $m\Delta f_{rep} < f_{rep}/2$, so from $\Delta\nu = mf_{rep}$ we obtain the limit on the optical spectrum width as:

$$\Delta\nu < \frac{1}{2}\frac{f_{rep}^2}{\Delta f_{rep}} \tag{7.4}$$

This gives a limit on the optical spectral width of 5 THz or ~40 nm for the above parameters.

When viewed in the time domain, as shown in Figure 7.8b, there is a short but increasing time delay between the consecutive pulses from the two mode-locked lasers, increasing in multiples of Δt where:

$$\Delta t = \left(\frac{1}{f_{rep}} - \frac{1}{f_{rep} + \Delta f_{rep}}\right) \cong \frac{\Delta f_{rep}}{f_{rep}^2} \tag{7.5}$$

The time delay is very short, typically $\Delta t \sim 0.1$ ps for the above parameters. From the perspective of the time domain, the pulse train of one comb effectively samples the pulse train of the other comb. The Fourier transform of the resulting digitised time domain signal or interferogram gives the RF spectrum and the process is directly analogous to FT spectroscopy with the scanning delay path replaced by the increasing delay time of the sampling pulse.

In practice there are a number of issues that need to be considered for achieving good performance with dual-comb spectroscopy [27]. There is clearly the need for two stabilised frequency combs with a high degree of mutual coherence. The complexity and challenges involved in the design of the combs are related to the level of performance required in terms of frequency resolution, accuracy and signal-to-noise ratio

(SNR). This can range from relatively low resolution and accuracy in the GHz region with free-running combs to very high resolution and accuracy in the kHz range with mutually coherent, fully referenced combs. Detector linearity is very important to ensure faithful recovery of the RF spectrum. Acquisition times can be very fast and a single-shot RF spectrum can be acquired in a matter of milliseconds, but in practice multiple spectra are required to improve the SNR through averaging, so the acquisition time is limited by the required SNR. The SNR scales as the square root of the acquisition time and inversely with the number of combs lines in the measurement. A detailed discussion on how to achieve a high SNR in dual-comb spectroscopy is given by Coddington et al. [27].

In order to avoid the need for sophisticated optical and electronic stabilisation schemes to meet the stringent stability requirements for dual-comb spectroscopy, Ideguchi et al. [28] have demonstrated an adaptive approach which uses free-running mode-locked lasers and compensates for laser instabilities by electronic signal processing. In this technique, as illustrated in Figure 7.9, two RF beat signals are derived from different line pairs of the two combs. This is achieved in practice by heterodyning the two free-running fibre laser combs with two free-running CW fibre lasers. The beat notes are then electronically mixed to eliminate the contribution from the CW fibre lasers to give signals f_a and f_a, typically at 30 MHz and 20 MHz, respectively, as shown in Figure 7.9. These RF beat signals thus report on the relative fluctuations between two lines of the two combs. A 50 MHz signal, generated electronically as $(3f_a - 2f_b)$, is multiplied with the interferogram signal before digitisation to mainly compensate for phase fluctuations of the interferogram signal. A second signal at 100 MHz, generated as $10(f_a - f_b)$, provides the adaptive clock signal triggering the data acquisition of the interferogram and accounting for timing fluctuation of the pulses. With this simplified system, high-quality molecular spectra down to Doppler-limited resolution were

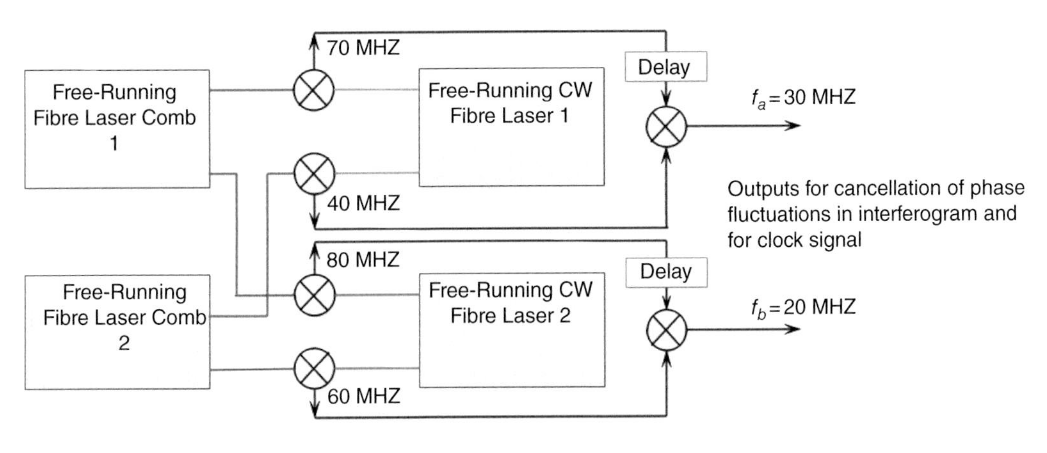

Figure 7.9 Generation of signals in adaptive dual-comb spectroscopy. Adapted from [28].

recorded, suitable for line intensity and profile measurements with accuracy of ~5%. However, the disadvantage of using free-running lasers is that self-calibration of the frequency scale is not possible and calibration must be performed against known molecular lines in the recorded spectra.

Because of the complexities associated with the operation of two mutually coherent mode-locked lasers, current research is investigating alternative techniques for dual-comb spectroscopy. For example, a simplified dual-comb system based on a single, dual-wavelength, mode-locked fibre laser has been experimentally demonstrated by Zhao et al. [29]. Mutual coherence of the combs is ensured, since the two pulse trains share the same laser cavity. The laser is passively mode-locked by a single-wall carbon nanotube mode-locker and generates two ultrashort pulse trains around 1533 nm and 1544 nm with similar repetition rates of ~52.74 MHz, but differing by 1250 Hz. The slight difference in the pulse repetition rates arises from group velocity dispersion in the fibre laser cavity due to the 11 nm wavelength difference of the pulses. The pulses are separated into the arms of a fibre interferometer by a bandpass filter, where they are amplified by an EDFA and spectrally broadened. The viability of the single free-running fibre laser system was demonstrated by measuring the absorption spectra of acetylene over the wavelength region of 1528–1543 nm, demonstrating its capability of resolving the comb teeth with a spectral resolution of a few picometres.

7.3.4 Cavity-Enhanced Dual-Comb Spectroscopy

While frequency comb spectroscopy allows fast and high-resolution acquisition of absorption spectra over a broad spectral region, extraction of the gas parameters is based on direct analysis of the recovered absorption line profiles and therefore lacks the signal-to-noise and sensitivity advantages associated with techniques such as WMS. In order to improve the sensitivity, especially with weak absorption lines, various forms of cavity-enhanced techniques have been explored, including cavity ring-down with a single comb [38] and cavity-enhanced dual-comb spectroscopy [30–33] as illustrated in Figure 7.10.

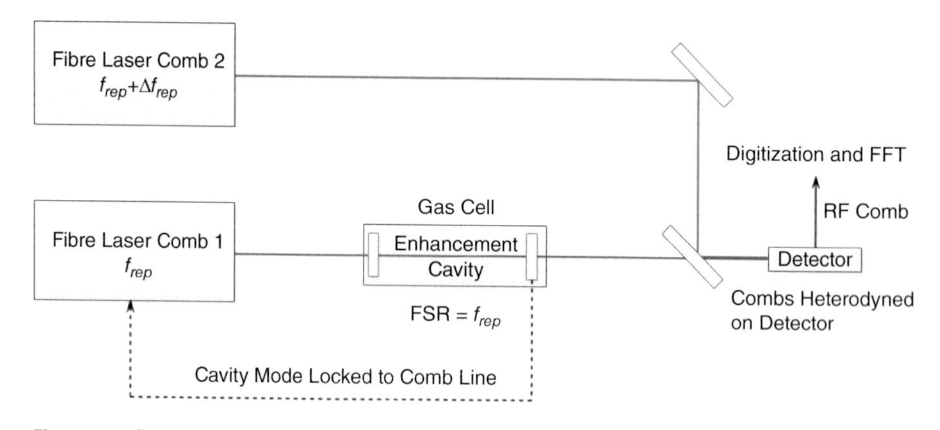

Figure 7.10 Schematic of system for cavity-enhanced dual-comb spectroscopy.

In Figure 7.10, the gas cell consists of a high-finesse optical cavity with the free spectral range (FSR) of the cavity matched to the pulse repetition frequency of mode-locked laser 1. This laser produces the comb for interrogation of the gas sample, with a comb line locked to one of the cavity modes of the cell. The output from the second laser comb is combined with the output from the gas cell and heterodyned on the detector to create the down-converted RF comb. The time-domain interferogram is then digitised and Fourier-transformed to obtain the absorption spectrum, as discussed in Section 7.3.3.

Considering the gas cell as a high-finesse Fabry–Perot etalon, then as shown in Chapter 5, see (5.20), the transmission of a comb line when it is lined up with a cavity mode is given by [43]:

$$\frac{P_o}{P_i} = \frac{T^2 \mathrm{e}^{-\alpha Cl}}{(1 - R\mathrm{e}^{-\alpha Cl})^2} \tag{7.6}$$

where T and R are the transmittance and reflectance of the similar mirrors, respectively, l is the cell length, α is the absorption coefficient and C is the gas concentration.

As derived in Chapter 5, if we assume that the reflectance of the mirrors is near unity and that the absorbance is small, the enhancement factor on the cell length is deduced from the expression for the absorbed power, see (5.21) and (5.22):

$$\frac{\delta P_o}{P_i} = \frac{(P_i - P_o)}{P_i} \simeq \frac{2\alpha Cl}{(1 - R)} \simeq \alpha Cl\left(\frac{2\mathfrak{F}}{\pi}\right) \tag{7.7}$$

where the finesse is given by $\mathfrak{F} = \pi\sqrt{R}/(1 - R) \simeq \pi/(1 - R)$.

We see from (7.7) that the effective path length of the cell is increased by a factor, $\approx 2\mathfrak{F}/\pi$. Since $\mathfrak{F}$ may typically be ~1000 or more, an enhancement factor of two to three orders of magnitude may be attained.

To be effective, however, the finesse and hence the mirror reflectivity must be maintained sufficiently high over the broad spectral range of interest from the comb. Also, dispersion from the mirrors must be minimised, since it causes the FSR of the cavity to vary with optical frequency, so the cavity modes 'walk off' from the uniformly spaced comb lines. The spectral range of the mode-matching is thus limited and reduces with increasing finesse. As an alternative to locking a cavity mode to a comb line, the FSR cavity may be scanned to effectively extend the spectral range of mode-matching, although this reduces the transmitted power.

7.3.5 Applications of Fibre Laser Combs for Spectroscopy

In recent years there have been a number of demonstrations of erbium fibre laser dual-comb systems applied to gas spectroscopy [34–39]. Some recent examples are as follows.

Rieker et al. [34] have demonstrated a dual-comb system for atmospheric trace gas monitoring over an open air path of 2 km. The comb outputs of two mode-locked erbium fibre lasers at a repetition rate of ~100 MHz and a difference frequency of

444 Hz are amplified, pulse-compressed in a large-mode-area fibre and spectrally broadened in non-linear fibre. The combined light output, limited to 1.5 mW in power, was launched over a horizontal atmospheric path of 1 km and returned to the detector by a plane mirror. Around 700 absorption lines were acquired at sub-1-kHz accuracy over the spectral region of 1600–1670 nm and fitted to HITRAN database models, with the measured atmospheric pressure as the only input to the fitting algorithm with temperature and gas concentration as the outputs. In this way, the atmospheric concentration of five gases, namely, CO_2, $^{13}CO_2$, CH_4, H_2O and HDO, were measured with a precision of <1 ppm for CO_2 and <3 ppb for CH_4 in 5 min and time-resolved measurements performed over a three-day period at 5 min averaging time. The fast acquisition time of the dual-comb system in acquiring broadband spectra gives immunity to intensity noise arising from atmospheric turbulence, while the high resolution and accuracy gives well-defined spectra for accurate retrieval of multiple gas concentrations.

Okubo et al. [35] have demonstrated dual-comb spectroscopy spanning a broad spectral region of 1–1.9 μm. The system uses two mode-locked erbium fibre lasers with a repetition rate of ~48 MHz to generate the combs, which are phase-locked to each other, with one of the comb lines locked to a stabilised 1.54 μm CW laser. To achieve the broad spectral range, the combs have a sub-Hz linewidth and the difference frequency between the laser combs is set at a very small value of 7.6 Hz. The spectrum of five rovibrational lines of C_2H_2, CH_4 and H_2O were measured with a 50 cm long single-pass cell for methane and a 15 cm long, 26-pass White cell (see Figure 5.1 in Chapter 5) for acetylene. Acetylene line positions were measured with a very high degree of accuracy, within ~1 MHz, as compared with previously obtained sub-Doppler resolution data.

In the dual-comb system demonstrated by Giorgetta et al. [36] a similar 2 km open air path was used for atmospheric measurements as described by Rieker [34]. However, in this case, instead of combining the combs prior to transmission over the atmospheric path, only one (signal) comb is transmitted over the open air path, while the other (reference) comb falls directly on the detector to give the heterodyne RF comb. As noted in Section 7.3.3, with this configuration, the Fourier transform of the time domain interferogram from the receiver gives the full phase and amplitude information for the absorption spectra. The frequency stability and high sampling speed possible with dual-comb spectroscopy, combined with use of adaptive compensation, allowed the acquisition of phase spectra over the turbulent air paths associated with atmospheric sensing. The atmospheric phase spectrum, spanning hundreds of rovibrational lines of CO_2, CH_4 and H_2O, was measured at each of the 70 000 comb teeth over a wavelength range of ~1.6 to 1.65 μm, with a sensitivity corresponding to 10^{-13} in refractive index and concentration uncertainty of 0.7 ppm for CO_2.

Of importance for practical applications is the design and operation of combs that can be deployed in the field. Such a system has been demonstrated by Coburn et al. [37] for continuous detection, location and quantification of methane emissions over areas of several square kilometres. To make the system field-deployable, a number of steps were taken in the design of the dual-comb lasers. A linear cavity design was used, with mode-locking by a SESAM and polarisation-maintaining fibre used throughout the system. Stabilisation was achieved by locking the carrier offset frequency, f_{ceo}, using *f*-to-2*f* locking, as discussed in Section 7.3.1, and phase locking an individual tooth from each comb to a common 1 kHz linewidth CW commercial diode laser. The fibre lasers were used to generate combs around 1.55 μm over a ~10 nm spectral range, which was then amplified and spectrally broadened to 1.0 to 2.2 μm in highly non-linear fibre (for the *f*-to-2*f* locking). The combined light from the combs was filtered to limit the spectral region to 1.62–1.69 μm and transmitted by single-mode fibre to a telescope transceiver. In a field demonstration of system operation, the dual frequency comb spectrometer was located in a centralised mobile trailer and the beam was directed sequentially over kilometre pathlengths to a number of retroreflector mirrors placed throughout a field site where controlled methane sources were dispersed to simulate emissions from natural gas production sites. In order to provide time-resolved location and quantification of the sources, a Bayesian inversion approach was applied to the array of integrated, open-path measurements obtained from the system. A sensitivity of ~2 ppb · km for CH_4 at 100 s averaging time was demonstrated, with the ability to continuously monitor emission rates down to 1.6 g min^{-1} from a distance of 1 km and to discern two leaks among a field of many potential sources. This level of performance means that the system is capable of detecting continuous or intermittent sources relevant to the oil and gas industry.

In a different application area, Schroeder et al. [38] have installed a dual-comb laser frequency spectrometer of a similar design to that discussed earlier [22, 37] in a power plant and demonstrated the simultaneous measurement, over a 5 hour period, of temperature, and H_2O and CO_2 concentrations in the exhaust of a 16 MW stationary gas turbine with a time resolution of 10–60 s. The comb teeth interrogated 16 000 wavelength points over the range 1435–1445 nm, spanning 279 absorption lines of H_2O and 43 lines of CO_2, although this range could be extended with different system parameters. The spectra were acquired at a rate of ~12 kHz, faster than beam fluctuations from turbulence, and with a round-trip optical pathlength in the exhaust duct of ~0.7 m. The uncertainties in the measurements over a 2.5 hour period were estimated to be ~1% for temperature and H_2O concentration and ~18% for CO_2 due the weak absorption lines of CO_2 in this wavelength region.

Finally, Chen et al. [39] have experimentally demonstrated the combination of dual-comb spectroscopy with evanescent-wave spectroscopy (see Chapter 5, Section 5.4.1) using a tapered fibre, with a taper length of 10 mm and waist diameter of 1 μm. The combs were generated by two commercial femtosecond erbium fibre lasers with a repetition rate of ~100 MHz and a rate difference of 155 Hz. The absorption and dispersion spectrum of methane at a pressure of 68.5 kPa was measured over the spectral region of 1.6–1.7 μm.

7.4 Ring-Down Spectroscopy with Passive and Active Fibre Cavities

Several authors have investigated the possibilities of using optical fibre cavities as a convenient way of performing ring-down spectroscopy [44–48] for enhancing the sensitivity, especially when using near-IR gas absorption lines. The basic concept is illustrated in Figure 7.11.

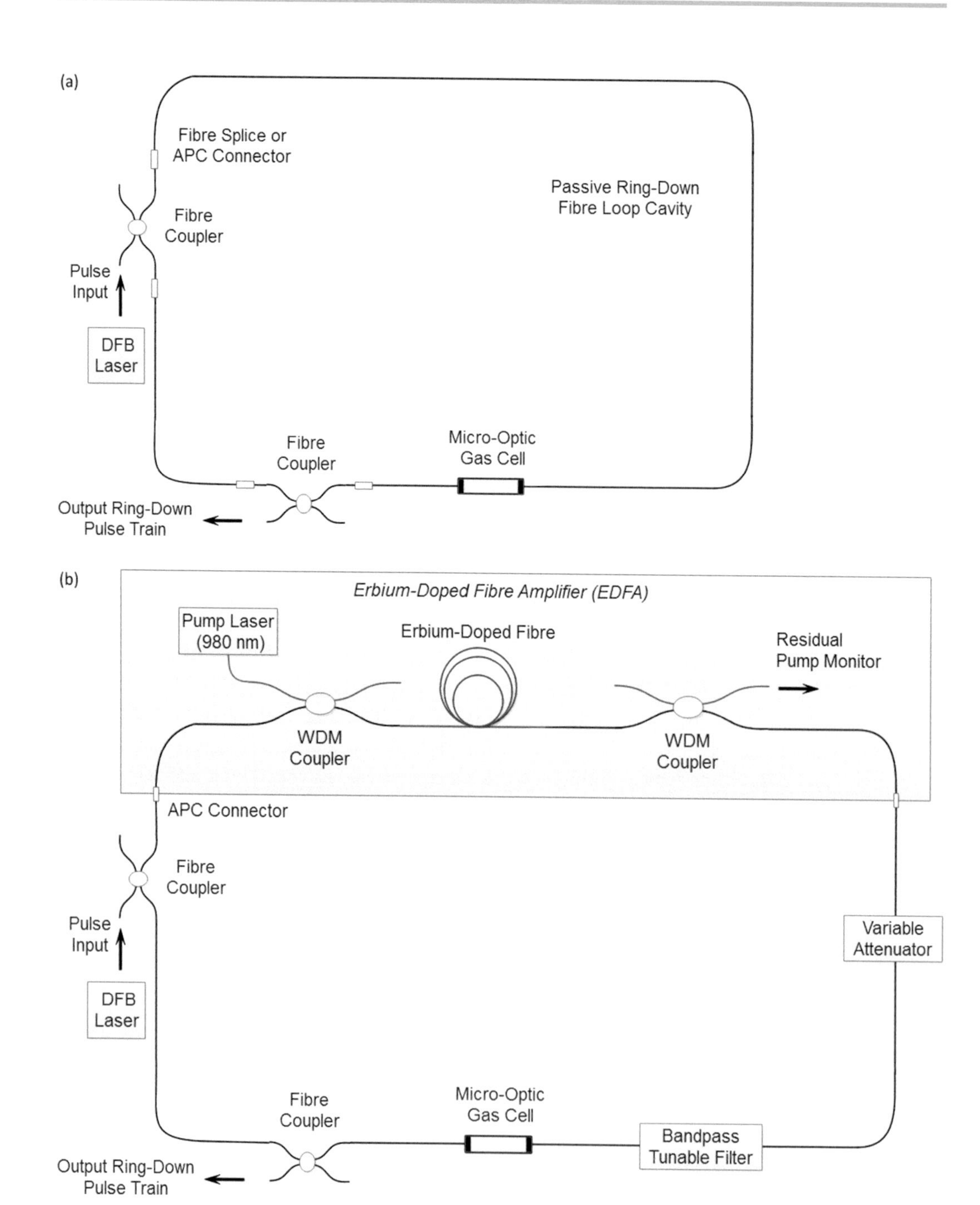

Figure 7.11 (a) Passive optical fibre loop cavity for ring-down. (b) Loop cavity with an EDFA to compensate for background loss. (c) Typical ring-down pulse train from the fibre loop cavity with the EDFA gain just below threshold. (d) Typical ring-down pulse train from the fibre loop cavity with the EDFA gain just above threshold. Adapted with permission from [46] © The Optical Society.

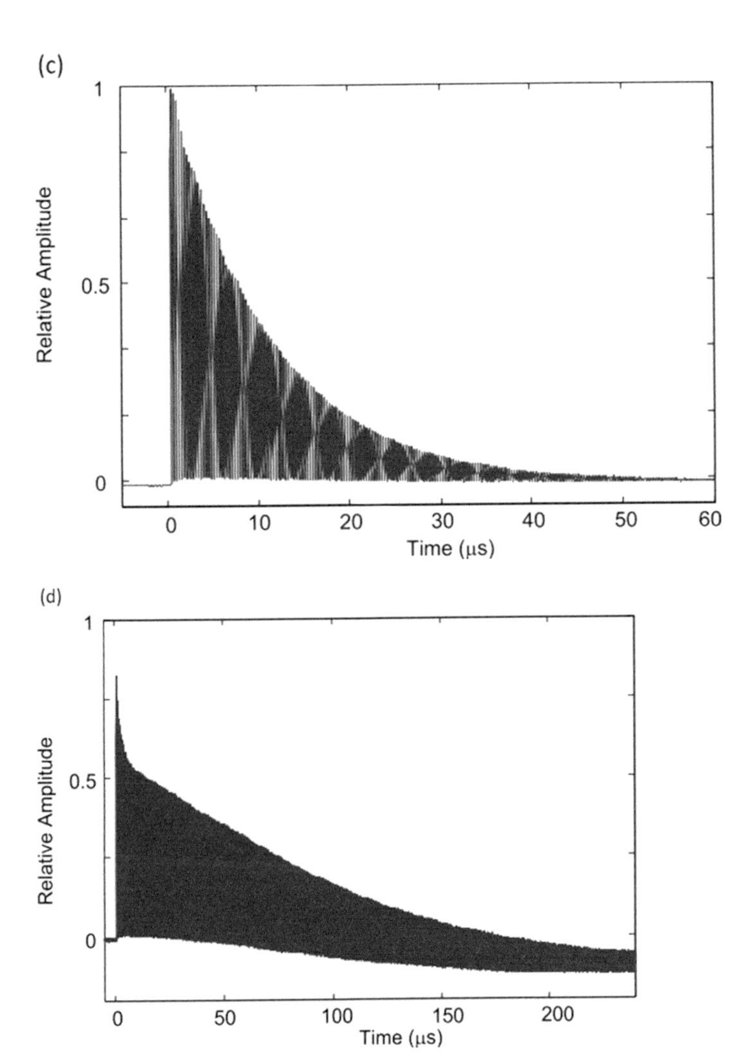

Figure 7.11 (cont.)

In Figure 7.11a, a pulsed DFB laser source injects pulses into the fibre loop cavity containing a micro-optic gas cell. The pulse width, typically a few tens of nanoseconds, is significantly less that the round trip time of the loop (τ ~100 ns for a 20 m loop) so successive pulses do not interfere with each other. Multi-circulation of pulses in the loop and through the gas cell create an exponentially decaying pulse train from the output coupler with a characteristic decay time or ring-down time of t_r, which is related to the number of circulations, N, in the loop over the ring-down time by:

$$N = \frac{t_r}{\tau} = \frac{4.34}{\Gamma_{\text{dB}}} \tag{7.8}$$

where Γ_{dB} is the round trip loop loss in units of decibels.

The presence of a gas in the cell of length L introduces a small additional loss in the loop of $\delta\Gamma_{dB} = 4.34\alpha CL$ so from (7.8) the fractional change in ring-down time from the gas absorption is:

$$\frac{\delta t_r}{t_r} = N \cdot (\alpha CL) \tag{7.9}$$

Similar to ring-down in a high-finesse bulk cell, as discussed in Chapter 5, Section 5.3.1, the principle of operation is to measure the gas concentration through the change in the ring-down time of the pulse train, as given by (7.9). The change in ring-down time may be directly measured in the time domain or, alternatively, through the phase shift of a sinusoidally modulated CW input to the cavity. In the latter case, the phase shift, ϕ_{mr}, is related to the ring-down time by: $\tan\phi_{mr} = \omega_m t_r$ where ω_m is the (angular) modulation frequency. For a measured 1% change in the ring-down time, from (7.9) we need $N > 100$ to detect an absorbance of less than 10^{-4}. However, to achieve even a modest cell-enhancement factor of $N \sim 100$, then, from (7.8), the background loop loss must be less than 0.05 dB, which is very hard to achieve in a passive fibre loop since the insertion loss of the micro-optic gas cell alone is much higher than this value. (Interaction with the external environment through the evanescent wave of a polished fibre device can give much lower insertion loss, but the advantage is negated for gas absorbers by the greatly reduced power in the evanescent field and hence evanescent-wave interaction is only really effective with liquids [47] where the refractive index is not too dissimilar from the fibre.)

One possibility to overcome this difficulty is to use an EDFA, as illustrated in Figure 7.11b, to compensate for the background losses of the fibre loop. If the EDFA gain is adjusted close to, but just below, the threshold for lasing action in the fibre loop, then ring-down times of around tens of microseconds can be achieved [46], as illustrated by the example shown in Figure 7.11c, which corresponds to a background loop loss of ~0.1 dB and an enhancement factor of $N \sim 50$. However, such an arrangement is critically dependent on tight control of gain fluctuations in the EDFA. A significant improvement in ring-down times can be realised by introducing automatic gain control through the use of lasing action in the loop since the gain in a lasing fibre loop is clamped at the threshold value, as explained in Chapter 6. For this situation, the same system as in Figure 7.11b is used, but now the gain is increased above threshold so that the loop is now in laser operation. The lasing action effectively creates a very high-finesse cavity and ring-down pulses are observed by injecting, as before, the external DFB pulse into the gain-clamped loop, probing the cavity loss profile around the fibre lasing wavelength. In this way ring-down times reaching several hundred microseconds can be demonstrated, as shown by the example in Figure 7.11d with $N \sim 500$. However, there are still a number of issues with this configuration. The ring-down time that can be attained depends critically on the relative positions of the input pulse wavelength from the DFB laser and the lasing wavelength of the fibre loop (the band-pass tuneable filter shown in Figure 7.11b can be used to adjust the fibre loop lasing wavelength). When off coincidence, the cavity loss and hence ring-down time is determined by the attenuation profile of the optical filter around its band-pass region and hence the longest ring-down times occur when the wavelengths are closely coincident, within the filter bandwidth. If, however, the pulse and fibre lasing wavelengths are at, or are very near to, coincidence, relaxation oscillations are induced by the injected pulses. In addition, drift and thermal or mechanical perturbations impose a limit on the accuracy and repeatability of the ring-down times, especially when they are very long, so that attempting to measure changes of the order of 1% in ring-down times is unreliable. Other configurations are possible, such as nested loops [44, 47], but suffer from much the same problems. In any case the system is now somewhat similar to the various versions of intra-cavity laser absorption spectroscopy (as discussed below in Sections 7.5 and 7.6), but with the additional complexity of requiring an external pulsed DFB laser source.

7.5 CW Fibre Lasers with an Intra-Cavity Gas Cell

We shall discuss in Section 7.6 the specific technique of intra-cavity laser absorption spectroscopy (ICLAS), where the spectral evolution of a laser is monitored during the transient start-up period, but here we first discuss the effect of an intra-cavity gas cell on the operation of a CW laser. Consider the loop cavity fibre laser system, as investigated by Liu et al. [50–53] and as illustrated in Figure 7.12, containing a tuneable filter, attenuator and the gas cell within the fibre cavity.

Using (6.22) to (6.25) for the CW output of a fibre laser, as given in Chapter 6, then by differentiation of the expressions for the slope efficiency and the threshold power,

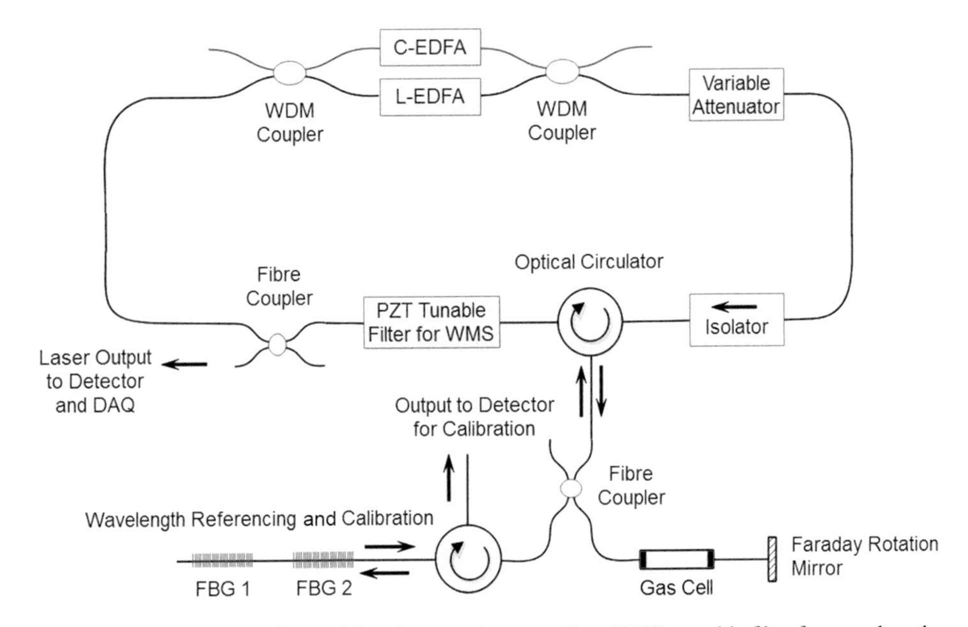

Figure 7.12 CW fibre loop laser with an intra-cavity gas cell and PZT tuneable filter for wavelength scanning and modulation.
Adapted from [50–53].

and neglecting the effects of spontaneous emission, we can write the effect of changes in the total cavity loss, α_c, on the output power as:

$$\frac{dP_o}{P_o} \cong -\left\{\frac{1}{e^{\alpha_c}-1}+\frac{1}{(\alpha_a l+\alpha_c)\left(\frac{P_p}{P_{th}}-1\right)}\right\}d\alpha_c \tag{7.10}$$

The sensitivity to small changes, $d\alpha_c$, in the overall cavity loss from gas absorption in an intra-cavity cell is defined by the bracketed expression in (7.10) and this must be greater than unity for any enhancement to take place. From consideration of the typical values for the various fibre laser parameters given in Table 6.4 in Chapter 6, it is clear that significant enhancement can only occur near threshold conditions, when the pump power is very close to the threshold power. Note that when the effects of spontaneous emission are included in the analysis [48, 49] the enhancement factor does not continue to increase to infinity as suggested by (7.10). However, operation near threshold means that the output power is low, with stability and repeatability issues since the enhancement factor is strongly dependent on small differences between the pump and threshold powers. Liu et al. [50–53] have investigated this approach for gas sensing with erbium fibre lasers operating in the C-band (1530–1560 nm) and L-band (1560–1620 nm). As illustrated in Figure 7.12, a PZT-driven tuneable filter within the cavity was used to generate both wavelength scanning and modulation for the application of WMS techniques, with the variable attenuator for selecting an appropriate operation point near threshold, resulting in an enhancement factor of ~50. Two FBGs are coupled to the cavity for wavelength referencing and calibration. Minimum detectable concentrations of ~0.6 ppm for acetylene, ~17 ppm for CO and ~19 ppm for CO_2 were obtained [52].

An alternative approach with an intra-cavity cell to achieve greater enhancement factors of ~500 with a simple and inexpensive system is to make use of the amplified spontaneous emission (ASE) within the fibre laser cavity [54–56]. As discussed in Chapter 6, Section 6.4.2, the ASE exists over the gain bandwidth of the erbium-doped fibre and provides a convenient broadband source for the interrogation of multiple absorption lines of the same or several gases. Also, as explained in Chapter 6, Section 6.5.1, the CW lasing action in a fibre loop effectively creates a very high-finesse fibre cavity with a small residual loss factor of δ_{ASE} at the lasing wavelength given by (6.20) and (6.21) and this determines the finesse of the cavity through (6.29) and the Schawlow–Townes limit in laser linewidth through (6.30) and (6.31). Away from the lasing wavelength, the fibre cavity still has a high finesse if the gain curve of the doped fibre is relatively flat over a range of wavelengths around the lasing wavelength (and if any optical filtering in the cavity for selection of the lasing wavelength also has a flat-top response). Hence, the ASE distribution in the vicinity of the lasing wavelength will undergo multiple circulations within the high-finesse cavity, which greatly enhances the effective pathlength of an intra-cavity gas cell. This process can be simply modelled as the incoherent summation of the individual contributions to the total ASE output from the simple fibre loop laser illustrated in Figure 7.13.

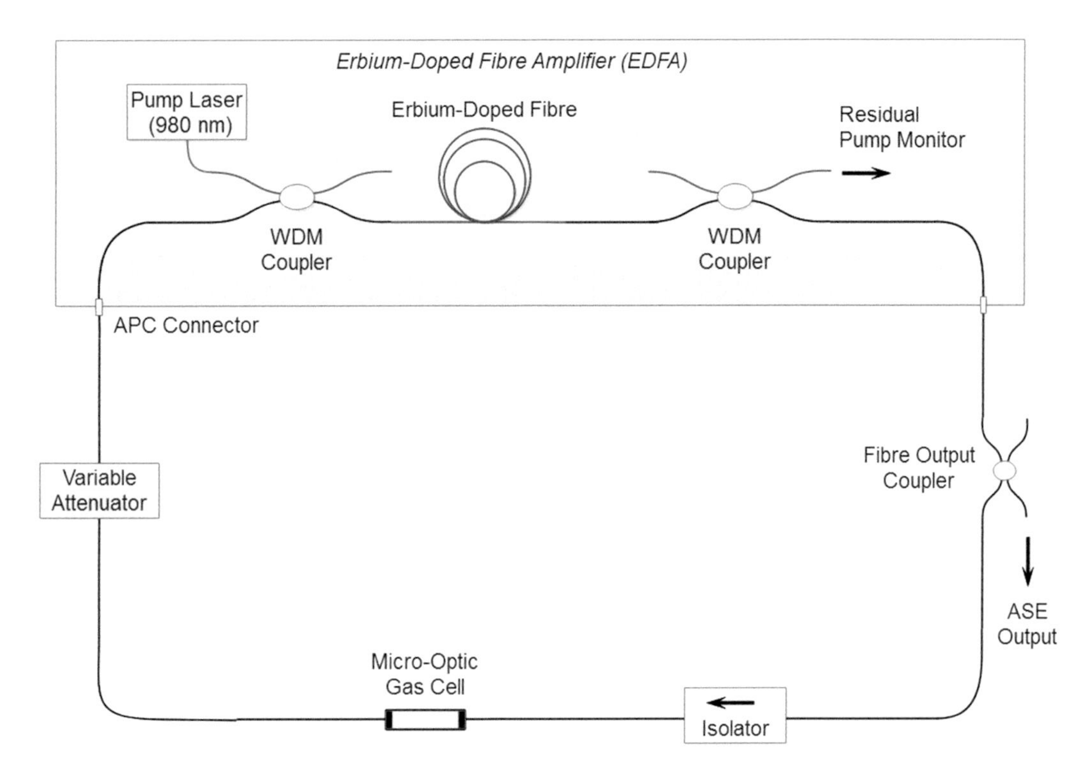

Figure 7.13 Fibre loop laser with intra-cavity cell for observation of ASE.

From Chapter 6, the ASE emerging from one end of the erbium-doped fibre over the optical frequency bandwidth, ν to $\nu + \mathrm{d}\nu$, is given by (6.16) in units of photons per second as:

$$P_{\mathrm{ASE}}(\nu) = 2\left(\overline{N_2}l\right)\gamma_e(\nu)A_{\mathrm{SE}}(\nu)\mathrm{d}\nu \tag{7.11}$$

In the vicinity of the lasing wavelength, the residual loss factor of the fibre loop cavity, $\delta(\nu)$, is defined as the difference between the total background loss of the cavity and the erbium fibre gain at the optical frequency ν. Hence, by analogy with (6.20), we can write:

$$\delta(\nu) = \left\{\alpha_c(\nu) - \overline{g_{ss}(\nu)\cdot l}\right\} \geq \delta_{\mathrm{ASE}} \tag{7.12}$$

The summation of multiple circulations of the ASE around the fibre cavity containing a cell with gas absorption of $\delta\Gamma_{\mathrm{dB}} = 4.34\alpha CL$ is then represented by:

$$P_{\mathrm{sum}}(\nu) = P_{\mathrm{ASE}}(\nu)[1 + \{\mathrm{e}^{-\delta(\nu)} \cdot \mathrm{e}^{-\alpha CL}\} + \{\mathrm{e}^{-\delta(\nu)} \cdot \mathrm{e}^{-\alpha CL}\}^2 + \cdots] \tag{7.13}$$

Hence the total output ASE power is:

$$P_{\mathrm{out}}(\nu) = \frac{(R\mathrm{e}^{-\alpha_{23}})P_{\mathrm{ASE}}(\nu)}{1 - \mathrm{e}^{-\delta(\nu)}\cdot\mathrm{e}^{-\alpha CL}} \tag{7.14}$$

where α_{23} is the attenuation in the loop between the erbium fibre output and the output fibre coupler and R is the out-coupled power fraction of the coupler.

For the case of small residual loss and gas absorption we can replace the exponentials in (7.14) by the usual approximation, $e^x \cong 1 + x$, and we can write a normalised output ASE signal as:

$$P_n(\nu) \cong \frac{1}{1 + \frac{\alpha CL}{\delta(\nu)}} \approx 1 - \frac{\alpha CL}{\delta(\nu)} \tag{7.15}$$

where $P_n(\nu) = P_g/P_{bg}$ is the ratio of the output ASE with gas in the cell to the background level without gas and the linear approximation applies for weak absorption lines when $\alpha CL \ll \delta(\nu)$.

Equation (7.15) shows that absorption line profiles are enhanced in depth over the spectral range of the ASE where $\delta(\nu) < 1$ and may be measured from the normalised ASE output, knowing the enhancement factor. Typically, for $\delta(\nu)$~0.01 dB, the enhancement is ~400. A necessary condition for low values of $\delta(\nu)$ is to ensure a broad, flat gain spectrum around the lasing wavelength and preferably with the ability to select spectral regions of interest for gas absorption lines. The gain distribution for erbium-doped fibre at different inversion levels is illustrated in Chapter 6, Figure 6.7, where it can be seen that the gain is relatively flat around the spectral regions of 1530 nm and 1560 nm when the inversion level is ~60%. Also, as explained in Chapter 6, Section 6.5.1, the inversion level in a CW fibre laser can be selected by adjustment of the cavity loss so a variable attenuator in the fibre loop can be used to select an appropriate flat gain region. Figure 7.14 shows an example [54] of the normalised experimental ASE output that is obtained around the 1530 nm spectral

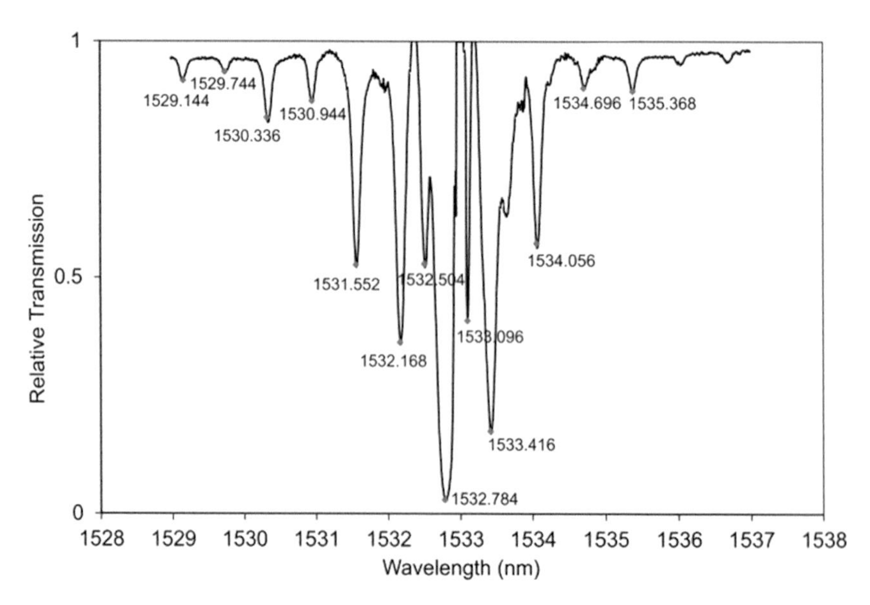

Figure 7.14 Normalised ASE output from a fibre loop laser with an intra-cavity gas cell containing 1% acetylene. © [2011] IEEE. Reprinted, with permission, from [54].

region under these conditions, with a 6 cm pathlength intra-cavity gas cell containing 1% acetylene. For Figure 7.14, the erbium fibre loop laser has a total cavity length of 13 m with a 20% output coupling ratio and the lasing peak at 1533 nm has been removed for clarity. Note that the cavity attenuation can be adjusted to set the inversion level at ~60% by observing the point where dual lasing occurs at 1530 nm and 1560 nm, as explained in Chapter 6.

Around 16 acetylene lines can be observed in Figure 7.14 with their relative depths reflecting both the absorption line strengths and the degree of enhancement. It is clear from Figure 7.14 that, as expected, the degree of enhancement is greatest in the vicinity of the central lasing wavelength, attaining a value of ~500, and falls off according to the curvature of the gain and/or the filter response curve. For quantitative analysis using (7.15), it is necessary to know $\delta(\nu)$ and this may be calculated theoretically from (7.12) with knowledge of the erbium fibre parameters, as discussed in Chapter 6. In practice, it is advantageous to have some form of calibration for $\delta(\nu)$ and this may be done by introducing into the loop a small, uniform etalon fringe pattern which will also experience the same enhancement factor over the spectral range of interest. This may be done, for example, through an optical fibre Mach–Zehnder interferometer formed from two fibre optic couplers with small coupling ratios to generate a small fringe ripple and a length difference of 1–10 cm between the two fibre arms to give a fringe spacing in the range of 2–20 GHz. Figure 7.15 illustrates the enhancement effect on the fringes of a fibre optic Mach–Zehnder with coupling ratios of 90:10 and 99:1 and a length difference of 1.25 cm [54].

Valiunas et al. [55, 56] have used this ASE method for the detection of nitrous oxide (N_2O), at sub-ppmv concentration levels using the P(12) rotational line of N_2O at ~1522.2 nm. N_2O is a minor constituent in Earth's atmosphere but, as a greenhouse gas, is around 300 times more destructive than CO_2 and is likely to increase with

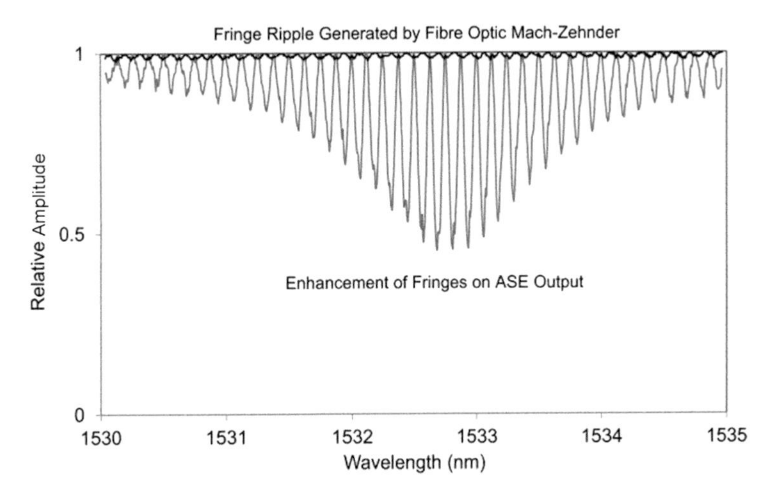

Figure 7.15 Experimental example of the enhancement of fringes observed in the ASE output with a fibre optic Mach–Zehnder interferometer in a fibre laser cavity. © [2011] IEEE. Reprinted, with permission, from [54].

increased use of agricultural fertilisers, which are a major source of N_2O. An intra-cavity multi-pass Herriott cell (see Figure 5.1b in Chapter 5) with an effective path-length of 30 m [55] or a hollow-core photonic crystal fibre [56] was used for the gas cell, with the sensitivity enhanced by the ASE circulation in the loop. The erbium fibre laser system was stabilised by a 0.5 m length of unpumped erbium-doped fibre to act as a saturable absorber with an FBG to select the wavelength of operation around the 1522 nm region.

In summary, the ASE intra-cavity method is similar to the ring-down technique in a fibre laser cavity, as discussed in Section 7.4, but is more versatile since it uses the existing broadband ASE in the cavity to probe the cavity loss around the lasing region rather than an external pulse source at a single wavelength. It also has the advantages of being relatively simple and inexpensive and may be applied in principle to most rare-earth-doped fibre lasers with ASE to cover a broad spectral range. Direct recovery of absorption lines makes for easier lineshape recovery for extraction of gas parameters. It does, however, require careful calibration and lacks the signal-to-noise advantages that are possible with WMS methods.

A different approach with an intra-cavity gas cell in a CW fibre laser is to use photoacoustic spectroscopy (PAS), as discussed in Chapter 4, to derive the gas signal. Potential advantages include a large PAS signal due to the high intra-cavity optical power (only a small optical tap from the cavity is required for monitoring purposes) and the ability to generate intensity modulation, at a large modulation index and without wavelength modulation, through pump current modulation around the relaxation oscillation frequency. Wang et al. [57] have demonstrated such an intra-cavity PAS system with an erbium fibre loop laser, as shown in Figure 7.16, using an intra-cavity gas cell with an acoustic resonant frequency of 2.68 kHz and low Q-factor of ~5. The FBG-based modulator shown in Figure 7.16, connected into the cavity via a circulator, was used to scan across the absorption line and modulate the laser wavelength at half the acoustic resonant frequency so that the second harmonic of the gas absorption could be observed in the PAS signal. The intra-cavity laser power was ~100 mW. A minimum detection limit of ~400 ppbv with a 2 s response time was demonstrated for acetylene on the 1531.6 nm absorption line, corresponding to a noise-equivalent absorption coefficient of $\sim 3\times10^{-7}$ cm^{-1} $Hz^{-1/2}$.

7.6 Intra-Cavity Laser Absorption Spectroscopy with Fibre Lasers

As noted in Chapter 6, Section 6.5.2, the technique generally referred to as 'intra-cavity laser absorption spectroscopy' or ICLAS involves placing a gas cell within a laser cavity and monitoring the spectral evolution of the laser output during the transient period of laser start-up, when light is building up within the cavity. At start-up, with a step input applied to the pump power, the inversion level initially spikes above the steady-state value so a large number of modes are initially excited across the spectral width of the gain, but as the laser output approaches steady-state, spectral narrowing occurs until the final state of a single or a few modes is reached, as determined by the

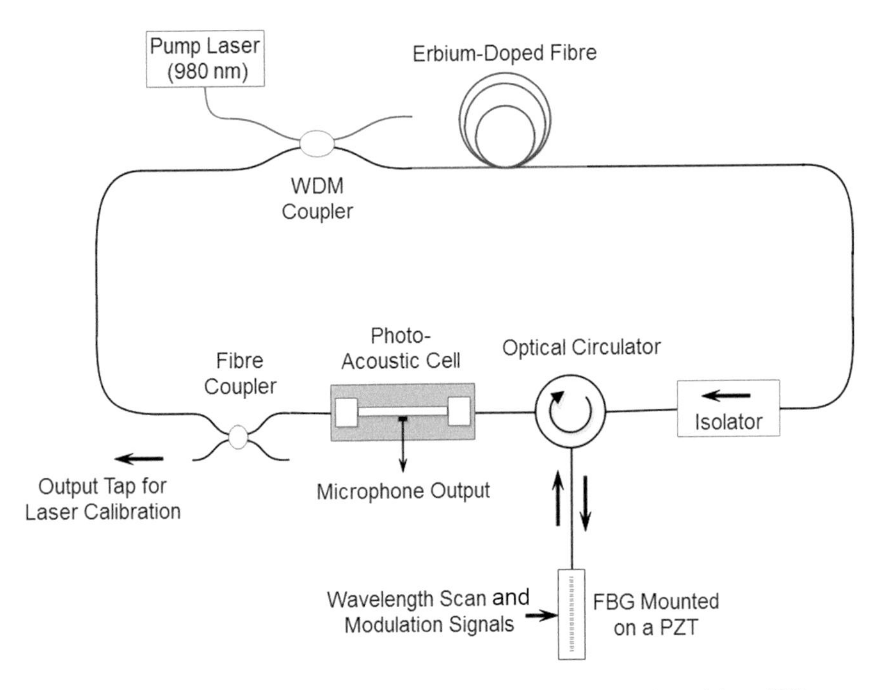

Figure 7.16 Fibre laser with an intra-cavity photoacoustic gas cell. Adapted from [57].

optical filtering in the cavity. For fibre lasers, this spectral narrowing occurs over a typical period of a few milliseconds, during which there are many thousands of circulations of light within the cavity. With a gas cell within the cavity, the spectral evolution is modified by the spectral absorption features of the gases present and the absorption lines are imprinted, with increasing depth, on the laser output due to the multiple circulations through the cell during the transient period. This process effectively enhances the length of the gas cell by several orders of magnitude, depending on the time of observation, t_o, of the output spectrum, and continues until spectral saturation is reached at the spectral saturation time denoted by t_s. For $t_o < t_s$ the enhancement factor on the cell length (number of circulations during the time period, t_o) for a loop cavity is given by $t_o/\tau = ct_o/L_{opt}$ where τ is the round-trip time of the laser cavity and L_{opt} is the total optical path length of the cavity containing the cell. For a Fabry–Perot type cavity, the round trip time is $2L_{opt}/c$, but there are two passes per round trip, so the enhancement factor is still t_o/L_{opt}. Hence a cavity of 3 m in optical length gives an enhancement factor of 1000 at an observation time of 10 μs.

Essentially, as noted in Sections 7.4 and 7.5, the lasing action creates a very high-finesse optical cavity, but here the degree of enhancement can be selected through the observation time. Ultimately, the enhancement factor is limited by other factors, such as the perturbations on light coherence through spontaneous emission and by Rayleigh scattering in fibres. In practice, intra-cavity reflections and etalon fringes may be present and these also are imprinted on the output spectra with increasing severity at longer times of spectral evolution. As discussed in Chapter 6, the fundamental process of spectral narrowing can be modelled for fibre lasers by numerical analysis of the rate equations (6.17) and (6.18). An important requirement for high sensitivity with ICLAS is that the absorber linewidth is smaller than the homogeneous gain bandwidth, which is true for fibre lasers. ICLAS has been extensively studied by Baev et al. [58] for a number of different types of lasers and has been demonstrated for various fibre laser systems, including erbium-, neodymium-, thulium-, tellurium- and holmium-doped fibre lasers [59–65]. Löhden et al. [62] have demonstrated an erbium-doped fibre ICLAS system, tuneable over the spectral range 1.53–1.61 mm, for the measurement of several gas species, such as CO_2, CO, H_2O, H_2S, C_2H_2 and OH, in combustion environments. Their system is illustrated in Figure 7.17. The fibre laser in Figure 7.17 consists of a linear erbium-doped fibre laser cavity, of total optical length of 2.35 m (round-trip time of 15.7 ns), including the gas cell, with end mirrors and pumped with a 980 nm laser diode. The bulk gas cell within the laser cavity has an optical path length of 48 cm for measurement of selected gas species or a 6 cm flame within the cell for measurement of flame spectra. Step modulation of the pump power above and below threshold was used to allow only the first peak in the relaxation oscillations to occur at the laser output and the pump power adjusted to set the observation time of spectral evolution at 10 μs. Over this 10 μs period, the number of round-trip circulations in the laser cavity is 10 μs/15.7 ns = 637 so, with two cell passes per round trip, the cell or flame length is enhanced by a factor of ~1270 for absorption measurements. A CCD camera on the spectrometer was used to record the

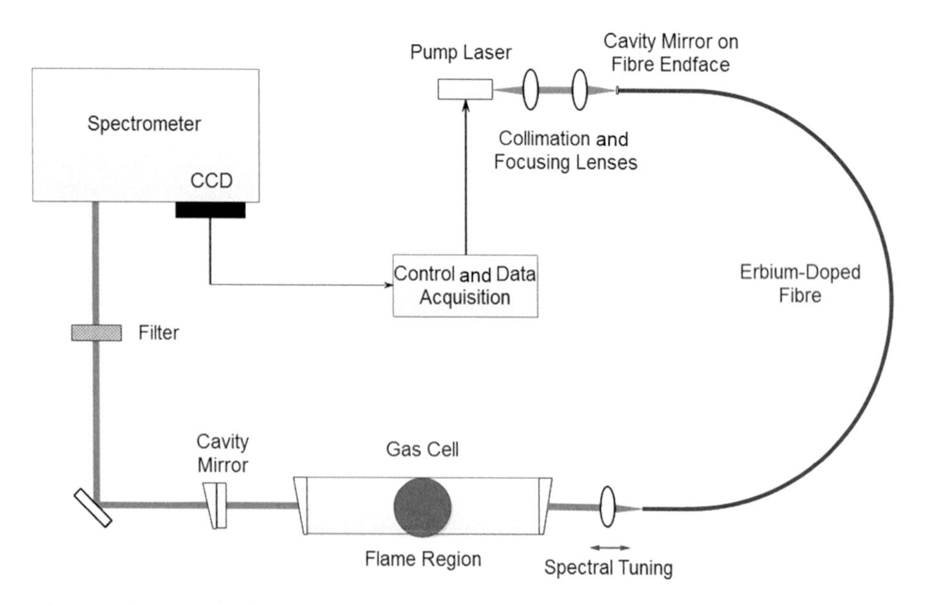

Figure 7.17 Intra-cavity laser spectroscopy system for flame and gas spectra. Adapted from [62-65].

output spectra, averaged over 200 ms, corresponding to 4000 averaged spectra at a laser repetition rate of 20 kHz and normalised to spectra without gas in the cell. The absorption spectra of CO_2, H_2S and C_2H_2 molecules introduced into the gas cell were measured with a minimum detectable concentration of ~25 ppm for CO_2 on the 1578.7 nm absorption line, 5 ppm for H_2S and ~10 ppb for C_2H_2 on the 1532.8 nm line. The set-up was also used for in situ measurements of the absorption spectra within methane- and propane-based low-pressure flames at different locations in the flame. Concentrations and the distribution within the flame of three different reaction products were monitored simultaneously, including CO, CO_2, H_2O and OH. The local flame temperature was estimated from measurements of the relative strengths of OH absorption lines. As noted earlier, a much larger enhancement factor is, in principle, possible by monitoring the spectral evolution for a longer period, over several peaks of the relaxation oscillations, but requires increased attention to other potential sources of reflection within the cavity such as etalon fringes, which are also imprinted in the spectral output.

Fjodorow et al. [63–65] have used a similar intra-cavity fibre laser system to that shown in Figure 7.17 for the quantification of various species in different environments. Fjodorow [63] reports on the time-resolved detection of temperature, total pressure and concentration of acetylene during the propagation of shock waves in an acetylene/argon gas mixture in a shock tube of inner diameter 80 mm contained within the fibre laser cavity. For the purpose of noise reduction in single-shot spectra, a 30 m length of fibre was added to the erbium-doped fibre shown in Figure 7.17, giving the total optical length of the laser cavity as 45 m. The measurement time for observation of the spectral evolution was set at 28 μs but, with the increased laser cavity length, the enhancement factor on the gas cell length is reduced to ~188. Output spectra for C_2H_2 were recorded over the range 1526–1536 nm and for a duration of 10 ms at a CCD frame rate of 10 kHz with a spectral resolution of 3.6 GHz (~0.03 nm). The normalised spectra were fitted to spectra from the HITRAN database for calibration purposes and to extract the relevant gas parameters of concentration, pressure and temperature. Noise-equivalent detection limits for acetylene were ~10 ppm at 100 mbar pressure and 300 K, or ~200 ppm at 1 bar pressure and 1000 K. In Fjodorow [64], the absorption spectrum of formaldehyde (CH_2O) was measured over the range of 1558–1605 nm at a resolution of ~3 GHz with a detection limit of ~5 ppm for the strongest absorption line in this spectral region. In this case, a 50 m length of fibre was added to the erbium-doped fibre, giving the total optical length of the laser cavity as 77 m. The intra-cavity cell had a path length of 25 cm and so, with a spectral evolution time of 50 μs, chosen at the first peak of the relaxation oscillations, the cell length enhancement factor was ~195. Calibration was performed by measuring the absorption line spectra over the ~1570–1575 nm band of CO_2 introduced into the cell at 50 mbar and fitting to the corresponding spectra from the HITRAN database. The noise-equivalent detection limit for CO_2 was ~50 ppm. In Fjodorow [65], the same type of fibre laser intra-cavity system was used, but with a 30 cm length of thulium/holmium co-doped fibre (replacing the erbium-doped fibre in Figure 7.17), operating in the 1.8–2.09 μm spectral region and pumped by a laser diode at 790 nm. The total laser cavity length was 1 m with an intra-cavity cell of 40 cm length. The spectrometer and line scan camera for capturing the output spectra were operated at a resolution of 3.6 GHz and experimental observations were made at the first peak of the relaxation oscillations, corresponding to a spectral evolution time of ~7 μs and cell length enhancement factor of ~2000. The capabilities of the system were demonstrated by sensitive measurements of NH_3 and atmospheric water vapour, with detection limits at ppb levels and simultaneous measurements of three stable isotopes of CO_2 in human breath at ppm levels. Typically, the spectra were acquired at a rate of 10 kHz, with 2000 spectra averaged over a period of 200 ms. Calibration was performed by fitting recorded spectra of H_2O and CO_2 to spectra from the HITRAN database.

7.7 Conclusion

This chapter has reviewed the potential applications of fibre amplifier and fibre laser systems for near-IR spectroscopy. There are a number of advantages that can be derived by the use of these systems, but there are also a number of challenges for practical deployment. Although fibre laser sources offer the potential of a wide bandwidth, they are more awkward for WMS spectroscopy techniques, as compared with DFB laser sources. Enhanced sensitivity may be achieved in other ways by using a fibre laser to effectively create a high-finesse cavity for ring-down or cavity-enhanced spectroscopy, but stability and repeatability are key problems. Dual-comb spectroscopy with fibre lasers offers simultaneous fast, high-resolution and broad spectral coverage with direct recovery of absorption line profiles (and hence recovery of the gas parameters of concentration, pressure or temperature) without the need for prior knowledge of modulation parameters, as in WMS. However, it lacks the signal-to-noise and hence sensitivity advantages associated with WMS lock-in detection methods. Intra-cavity laser absorption spectroscopy with fibre lasers gives a very high sensitivity over a medium spectral range, but, to date, the resolution is low and the gas sample must be contained within the fibre laser cavity. Further research and development will no doubt improve the opportunities of fibre laser systems for application in spectroscopy as well as extend the available spectral range through the use of other rare-earth-doped fibres. Chapter 8 will also consider how near-IR fibre laser sources may be used for the generation of coherent mid-IR light, as well as for the generation of mid-IR frequency combs for dual-comb spectroscopy.

References

1. S. D. Agger and J. H. Povlsen, Emission and absorption cross section of thulium doped silica fibers, *Opt. Express*, 14, (1), 50–57, 2006.
2. Z. Li, A. M. Heidt, J. M. O. Daniel, et al., Thulium-doped fiber amplifier for optical communications at 2μm, *Opt. Express*, 21, (8), 9289–9297, 2013.
3. Y. Feng, J. Nilsson, S. Jain, et al., LD-seeded thulium-doped fibre amplifier for CO_2 measurements at 2μm, *6th EPS QEOD Europhoton Conference* (Europhoton 2014), Neuchatel, Switzerland, Poster TuP-T1-P-12, 24–29 August 2014.

4. Y. Ma, Y He, L. Zhang, et al., Ultra-high sensitive acetylene detection using quartz-enhanced photoacoustic spectroscopy with a fiber amplified diode laser and 30.72kHz quartz tuning fork, *Appl. Phys. Lett.*, 110, 031107-1–031107-5, 2017.
5. R. Bauer, T. Legg, D. Mitchell, et al., Miniaturized photoacoustic trace gas sensing using a Raman fiber amplifier, *IEEE J. Lightwave Technol.*, 33, (18), 3773–3780, 2015.
6. J. Cousin, L. P. Masselin, W. Chen, et al., Application of a continuous-wave tunable erbium-doped fiber laser to molecular spectroscopy in the near infrared, *Appl. Phys. B*, 83, 261–266, 2006.
7. K. Bremer, A. Pal, S. Yao, et al., Sensitive detection of CO_2 implementing tunable thulium-doped all-fiber laser, *Appl. Opt.*, 52, (17), 3957–3963, 2013.
8. A. Ghosh, A. S. Roy, S. D. Chowdhury, R. Sen and A. Pal, All-fiber tunable ring laser source near 2μm designed for CO_2 sensing, *Sens. Actuators B: Chem.*, 235, 547–553, 2016.
9. IPG Photonics Corporation. Low power CW fibre lasers. 2019. [Online]. Available: www.ipgphotonics.com/en/products/lasers/low-power-cw-fiber-lasers (accessed April 2020)
10. NP Photonics. Single frequency fibre laser systems. 2019. [Online]. Available: www.npphotonics.com/single-frequency-lasers (accessed April 2020)
11. M. A. Mirza and G. Stewart, Multi-wavelength operation of erbium-doped fibre lasers by periodic filtering and phase modulation, *IEEE J. Lightwave Technol.*, 27, (8), 1034–1044, 2009.
12. G. Whitenett, G. Stewart, H. Yu and B. Culshaw, Investigation of a tuneable mode-locked fibre laser for application to multi-point gas spectroscopy, *IEEE J. Lightwave Technol.*, 22, (3), 813–819, 2004.
13. D. Tosi, Review of chirped fiber Bragg grating (CFBG) fiber-optic sensors and their applications, *Sensors*, 18, (2147), 1–32, 2018.
14. H. A. Haus, Mode-locking of lasers, *IEEE J. Sel. Top. Quantum. Electron.*, 6, (6), 1173–1185, 2000.
15. N. R. Newbury and W. C. Swann, Low-noise fiber-laser frequency combs (invited), *J. Opt. Soc. Am. B*, 24, (8), 1756–1770, 2007.
16. S. A. Diddams, The evolving frequency comb (invited), *J. Opt. Soc. Am. B*, 27, (11), 51–62, 2010.
17. A. Foltynowicz, P. Maslowski, T. Ban, et al., Optical frequency comb spectroscopy, *Faraday Discuss.*, 150, 23–31, 2011.
18. J. Kim and Y. Song, Ultralow-noise mode-locked fiber lasers and frequency combs: principles, status, and applications, *Adv. Opt. Photonics*, 8, (3), 465–539, 2016.
19. D. Fehrenbacher, P. Sulzer, A. Liehl, et al., Free-running performance and full control of a passively phase-stable Er:fiber frequency comb, *Optica*, 2, (10), 917–923, 2015.
20. N. Kuse, J. Jiang, C.-C. Lee, T. R. Schibli and M. E. Fermann, All polarization-maintaining Er fiber-based optical frequency combs with nonlinear amplifying loop mirror, *Opt. Express*, 24, (3), 3095–3102, 2016.
21. L. C. Sinclair, I. Coddington, W. C. Swann, et al., Operation of an optically coherent frequency comb outside the metrology lab, *Opt. Express*, 22, (6), 6996–7006, 2014.
22. L. C. Sinclair, J.-D. Deschênes, L. Sonderhouse, et al., Invited article: a compact optically coherent fiber frequency comb, *Rev. Sci. Instrum.*, 86, 081301-1–081301-15, 2015.
23. C Gohle, B. Stein, A. Schliesser, T. Udem and T. W. Hansch, Frequency comb Vernier spectroscopy for broad-band, high-resolution, high-sensitivity absorption and dispersion spectra, *Phys. Rev. Lett.*, 99, 263902-1–263902-4, 2007.
24. S. K. Scholten, C. Perrella, J. D. Anstie, et al., Number-density measurements of CO_2 in real time with an optical frequency comb for high accuracy and precision, *Phys. Rev. Appl.*, 9, (054043), 1–8, 2018.

25. T. Ideguchi, Dual-comb spectroscopy, *Opt. Photonics News*, 32–39, 2017.
26. I. Coddington, W. C. Swann and N. R. Newbury, Coherent multi-heterodyne spectroscopy using stabilised optical frequency combs, *Phys. Rev. Lett.*, 100, 013902-1–013902-4, 2008.
27. I. Coddington, N. R. Newbury and W. C. Swann, Dual-comb spectroscopy, *Optica*, 3, (4), 414–426, 2016.
28. T. Ideguchi, A. Poisson, G. Guelachvili, N. Picqué and T. W. Hänsch, Adaptive real-time dual-comb spectroscopy, *Nature Comm.*, 5, (3375), 1–8, 2014.
29. X. Zhao, G. Hu, B. Zhao, et al., Picometer-resolution dual-comb spectroscopy with a free-running fiber laser, *Opt. Express*, 24, (19), 21833–21845, 2016.
30. M. J. Thorpe and J. Ye, Cavity-enhanced direct frequency comb spectroscopy, *Appl. Phys. B*,. 91, 397–414, 2008.
31. B. Bernhardt, A. Ozawa, P. Jacquet, et al., Cavity-enhanced dual-comb spectroscopy, *Nat. Photon.*, 4, 55–57, 2009.
32. F. Adler, M. J. Thorpe, K. C. Cossel, and J. Ye, Cavity-enhanced direct frequency comb spectroscopy: technology and applications, *Annu. Rev. Anal. Chem.* 3, 175–205, 2010.
33. A. Foltynowicz, T. Ban, P. Masłowski, F. Adler and Jun Ye, Quantum-noise-limited optical frequency comb spectroscopy, *Phys. Rev. Lett.*, 107, 233002-1–233001-5, 2011.
34. G. B. Rieker, F. R. Giorgetta, W. C. Swann, et al., Frequency-comb-based remote sensing of greenhouse gases over kilometre air paths, *Optica*, 1, (5), 290–297, 2014.
35. S. Okubo, K. Iwakuni, H. Inaba, et al., Ultra-broadband dual-comb spectroscopy across 1.0–1.9μm, *Appl. Phys. Express*, 8, 082402-1–082402-4, 2015.
36. F. R. Giorgetta, G. B. Rieker, E. Baumann, et al., Broadband phase spectroscopy over turbulent air paths, *Phys. Rev. Lett.*, 115, 103901-1–103901-5, 2015.
37. S. Coburn, C. B. Alden, R. Wright, et al., Regional trace-gas source attribution using a field-deployed dual frequency comb spectrometer, *Optica*, 5, (4), 320–327, 2018.
38. P. J. Schroeder, R. J. Wright, S. Coburn, et al., Dual frequency comb laser absorption spectroscopy in a 16MW gas turbine exhaust, *Proc. Combust. Inst.*, 36, 4565–4573, 2017.
39. Z. Chen, M. Yan, T. W. Hansch and N. Picque, Evanescent wave gas sensing with dual-comb spectroscopy. In *OSA Conference on Lasers and Electro-Optics (CLEO), San Jose, CA, USA, 14–19 May, 2017*, paper SF1M.7.
40. M. J. Thorpe, D. D. Hudson, K. D. Moll, J. Lasri and J. Ye, Cavity-ringdown molecular spectroscopy based on an optical frequency comb at 1.45-1.65μm, *Opt. Lett.*, 32, (3), 307–309, 2007.
41. Menlo Systems GmbH. Optical frequency combs. 2019. [Online]. Available: www.menlosystems.com/products/optical-frequency-combs/ (accessed April 2020)
42. Toptica Photonics. Compact low noise frequency comb. 2019. [Online]. Available: www.toptica.com/products/frequency-combs/dfc-core/ (accessed April 2020)
43. E. Hecht and A. Zajac, *Optics*, Reading, MA, Addison-Wesley, 306–309, 1974.
44. G. Stewart, K. Atherton, H. Yu and B. Culshaw, Investigation of optical fibre amplifier loop for intra-cavity and ring-down cavity loss measurements, *Meas. Sci. Technol.*, 12, (7), 843–849, 2001.
45. G. Stewart, P. Shields and B. Culshaw, Development of fibre laser systems for ring-down and intra-cavity gas spectroscopy in the near-IR, *Meas. Sci. Technol.*, 15, (8), 1621–1628, 2004.
46. G. Stewart, K. Atherton and B. Culshaw, Cavity-enhanced spectroscopy in fibre cavities, *Opt. Lett.*, 29, (5), 442–444, 2004.
47. H. Waechter, J. Litman, A. H. Cheung, J. A. Barnes and H.-P. Loock, Chemical sensing using fibre cavity ring-down spectroscopy, *Sensors*, 10, 1716–1742, 2010.

48. K. Liu, T. G. Liu, G. D. Peng, et al., Theoretical investigation of an optical fibre amplifier loop for intra-cavity and ring-down cavity gas sensing, *Sens. Actuators B: Chem.*, 146, 116–121, 2010.
49. Y. Zhang, M. Zhang, W. Jin, et al., Investigation of erbium-doped fiber laser intra-cavity absorption sensor for gas detection, *Optics Comm.*, 232, 1–6, 295–301, 2004.
50. K. Liu, T. Liu, J. Jiang, et al., Investigation of wavelength modulation and wavelength sweep techniques in intra-cavity fibre laser for gas detection, *IEEE J. Lightwave Technol.*, 29, (1), 15–21, 2011.
51. L. Yu, T. Liu, K. Liu, et al., Development of an intra-cavity gas detection system based on L-band erbium-doped fibre ring laser, *Sens. Actuators B: Chem*, 193, 356–362, 2014.
52. L. Yu, T. Liu, K. Liu, J. Jiang and T. Wang, Intra-cavity multi-gas detection based on multiband fibre ring laser, *Sens. Actuators B: Chem*, 226, 170–175, 2016.
53. L. Yu, T. Liu, K. Liu, J. Jiang and T. Wang, A method for separation of overlapping absorption lines in intra-cavity gas detection, *Sens. Actuators B: Chem*, 228, 10–15, 2016.
54. N. Arsad, M. Li, G. Stewart and W. Johnstone, Intra-cavity spectroscopy using amplified spontaneous emission in fibre lasers, *IEEE J. Lightwave Technol.*, 29, (5), 782–788, 2011.
55. J. K. Valiunas, G. Stewart and G. Das Detection of nitrous oxide (N_2O) at sub-ppmv using intra-cavity absorption spectroscopy (ICAS), *IEEE Photon. Technol. Lett.*, 28, (3), 359–362, 2015.
56. J. K. Valiunas, M. Tenuta and G. Das, A gas cell based on hollow-core photonic crystal fiber (PCF) and its application for the detection of greenhouse gas (GHG): nitrous oxide (N_2O), *J. Sens.*, 7678315, 1–9, 2016.
57. Q. Wang, Z. Wang, J. Chang and W. Ren, Fibre-ring laser-based intra-cavity photoacoustic spectroscopy for trace gas sensing, *Opt. Lett.*, 42, (11), 2114–2117, 2017.
58. V. M. Baev, T. Latz and P. E. Toschek, Laser intra-cavity absorption spectroscopy, *Appl. Phys. B*, 69, 171–202, 1999.
59. R. Bohm, A. Stephani, V. M. Baev and P. E. Toschek, Intra-cavity absorption spectroscopy with a Nd^{3+}-doped fibre laser, *Opt. Lett.*, 18, (22), 1955–1957, 1993.
60. A. Stark, L. Correia, M. Teichmann, et al., Intra-cavity absorption spectroscopy with thulium-doped fibre laser, *Opt. Comm.*, 215, 113–123, 2003.
61. G. Stewart, G. Whitenett, S. Sridaran and V. Karthik, Investigation of the dynamic response of erbium fibre lasers with potential application for sensors, *IEEE J. Lightwave Technol.*, 25, (7), 1786–1796, 2007.
62. B. Löhden, S. Kuznetsova, K. Sengstock, et al., Fiber laser intracavity absorption spectroscopy for in situ multicomponent gas analysis in the atmosphere and combustion environments, *Appl. Phys. B.*, 102, (2), 331–344, 2011.
63. P. Fjodorow, M. Fikri, C. Schulz, O. Hellmig and V. M. Baev, Time-resolved detection of temperature, concentration and pressure in a shock tube by intra-cavity absorption spectroscopy, *Appl. Phys. B*, 122, 159, 2016.
64. P. Fjodorow, O. Hellmig, V. M. Baev, H. B. Levinsky and A. V. Mokhov, Intracavity absorption spectroscopy of formaldehyde from 6230 to 6420 cm^{-1}, *Appl. Phys. B.*, 123, 147, 2017.
65. P. Fjodorow, O. Hellmig, V. M. Baev, A broadband Tm/Ho-doped fiber laser tunable from 1.8 to 2.09 μm for intracavity absorption spectroscopy, *Appl. Phys. B*, 124, 62, 2018.

8 Mid-IR Systems and the Future of Gas Absorption Spectroscopy

8.1 Introduction

The focus of this book has been on spectroscopy in the near-IR region, at wavelengths less than ~2 μm, making use of overtone gas absorption lines. The major advantage of this approach is the availability, maturity and relatively low cost of a wide range of optical components for spectroscopic measurement systems. However, this comes at a price, the relatively weak absorption line strengths of the overtone lines. There is therefore a strong case for mid-IR systems, at wavelengths beyond 2 μm, to take advantage of stronger or fundamental absorption lines where line strengths may be two or three orders of magnitude greater than the near-IR, especially for applications where gas concentrations may be low, or high sensitivities are required. Strong absorption lines also mean that for certain applications shorter path lengths may be employed or the use of multi-pass cells or cavity-enhanced techniques may be avoided. To illustrate, Figure 8.1 shows fundamental absorption lines in the 3–5 μm atmospheric window region for methane, carbon dioxide and carbon monoxide plotted using *HITRAN* on the Web (see reference [18] in Chapter 1); these may be compared with the near-IR lines for the same gases shown in Chapter 1, Figures 1.6, 1.7 and 1.9. However, in Chapter 1 the lines are plotted for pure gas with a 1 cm pathlength, equivalent to a concentration length product of Cl = 1 cm in (1.8), but for the plot in Figure 8.1 this product has been reduced by a factor of 100, clearly showing the vast improvement in absorption strength in the mid-IR.

It is not surprising then that the development of mid-IR gas sensors has gained much momentum in recent years, with the demonstration of new tuneable mid-IR laser sources and with the ongoing research on mid-IR fibre and detectors. In this chapter, we shall review some of the advances in mid-IR gas absorption spectroscopy and their impact in relation to the future of near-IR systems [1–4]. However, the various techniques and principles discussed throughout this book in the context of near-IR spectroscopy, such as wavelength modulation spectroscopy, photoacoustic spectroscopy, cavity-enhanced spectroscopy, comb spectroscopy, evanescent-wave spectroscopy, etc., are all directly transferable to the mid-IR region, albeit with suitable modification for the longer wavelength of operation.

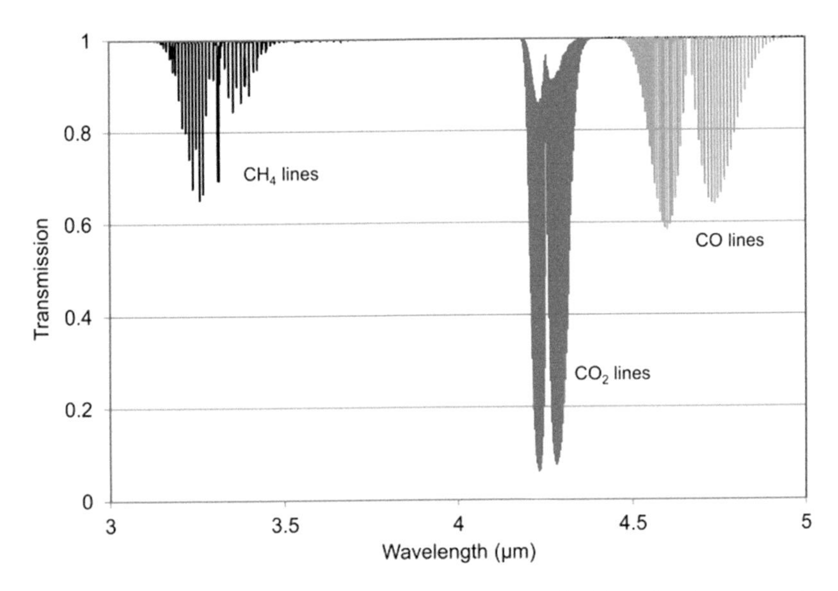

Figure 8.1 Mid-IR absorption lines for methane, carbon dioxide and carbon monoxide plotted using *HITRAN* on the Web with $Cl = 0.01$ cm.

8.2 Mid-IR Sources

A variety of laser sources are available to cover the mid-IR region at wavelengths beyond 2 μm. These include diode laser sources such as quantum well (QW) lasers, inter-band cascade lasers (ICLs), quantum cascade lasers (QCLs) and lead salt lasers. Several fibre laser sources are also available for the mid-IR region. Alternatively, wide coverage of the mid-IR region can be achieved through non-linear down-conversion of near-IR light from diode or fibre laser sources based on the principles of difference frequency generation (DFG) or optical parametric oscillation (OPO), but these systems are generally more complex and expensive. An important current area of research is the demonstration of mid-IR frequency comb sources. Finally, some gas lasers can generate light at a fixed mid-IR wavelength, for example, He–Ne at ~3.39 μm, CO at ~5 μm and CO_2 at ~10.6 μm. Figure 8.2 shows the approximate range of wavelengths available from the most common mid-IR sources, which we shall briefly review in the following sections.

8.2.1 Mid-IR Diode Laser Sources

For the wavelength range of ~1.8 μm to 3 μm, QW diode lasers, operating at room temperature, may be engineered based on the III-V semiconductor material system AlGaIn, AsSb, making use of inter-band (conduction to valence) electron–hole transitions [5]. Although absorption lines may not be as strong as those in the 3–5 μm region, the 2–3 μm region does include stronger CO_2 and CO absorption lines at 2.04 μm and

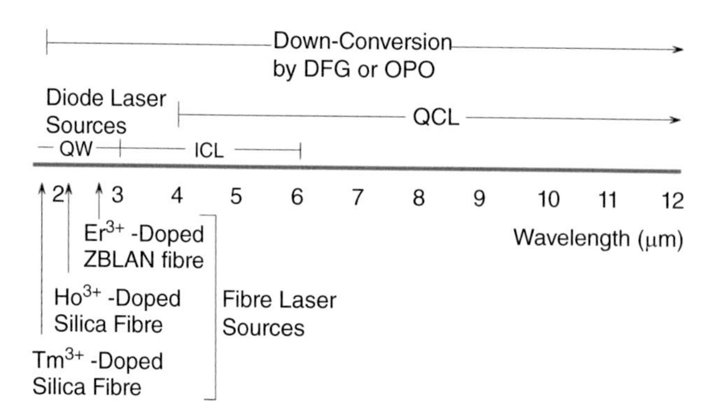

Figure 8.2 Common mid-IR sources with range of wavelengths available.

2.33 μm, and a range of DFB or discrete mode QW lasers are available commercially from several companies [6–8] in this wavelength range to target these lines.

For the 3–6 μm wavelength range, DFB ICLs are commercially available [7], which region includes strong fundamental absorption lines of CO_2, CO and NO and the stretching mode of the C–H bond around 3.3–3.4 μm, which is common to all hydrocarbons, as shown in Figure 8.1. Similar to conventional laser diodes, ICLs use electron–hole transitions between the conduction and valence bands, but do so through a cascade of quantum well structures as in a QCL. However, in comparison with QCLs, the use of the inter-band transitions results in a lower threshold current and input electrical power. ICLs based on GaSb typically operate at room temperature with a threshold current of ~50 mA, slope efficiency of ~0.06 mW mA^{-1}, a temperature tuning coefficient of 0.3 nm $°C^{-1}$ and current tuning coefficient of 0.2 nm mA^{-1} [7]. These parameters mean that ICLs may be operated in a similar manner to conventional near-IR laser diodes, as discussed in Chapters 2 and 3. Heterogeneous integration of an ICL onto a silicon chip has also been demonstrated [9], paving the way for chip-based gas sensor systems.

QCLs can be engineered to operate over a large range of wavelengths from ~4 μm to the far infrared [10, 11], as a result of their unique design where the operation wavelength is not dependent on the band-gap, but on the chosen thicknesses of the quantum wells in the active regions. QCLs make use of inter-sub-band transitions in a superlattice arrangement of quantum wells, consisting of alternating electron injector and active regions, as illustrated in Figure 8.3. The electric field applied across the structure gives the sloped energy distribution which is necessary for the electron cascade through the structure, from left to right, as shown in Figure 8.3. The injector regions consist of a series of quantum wells separated from the active regions by very thin barrier layers. Electrons from the first injector region tunnel through the thin barrier into the first active region, where the laser transition and photon emission occurs. The electron, now in a lower ground state, tunnels through the thin exit barrier into the next injector region and the process continues to the next active region. Figure 8.3 shows only two stages, but the electron may cascade through 20 or more stages of the structure.

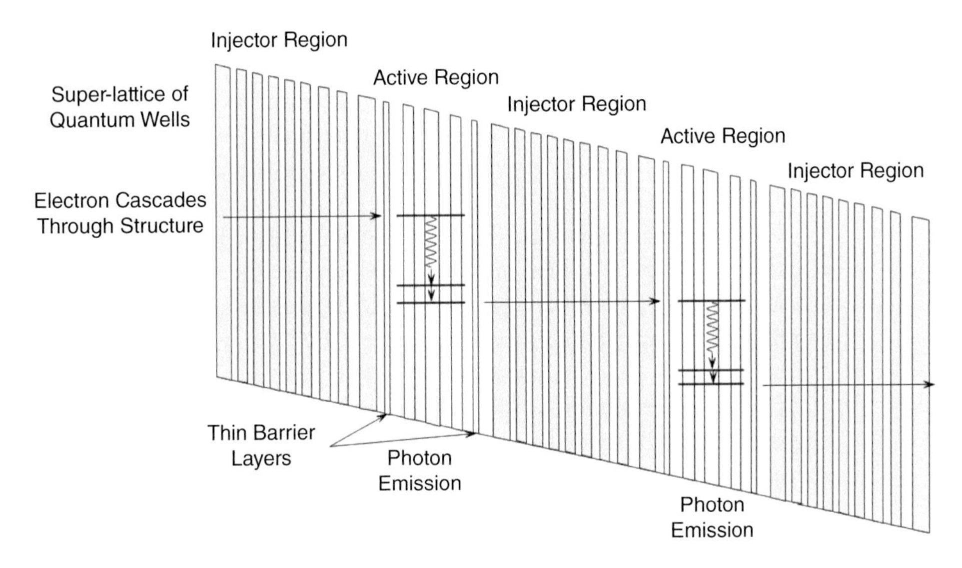

Figure 8.3 Schematic energy band diagram for two stages of a quantum cascade laser.

Since the first demonstration of a QCL at Bell Labs in 1994 by Faist and colleagues [11–13], extensive research and development over the last 20 years and more [14] has led to the QCL becoming one of the most important mid-IR sources for spectroscopic sensing, with mid-IR coverage demonstrated up to 25 μm and even extended into the THz region. DFB QCLs are commercially available [15–17] with room temperature operation, thermal tuning over a typical range of 50 GHz or more and high-output optical powers of tens or hundreds of milliwatts due to the cascade effect and the ability of QCLs to handle large currents. However, as already noted, they have relatively high threshold currents (typically several hundreds of mA to 1 A). Not surprisingly, there have been numerous applications of QCLs to mid-IR gas spectroscopy over the last decade for atmospheric and greenhouse gas sensing or combustion and emissions monitoring [1–4]. Multi-species spectroscopy has also been demonstrated using QCL arrays or dual wavelength operation of QCLs. For example, an array of nine DFB QCLs, with the beams spatially combined using a SiGe arrayed waveguide grating (AWG) device, has been used to demonstrate multi-species spectroscopy on CO, C_2H_2 and CO_2 [18]. Each QCL in the array had a different grating for a different operation wavelength with each thermally tuneable over ~90 GHz through temperature adjustment over 10–30 °C, giving a total wavelength span of ~60 nm around a wavelength of 4.5 μm. Another example is the use of two dual-wavelength QCLs along with a single wavelength DFB QCL for simultaneous, high-precision measurements at ppb levels of a number of gases, including CO, CO_2, NH_3, NO, NO_2, N_2O and O_3, in a compact multi-pass cell with an effective pathlength of 75 m [19]. Alternatively, the wavelength tuning range of a single QCL may be extended with the use of an external cavity (EC), and EC-QCLs in the 4.5–10 μm region are available commercially [20, 21] with a large mode-hop-free tuning range of 100 cm^{-1} (3 THz) or more. For example, Ghorbani et al. [22] used a water-cooled EC-QCL operating around 4.7 μm for the real-time analysis of CO and CO_2 in exhaled human breath in a low-volume multi-pass cell with wavelength modulation spectroscopy. Heinrich et al. [23] designed a laboratory reference system based on four EC-QCLs to provide a gapless wavelength coverage over the 6–11 μm spectral region for accurate, high-resolution measurements of hydrocarbon spectra under conditions where spectroscopic data is not available in current spectroscopic databases.

For more information on the many varied applications of QCLs in research and industry, the reader is referred to the series of conferences on the topic of Field Laser Applications in Industry and Research (FLAIR) and the special journal papers from the conference series published in *Applied Physics B, Lasers and Optics,* vol. 110, no. 2, February 2013, vol. 119, no. 1, April 2015 and vol. 123, 2017.

Finally, in this section on mid-IR diode lasers, we note that lead salt diode lasers [24] based on IV-VI semiconductors such as ternary lead compounds $Pb_xSn_{1-x}Te$ or quaternary compounds $Pb_xEu_{1-x}Se_yTe_{1-y}$ have been available since the 1970s for mid-IR spectroscopy and can be designed to operate over a wide range of mid-IR wavelengths beyond 3 μm. However, reliability issues, low optical power outputs and the need for cryogenic operation have limited their application and they have been largely superseded by ICLs and QCLs.

8.2.2 Mid-IR Fibre Laser Sources

As an alternative to diode lasers, which are generally single-wavelength operation, there are several doped-fibre laser based sources for the mid-IR region [25, 26] which, in principle, can be designed to operate over the range of wavelengths corresponding to their emission band. As noted in Table 6.1, Chapter 6, Tm^{3+}-doped silica fibre may be used to create fibre laser sources within the approximate emission range 1.7–2.1 μm and Ho^{3+}-doped silica fibre within the range 2–2.2 μm. Custom-made Tm^{3+}-doped fibre lasers for the 1.8–2.05 μm range are available commercially [27, 28], with fibre Bragg gratings to select the operation wavelength and to form the Fabry–Perot laser cavity. However, for fibre lasers beyond 2.2 μm, doped silica fibre is not suitable. This is because the non-radiative transition rate from phonon interactions increases since the energy gap between the electronic states involved in a mid-IR lasing transition must be smaller than that required for the shorter-wavelength near-IR (for mid-IR wavelengths >2.5 μm, the required energy gap is <4000 cm^{-1}). For silica glass, the lattice phonon energy can be as high as ~1100 cm^{-1}, so the energy gap between states for mid-IR laser transitions can be bridged by a few phonons, giving relatively fast, non-radiative transition rates. Hence, for mid-IR fibre lasers, glasses with lower phonon energies are required, which also has the beneficial effect that the phonon absorption edge is shifted to longer wavelengths, giving better mid-IR transmission characteristics. One important category is heavy-metal fluoride glass, where the heavier elements and weaker ionic bonds making up the glass result in lower phonon energies, but unfortunately also have a detrimental effect on the durability and hardness of the glass. The heavy-metal fluoride glass, ZBLAN, with the composition 53% ZrF_4, 20% BaF_2, 4% LaF_3, 3%AlF_3 and 20% NaF has a maximum phonon energy of ~500 cm^{-1}, less than half that of silica, and has been used to demonstrate fibre lasers over the 2–4 μm region with various rare-earth dopants, as shown in Table 8.1.

The fibre lasers listed in Table 8.1 are at various stages of research and development, and some, such as erbium at 3.45 μm and holmium at 3.95 μm, require nitrogen cooling for operation. One of the most important mid-IR fibre lasers is the erbium-doped fibre laser at ~2.7 μm, based on the $^4I_{11/2}-^4I_{13/2}$ electronic transition, as shown in Figure 8.4 (see also Figure 6.1 in Chapter 6). High-power, diode-pumped versions are available commercially [28] using erbium-doped ZrF_4 fibre with single-mode ZrF_4 fibre patch cables and FC/APC-type connectors for coupling the fibre laser output to a system.

To avoid the durability problems of fluoride glass fibre lasers and to extend the operation range further into the mid-IR, another approach is to use hollow-core fibres filled with a gas as the lasing medium. Cui et al. [29] have demonstrated a 4.3 μm fibre laser based on

Table 8.1 Mid-IR fibre lasers demonstrated in rare-earth-doped ZBLAN fibre

Rare-earth ion	Electronic transition	Lasing wavelength (μm)
Er^{3+}	$^4I_{11/2}-^4I_{13/2}$	2.7
Er^{3+}	$^4F_{9/2}-^4I_{9/2}$	3.45
Tm^{3+}	$^3H_4-^3H_5$	2.31
Ho^{3+}/Pr^{3+}	$^5I_6-^5I_7$	2.86
Ho^{3+}	$^5S_2-^5F_5$	3.22
Ho^{3+}	$^5I_5-^5I_6$	3.95
Dy^{3+}	$^6H_{13/2}-^6H_{15/2}$	2.9

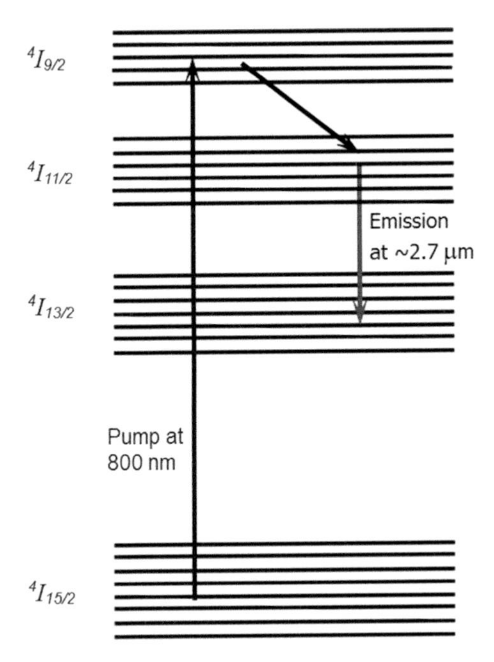

Figure 8.4 Energy levels in erbium-doped ZBLAN fibre for 2.7 μm laser operation.

hollow-core silica fibre filled with CO_2 and pumped at 2 μm. The fibre laser operated at room temperature with an output power of ~80 mW and efficiency of ~20%.

8.2.3 Mid-IR Sources Based on Near-IR Down-Conversion

A wide range of mid-IR wavelengths can be generated by the use of non-linear crystals [30], where near-IR light from diode lasers, fibre lasers or high-power solid-state lasers is down-converted to the mid-IR by non-linear processes. Down-conversion systems are generally more complex and expensive compared with mid-IR laser diodes, but they do offer advantages in some situations.

8.2.3.1 Difference Frequency Generation of Mid-IR Light

In the method of difference frequency generation (DFG), two near-IR laser outputs, as illustrated in Figure 8.5, consisting of a high-power pump source of wavelength λ_p and a second (tuneable) signal source of wavelength λ_s, are mixed in a non-linear crystal, such as periodically poled lithium niobate (PPLN).

The non-linear crystal generates a difference or idler frequency of wavelength λ_i given (through energy conservation) by:

$$\frac{1}{\lambda_i} = \frac{1}{\lambda_p} - \frac{1}{\lambda_s} \tag{8.1}$$

For example, if the pump source is a Nd:YAG laser at 1064 nm and the signal source is a near-IR DFB diode laser at 1547 nm, the idler generated is at a mid-IR wavelength of 3.4 μm suitable for methane detection [31]. Furthermore, if a sinusoidal modulation and linear ramp are applied to the diode laser current for WMS, as described in Chapter 3, then this wavelength modulation and scan are transferred to the mid-IR output. Also, standard silica optical fibres may be used to transport the near-IR light to the PPLN device for remote generation of the mid-IR light. Alternatively, if the DFB diode laser signal source is replaced by a tuneable source, such as an external-cavity diode laser, or an erbium-doped fibre laser tuneable over the range 1.5–1.6 μm, as discussed in Chapter 6, then a tuneable mid-IR output over the range 3.2–3.6 μm may be generated [32]. However, the non-linear DFG conversion efficiency is low, so with CW laser inputs the mid-IR output power is generally low, but much higher output powers are possible with pulsed lasers.

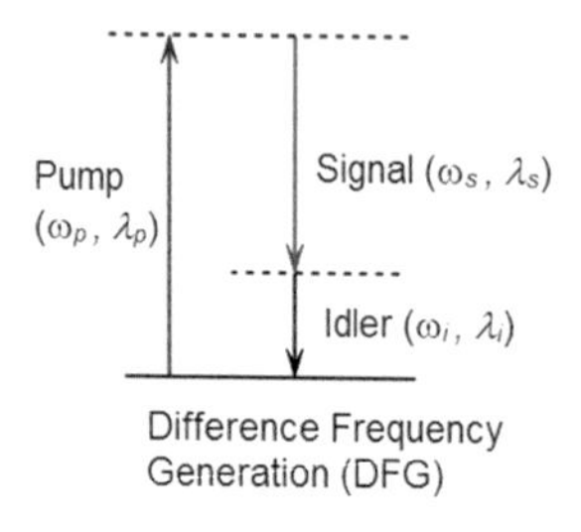

Figure 8.5 Principle of difference frequency generation in a non-linear crystal.

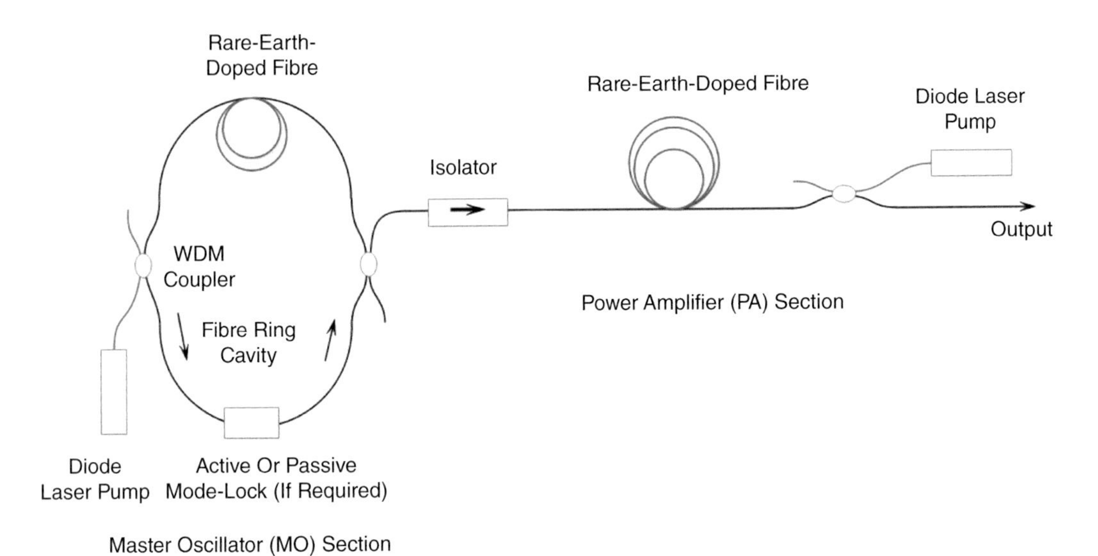

Figure 8.6 Typical master oscillator power amplifier (MOPA) configuration.

Several DFG-based mid-IR sources are available commercially [33–36]. An example is the Genia Photonics source [33, 34], which uses fibre-based lasers to provide both the pump and signal inputs. The pump input is from an erbium fibre laser source operating at a wavelength of either 1542 nm or 1597 nm in a master oscillator power amplifier (MOPA) configuration, as shown in Figure 8.6. The signal input is from a programmable laser based on thulium-doped fibre which is actively mode-locked to produce ~25 ps duration pulses and dispersion-tuned with a tuning range of 1893–2000 nm. The non-linear crystal for the DFG is orientation-patterned GaAs, which allows quasi-phase-matching over a wide bandwidth and has a wider transmission window (0.9–17 μm), as compared with lithium niobate (transparent up to 5 μm). Output power levels of ~0.4–1 mW are generated over a broad mid-IR spectral range of 6.7–10.2 μm.

Toptica [35] market the FemtoFiber dichro mid-IR femtosecond laser source, which provides an output power of ~1 mW over a tuneable wavelength range of ~5–15 μm. The core of the system is an erbium-doped fibre ring laser, operating at 1.5 μm and mode-locked with a saturable absorber mirror (SAM). The SAM ring laser generates seed pulses at a repetition rate of 80 MHz which are fed to two erbium-doped fibre amplifiers (EDFAs) in a similar way to the MOPA of Figure 8.6. The pulsed output of one EDFA is used to generate a supercontinuum from a length of highly non-linear fibre over the spectral region ~1.7–2 μm to provide the seed signal for the DFG process in a non-linear crystal. The 1.5 μm pulsed output from the other EDFA provides the pump signal, which is first passed through a delay stage to ensure perfect temporal overlap with the signal pulses.

8.2.3.2 Optical Parametric Oscillator for Generation of Mid-IR Light

The use of an optical parametric oscillator (OPO) is another way of generating mid-IR light by down-conversion from the near-IR. With an OPO, as illustrated in Figure 8.7,

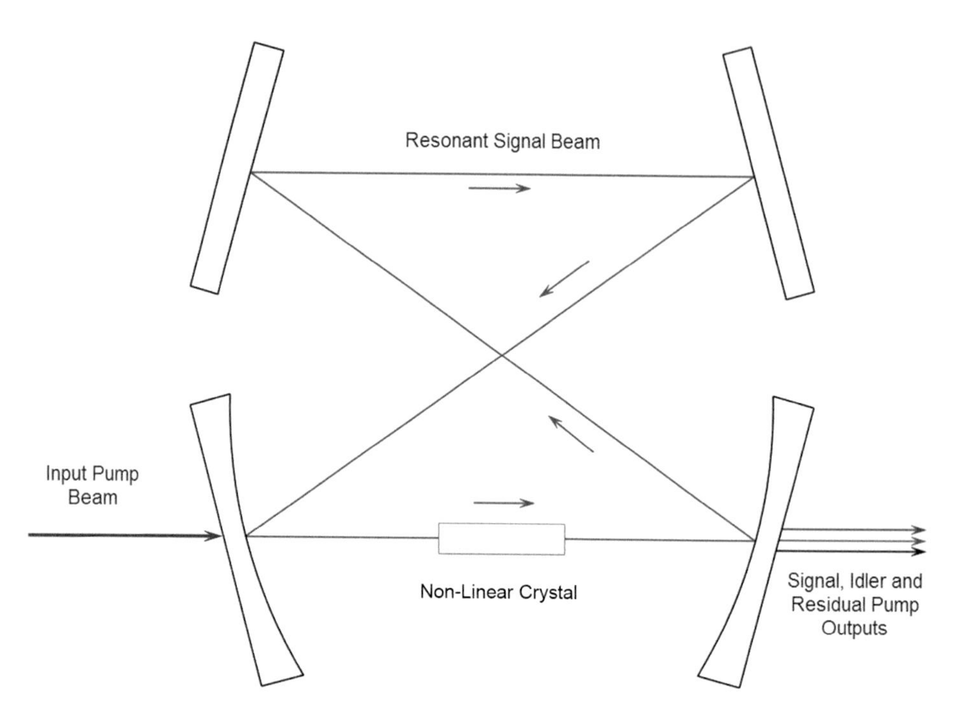

Figure 8.7 Schematic of a ring-cavity optical parametric oscillator (OPO).

a single high-power laser pump, when above a certain threshold intensity level, generates both signal and idler wavelengths in a non-linear crystal. The cavity containing the non-linear crystal is designed to be resonant at either the signal frequency or the idler frequency (or both in a doubly resonant OPO).

An OPO can give much higher output powers, of the order of watts, as compared with DFG, and also the mid-IR wavelength can be tuned over a large range. This is possible because the pair of wavelengths generated through parametric pumping must satisfy the phase-matching (or conservation of momentum) condition that $k_p = k_s + k_i$ so:

$$\frac{n_p}{\lambda_p} = \frac{n_i}{\lambda_i} + \frac{n_s}{\lambda_s} \tag{8.2}$$

where n represents the refractive index of the crystal at the corresponding wavelength. In general, the indices depend on the orientation of the (birefringent) non-linear crystal and its temperature, so rotation of the crystal or change in its temperature modifies the output signal and idler wavelengths in order that condition (8.2) remains satisfied. Several commercial OPO sources are now available [37, 38], such as the tuneable optical parametric oscillator (TOPO) from Toptica [37], which uses a DFB seed laser and fibre amplifier for the pump source. It gives a coherent CW output of 1–2 W power with linewidth of ~2 MHz, tuneable across the broad wavelength range of 1.45–4.0 μm (1.45–2.07 μm from the signal beam and 2.19–4.00 μm from the idler beam). The coarse wavelength tuning is achieved through crystal translation and temperature, while mode-hop-free fine tuning over ~300 MHz is achieved through pump or PZT tuning. A wider fine tuning/scan range is generally desirable for WMS applications, but an example of using an OPO for ethane detection at parts per trillion levels is given in [39]. Both DFG and OPOs are finding important application in the generation of mid-IR frequency combs, as discussed in the next section.

8.2.4 Mid-IR Laser Combs

As discussed in Chapter 7, the generation of frequency combs and their application to dual comb spectroscopy (DCS) have been extensively investigated in the near-IR and there is much current research on extending the techniques to the mid-IR [40–60]. Mid-IR combs can, in principle, be generated directly from mode-locked mid-IR laser sources, or indirectly through down-conversion of a near-IR comb by an OPO or by DFG. The comb spacing from mid-IR sources such as QCLs is relatively large, in the GHz range, due to the short millimetre lengths of laser cavities, whereas with fibre lasers, where cavity lengths are of the order of several metres, the comb spacing is in the MHz range. The advantage of down-conversion is that many of the underlying properties of mature near-IR combs can be directly translated into the mid-IR. In particular, passively mode-locked, near-IR femtosecond ytterbium (Yb) fibre laser combs, operating at ~1 μm wavelength, give high power outputs with robust performance and provide a very important basis for the generation of mid-IR combs by down-conversion in a non-linear crystal [42]. On the other hand, direct generation of combs with a QCL offers the possibility of compact, chip-scale, dual-comb sources for spectroscopy.

Several examples of mid-IR combs based on an OPO with MgO-doped PPLN as the non-linear crystal have been demonstrated. Adler et al. [43] used a 10 W femtosecond Yb:fibre laser with a pulse repetition rate of 136 MHz to generate a mid-IR comb by pumping a singly resonant OPO. The mid-IR comb was generated on the idler output and could be tuned over a range of 2.8–4.8 μm by translating the non-linear crystal, designed with a variable poling period. A stable mid-IR comb with $>10^4$ comb lines covering a spectral bandwidth of up to 0.3 μm at 3.26 μm was demonstrated, with >30 μW power per comb line. Zhang et al. [44] used two independent mode-locked Yb:KYW (ytterbium-doped potassium yttrium tungstate crystal) lasers at 1058 nm with different pulse repetition rates seeding fibre amplifiers to simultaneously pump the OPO for the generation of a dual mid-IR comb output, as illustrated in Figure 8.8. The pump lasers generated 3 ps pulses at ~2.5 W average power at a repetition rate of ~100 MHz and a rate difference of Δf_{rep} = 96 Hz. The mid-IR dual comb generated on the idler output was centred at ~3.3 μm with $>10^4$ comb lines (spectral bandwidth of ~170 nm or 4.7 GHz) and ~1 μW power per comb line. Dual-comb spectroscopy was demonstrated with a 20 cm cell containing 0.7% CH_4 gas, giving a good fit to the P-branch absorption lines over the 2900–3000 cm^{-1} region, with resolution of 0.2 cm^{-1} (the fit to the Q-branch was not so good, possibly due to the fact that the carrier-envelope-offset frequency, f_{ceo}, for both lasers was free-running and only one laser had its repetition frequency locked so that Δf_{rep} was also free-running).

Similarly, Jin et al. [45] used two independent Yb:fibre laser systems (consisting of a seed laser, amplifier and compressor) at 1040 nm with 80 fs pulses to pump the OPO.

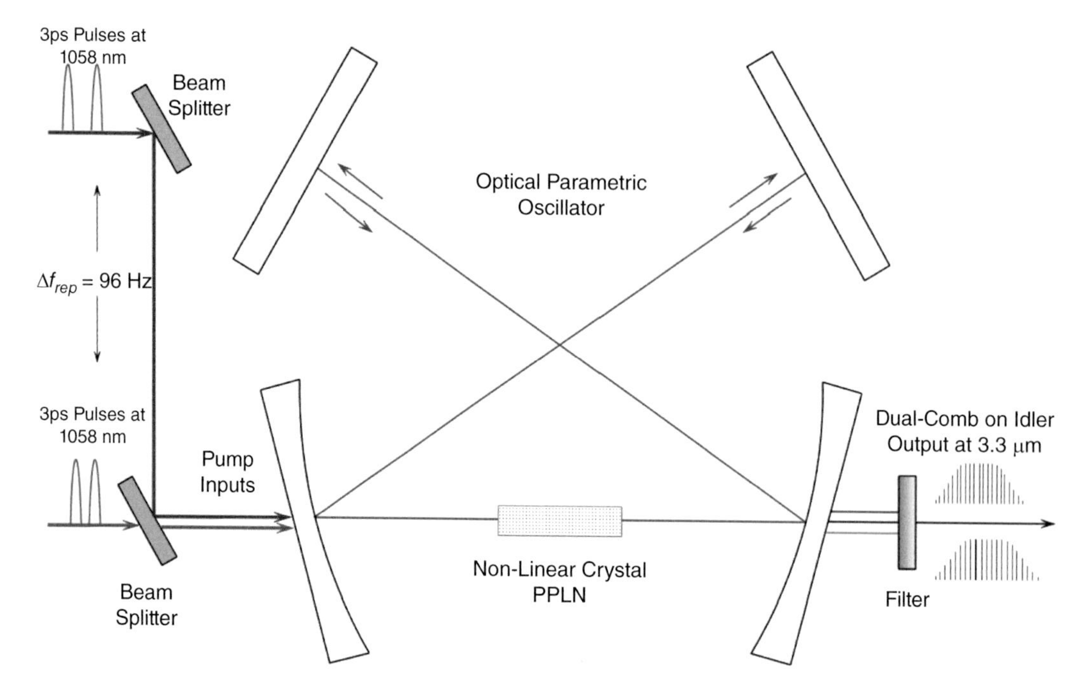

Figure 8.8 Generation of a mid-IR dual-comb output using an OPO. After [44].

The pulse repetition rates of the two lasers were stabilised at 90 MHz, with a difference of 185 Hz. The dual comb generated on the idler output could be tuned over the range 2.7–4.7 μm and was used to measure the methane spectrum over a 300 cm^{-1} bandwidth, with a resolution of 0.4 cm^{-1}. An alternative system was also demonstrated which used two MgO-doped PPLN crystals in the same OPO cavity so that the two idler combs were spatially separated, giving an improved resolution of 0.2 cm^{-1}.

Generation of mid-IR combs through DFG is also important and Menlo Systems [46] market a product with a spectral range of 3.1–3.4 μm or 6.5–7.8 μm, a spectral bandwidth of greater than 200 nm and a comb mode spacing of either 100 or 250 MHz. Generation of dual combs by DFG for spectroscopy has been investigated by several researchers. Baumann et al. [47] used two near-IR femtosecond laser combs operating around 1560 nm (with repetition rates of ~100 MHz and a rate difference of ~1.5 kHz) combined with ~0.5 W of amplified light from a CW fibre laser at 1064 nm to generate a dual mid-IR comb around 3.4 μm by DFG in PPLN. Dual-comb spectroscopy was performed to measure the P, Q and partial R branches of the methane spectrum, with a point spacing of 25–100 MHz and a resolution of less than 10 kHz. In later work, Ycas et al. [48] made use of a highly stabilised near-IR laser comb operating around 1560 nm, with a repetition rate of 200 MHz, as the basis for the production of both the pump light at 1070 nm and the signal light at 1350–1750 nm for the generation of a mid-IR comb with the DFG process in PPLN. This was accomplished through the use of highly non-linear optical fibre (HNLF) with appropriate optical filtering, compression and amplification with Yb-doped and erbium-doped fibre amplifiers to boost the pump and signal powers. Using two such linked systems of different repetition rates, dual-comb spectroscopy spanning the mid-IR region of 2.6–5.2 μm was demonstrated, with sub-MHz frequency precision and accuracy. The complex spectra of several gases, including propane, carbonyl sulfide and a mixture of methane, acetylene and ethane [48] were measured, as well as various volatile organic compounds and atmospheric ethane in the presence of interfering species over long open-air pathlengths of up to 1 km [49].

As noted earlier, mid-IR combs may be generated directly from mid-IR laser sources such as QCLs [50–56]. Compact, electrically pumped, chip-based QCL combs have the advantages of simplicity, robustness and the potential for mass production based on standard semiconductor technologies. Comb generation in a QCL takes place in the laser chip through four-wave mixing and mode-locking as a result of the large third-order non-linearity arising from the inter-sub-band transitions, provided the dispersion has been minimised. Unlike conventional mode-locked lasers, since the gain recovery time in a QCL is much shorter than the cavity round-trip time, a QCL comb source does not emit a train of pulses, but the output intensity remains relatively constant with phase-coherent comb lines, similar to a frequency-modulated source. Development of chip-based QCL comb sources for spectroscopy is an active research area. Villares et al. [51] have demonstrated an on-chip dual-comb source where two QCL combs are formed on a single chip and the comb repetition and offset frequencies are controlled through temperature tuning of the refractive index by integrating micro-heaters next to each QCL. Several authors have addressed the issues associated with the application of chip-based QCL combs for spectroscopy. One such issue is dispersion compensation. For comb operation in the 7–9 μm range, the negative material dispersion of InP is compensated by the positive dispersion of the active material and the waveguide so that the overall dispersion can be close to zero, but this is not true outside this wavelength range. Lu et al. [52] have addressed this problem of dispersion compensation at shorter mid-IR wavelengths and demonstrated a QCL comb at 5 μm with 95 comb lines spanning a spectral range of 100 cm^{-1} with total output power up to 1 W. Other issues have been addressed by Hillbrand et al. [53] who noted that, depending on the bias current, a high phase-noise regime may occur in the operation of a QCL comb caused by the finite laser dispersion, limiting the application of QCLs to dual-comb spectroscopy. Also, optical feedback, which is difficult to eliminate in integrated chip-based combs, is fatal to comb operation. To address both these issues, Hillbrand et al. demonstrated the use of electrical injection-locking, where the laser round-trip frequency is locked to an external stabilised RF oscillator. This ensures the generation of a coherent, phase-locked QCL spectrum and mitigates the deleterious effects of optical feedback. A commercially available, mid-IR QCL dual-comb source, with a slightly different line spacing for the two combs is available from the company IRsweep [55]. Around 200 comb lines at ~7.5 GHz line spacing are generated in the ~6 mm long cavity of an InGaAs/InAlAs QCL, with centre wavelengths in the range 5–10 μm. The instrument provides a spectral coverage of ~60 cm^{-1}, spectral sampling interval of less than 0.5 cm^{-1} and a resolution of 10 MHz [56].

Compact laser combs have also been demonstrated with micro-resonators in silicon [57–60]. Silicon has a number of properties which make it an excellent platform for the generation of mid-IR combs and for the integration of several optical components on a single chip. These include a wide-transparency window over 1.2–8 μm, a high refractive index for tight confinement in ring resonators with high power densities and a large third-order optical non-linearity for the generation of comb lines through four-wave mixing. Low-loss micro-ring resonators of typically 100 μm radius using silicon nitride (Si_3N_4) deposition, compatible with complementary-metal-oxide-semiconductor (CMOS) technology, can be fabricated with very high *Q*-factors of $>10^7$ and low-threshold pump powers. The ring resonator may be pumped with a CW laser and power transferred through parametric four-wave mixing to the comb lines generated at the resonant mode positions, with a typical comb line spacing in the range of 20–400 GHz, depending on the radius of the micro-ring resonator. Tuning of the comb parameters over tens of MHz may be achieved through thermal control of the resonators and mode-locking by pumping in the anomalous dispersion regime. Using this technology, Stern et al. [58] have demonstrated a battery-operated near-IR comb chip, combining a micro-resonator and a pump laser on an integrated hybrid III-V/ Si_3N_4 platform. A III-V reflective semiconductor optical amplifier formed the gain section, which was coupled to a Si_3N_4 waveguide laser cavity to create a compact, low-power pump source at 1579 nm, requiring electrical power of less than 100 mW. Pumping the integrated micro-resonator generated a near-IR comb around 1550 nm, with a comb line spacing of 194 GHz and spanning a wavelength range of ~70 nm. For the mid-IR, Yu et al. [59, 60] have fabricated a mid-IR dual comb system on a silicon chip. The two mutually coherent mode-locked frequency combs were generated in two silicon, 100 μm radius, micro-ring resonators simultaneously pumped by the same external CW OPO source at 3 μm wavelength. The micro-ring resonators had a 12.8 MHz difference in FSR, which could be coarsely tuned by independent control of the micro-resonator temperatures. Around 305 comb lines were generated at a comb spacing of 127 GHz and spanning a wavelength range of 2.6–4.1 μm. Relatively low pump powers of 50–80 mW were required for the micro-resonators to generate high-power comb lines, varying from 2.5 μW to 2 mW over the limited range of 2.8–3.2 μm. The capability of the device was demonstrated by performing dual-comb spectroscopy on liquid acetone to acquire its spectrum over 2.9–3.1 μm at the comb line resolution of 127 GHz. Application to gas spectroscopy may be possible, but challenges arise due to the non-uniform power distribution across the comb lines, as well as their relatively wide spacing.

8.3 Mid-IR Spectroscopy Techniques

In general, the various spectroscopic techniques discussed earlier in this book, including wavelength modulation spectroscopy, ring-down spectroscopy, cavity-enhanced absorption spectroscopy, evanescent-wave spectroscopy and photoacoustic spectroscopy are all equally applicable in the mid-IR region, taking into account the longer wavelength of operation and any limitations imposed by the properties of mid-IR sources. We highlight a few examples in the following discussion.

8.3.1 Wavelength Modulation Spectroscopy (WMS) in the Mid-IR

There are a number of examples in the literature where standard WMS techniques, as discussed in Chapter 3, have been applied to mid-IR diode lasers. Tanaka et al. [61] used WMS with a DFB inter-band cascade laser (ICL) for ethylene (C_2H_4) detection in the exhaust of a portable power generator. The wavelength of the ICL was sinusoidally modulated at ~12 kHz and scanned at 1 Hz over the C_2H_4 absorption line at 3.356 μm. With a ~30 m pathlength Herriott cell and WMS based on the *2f* harmonic, a detection limit of ~100 ppbv at 3 kPa pressure was demonstrated. Similarly, Golston et al. [62] used WMS with a 3.27 μm GaSb laser for operation of a methane sensor fitted on board unmanned aerial vehicles for atmospheric measurements at various altitudes and over flexible flight paths. The modulation signals applied to the laser diode current consisted of a 200 Hz sawtooth ramp and a 20 kHz sinusoid for WMS with gas detection via a 24-pass Herriott cell of base length 11.2 cm and optical pathlength 2.7 m. Methane and water vapour concentrations were determined from the detected second harmonic WMS signal, normalised by the first harmonic, giving an in-flight precision of 5–10 ppbv $Hz^{-1/2}$ for methane. In Chapter 3, Section 3.5.1, it was noted that it is possible in certain circumstances to monitor in real time and in situ the laser and modulation parameters that are required for the simulation and analysis of the *2f/1f* signals obtained with WMS. Upadhyay et al. [63] have successfully investigated this approach for several laser types, including a mid-IR 5.25 μm QCL with sinusoidal current modulation at 8 kHz for concentration and pressure measurements on the R7 line of nitric oxide (NO) gas in a 10 cm pathlength cell. Compared with near-IR DFB lasers, the QCL had a smaller tuning range and a significantly larger *2f* no-gas background level arising from the non-linearity of the intensity-current (*LI*) curve. However, it was shown that the *3f/1f* ratio may be used for extraction of gas parameters, since the third harmonic background is almost zero for the QCL.

8.3.2 Cavity-Enhanced Absorption Spectroscopy (CEAS) in the Mid-IR

As discussed in Chapter 5, Section 5.3, the various techniques of CEAS are well established in the near-IR and research over the last several years has attempted to reach similar levels of enhancement for the mid-IR region [64–67]. The combination of QCLs with resonant cavities for attaining enhancement in gas absorption is reviewed in reference [64]. CEAS with optical feedback (OF-CEAS) is particularly appropriate for QCLs since it permits automatic self-locking of the laser output to cavity modes over a wavelength scan without the complexities of electronic locking and gives excellent line narrowing and signal-to-noise performance, even with lasers that may be spectrally broadened and affected by phase noise [65]. Ventrillard et al. [66] have used OF-CEAS with a QCL operating at 5.26 μm for trace gas analysis of NO, with a detection limit of 60 ppt and down to ~8 ppt for 10 s averaging time.

The combination of mid-IR-frequency comb sources with resonant enhancement cavities has also been demonstrated, where either the cavity mode spacing is locked to the comb spacing [68] or, as in Vernier spectroscopy [69], is slightly detuned from the comb spacing. For example, Foltynowicz et al. [68] have used CEAS with a mid-IR comb for trace detection of hydrogen peroxide in breath analysis. The mid-IR comb was generated by pumping an OPO with a high-power Yb:fibre laser at a pulse repetition rate of 136.6 MHz, giving a comb spectrum on the idler output at ~3.76 μm with ~200 nm bandwidth. The comb was locked to an external high-finesse gas cavity (finesse >3500) using the Pound–Drever–Hall method with both f_{rep} and f_{ceo} locked to the cavity to ensure stable operation. With a Fourier-transform spectrometer and auto-balancing detection, a sensitivity of 5.4×10^{-9} cm^{-1} Hz$^{-1/2}$ at a resolution of 800 MHz was achieved and a detection limit of 8 ppb hydrogen peroxide (in the absence of water) in a 1 s time interval.

8.3.3 Evanescent-Wave Spectroscopy in the Mid-IR

Early experimental work [70] on evanescent-wave gas spectroscopy in mid-IR used a standard step-index 125/50 μm silica fibre with a 10 mm length section tapered down to a diameter of 1.8 μm to increase the power fraction in the evanescent field outside the fibre to up to 40%. The 3.392 μm emission line from a helium–neon laser source was used to measure methane at a lower limit of 1% concentration, but the high loss of the silica fibre in the mid-IR and the fragile nature of the tapered fibre made the sensor rather impractical. A more practical approach is to make use of a waveguide evanescent-wave gas cell. As discussed in Chapter 5, Section 5.4.1, the high index contrast of a silicon waveguide formed on a SiO_2 substrate means that a significant fraction of the power can be carried in the superstrate evanescent field (typically ~25%), which has led to the realisation of a silicon-chip-based methane sensor in the near-IR at 1650 nm [71, 72]. The normalised curves of Figure A5.1.2 in Appendix 5.1 show that similar power fractions may be realised in the evanescent field at mid-IR wavelengths [73] with suitable adjustment of the waveguide dimensions according to the normalised thickness parameter. Hence, with further developments and novel waveguide architectures to avoid the absorption from the SiO_2 substrate, it may be possible to take advantage of the transparency of silicon up to 8 μm for the realisation of compact, silicon-chip-based evanescent-wave sensors in the mid-IR.

8.3.4 Photoacoustic Spectroscopy in the Mid-IR

The principles and methods of photoacoustic spectroscopy (PAS), as discussed in Chapter 4, are readily transferable to the mid-IR region, since the optical wavelength is not involved in the design of a resonant acoustic cell and, furthermore, detection is through the acoustic signal received by a microphone and not by a mid-IR optical detector. Provided intensity or wavelength modulation can be applied to the output of a mid-IR source, the advantages of PAS, such as a low acoustic background level and high sensitivity with high optical power can be attained in the mid-IR, with the additional benefit of strong acoustic signal generation from strong mid-IR gas absorption lines. Quartz-enhanced PAS (QEPAS) may be also be used with mid-IR sources and a review of the performance of QEPAS in combination with QCLs for the detection of a range of trace gases is given by Tittel [74]. An interesting example of the combination of PAS with a mid-IR frequency comb is given by Sadiek et al. [75] for the acquisition of broadband spectra of gaseous species in small sample volumes. The mid-IR comb was generated from a doubly resonant OPO pumped at 1.95 μm by a Tm: fibre femtosecond laser at a repetition rate of 125 MHz. The output signal comb from the OPO at ~3.3 μm and 250 nm bandwidth was first coupled to a fast-scan Fourier-transform spectrometer to apply intensity modulation on the comb lines at a frequency of $f_m = V_s/\lambda_{line}$, where V_s is the scan velocity of the optical path difference (0.16 cm s^{-1}) and λ_{line} is the wavelength of a comb line. Hence each comb line acquired a unique modulation frequency over the range ~470–510 Hz, which translates to an acoustic signal of the same frequency after interaction with the gas in a double-pass 10 cm long photoacoustic cell. A cantilever-enhanced broadband acoustic detector, operating in a non-resonant mode, acquired the photoacoustic spectrum, which was normalised by the intensity envelope of the comb. In this way the absorption spectrum of CH_4 around 3.3 μm in nitrogen at pressures between 400–1000 mbar was obtained with a detection limit of 0.8 ppm and a normalised noise-equivalent absorption (NNEA) of 8×10^{-10} W cm^{-1} $Hz^{-1/2}$.

8.4 Mid-IR Materials and Fibres

Table 8.2 shows a number of optical materials that may be used in the mid-IR for windows, lenses, etc., along with their refractive index and approximate transmission range. For harsh, high-temperature environments, such as in combustion analysis, quartz and sapphire are the best, since the other materials are generally less robust or, as in the case of Ge and Si, become opaque at high temperatures [3].

Compared with silica fibres, which have been extensively developed for the telecommunications market, mid-IR fibres [76] are in general much less mature and are more

Table 8.2 Common optical materials for mid-IR windows or lenses

Material	Refractive index	Transmission range (μm)
Quartz, SiO_2	1.45	0.2–3.5
Sapphire, Al_2O_3	1.7	0.2–5
Silicon, Si	3.5	1–10
Germanium, Ge	4.0	2–20
ZnS	2.3	0.5–10
ZnSe	2.4	0.5–15
MgF_2	1.35	0.2–7
CaF_2	1.4	0.2–8
BaF_2	1.45	0.2–12

expensive, exhibit higher attenuation, limiting the length over which they can be used, and in some cases are less durable physically and mechanically, with lower laser-damage thresholds. Mid-IR fibres fall into two basic categories, namely, those based on mid-IR transmitting glasses and hollow-core silica fibres.

The most common mid-IR fibres are based on heavy-metal fluorides and have a refractive index around 1.5, similar to that of silica fibres, but have lower tensile strengths and are susceptible to moisture and surface crystallisation. The fibre known as ZBLAN has a typical composition of 53% ZrF_4, 20% BaF_2, 4% LaF_3, 3%AlF_3 and 20% NaF and is commercially available in various forms, including single- or multi-mode versions, fibre patch cords and rare-earth-doped fibre for fibre amplifiers and lasers [77, 78]. The transmission window of ZBLAN fibre is ~0.4–4 μm with a loss of ~50–100 dB km^{-1} over this range. Efforts to reduce the loss (from scattering and micro-crystal formation) and approach the theoretical limits are being carried out by manufacturing ZBLAN fibre in the micro-gravity environment of space [79]. Heavy-metal-fluoride fibre based on AlF_3 with the composition AlF_3-BaF_2-SrF_2-CaF_2-MgF_2-YF_3 has better mechanical and hygroscopic properties and a higher laser-damage threshold compared with ZBLAN, but has higher losses beyond 3 μm [77]. Fluoride fibre based on InF_3 extends the transmission range to ~5 μm [78]. Another possible material for mid-IR fibre is chalcogenide glass, based on the chalcogen elements in the form of sulfides, selenides or tellurides [76]. Compared with fluoride fibres, chalcogenide fibres have the advantages of being more durable and less sensitive to moisture, but have a relatively high refractive index (>2) and may be brittle. They also exhibit higher losses in the mid-IR region than fluorides and are usually coloured, so transmission in the visible region is impaired. Examples of chalcogenide glasses that have been used for fibres include As_2S_3, As_2Se_3 and the binary system 70% Ga_2S_3:30% La_2S_3, with a transmission window extending to ~10 μm. However, absorption peaks may be present within the transmission window arising from the presence of impurities, notably the hydroxyl (OH) group, H_2O and S–H bonds. Fibres may also be formed from germanate glass (GeO_2) similar to silicate glass (SiO_2), but with the silicon replaced by germanium and PbO doping in the core to raise its refractive index. The transmission window of germanate-based fibres extends further into the IR to ~3 μm compared with silica fibres, but is dependent on the effective removal of hydroxyl (OH) group impurities [80].

Rather than using mid-IR transmitting glass, hollow-core fibres are made from conventional silicate glass, but the light is guided in a hollow core surrounded by a pattern of air holes forming an anti-resonant structure [2, 81]. Since most of the light is contained within the hollow core, mid-IR absorption from the silicate structure is minimal and hollow-core anti-resonant fibres have a broad mid-IR transmission window, with losses similar to fluoride fibres over the 3–5 μm range. Compared with fluoride fibres, they have the advantage that they do not suffer from hydroscopic and durability problems, but they must be protected from the ingress of contaminants at the end faces and may require purging to avoid attenuation arising from the presence of ambient CO_2 and H_2O absorption lines.

8.5 Mid-IR Detectors

The general principles and definitions outlined in Chapter 5, Appendix 5.2 for photodiode receiver circuits in the near-IR apply equally well in the mid-IR region. However, since the photon energy is inversely proportional to the wavelength, the longer mid-IR wavelengths require semiconductor photodiodes with smaller band-gap energies, E_g, as compared with the near-IR. Thermal transitions across the smaller band gap are more probable, so in general the noise performance of mid-IR detectors is inferior to those in the near-IR unless appropriate cooling of the detector is employed. The important thermal factor affecting noise performance is the Boltzmann distribution term:

$$\exp\left(-\frac{E_g}{kT}\right) = \exp\left(-\frac{hc}{kT\lambda_c}\right) \tag{8.3}$$

where λ_c is the longest wavelength of detection (cut-off wavelength for the detector).

It is clear from (8.3) that the product of temperature and cut-off wavelength must be constant for a given level of performance and the required detector temperature to achieve background-limited performance is approximated by [82, 83]:

$$T \approx \frac{300\,\text{K}}{\lambda_c(\mu\text{m})} \tag{8.4}$$

A range of mid-IR detectors are available commercially [84–86] and Table 8.3 lists the common materials used, along with the typical wavelength response range, depending on the exact manufacturer's specifications and the operation mode. Detectors may be operated in either photovoltaic or photoconductive mode, but photovoltaic operation is generally preferred due to its lower dark current. As discussed, some detectors have sufficiently low noise to be operated at room temperature, but others require cooling to improve the detectivity, D^*, which is typically in the range 10^9–10^{12} cm Hz^{½} W^{-1}. Peltier thermo-electric cooling is desirable since it is relatively simple to implement and may be sufficient in some cases, but cryogenic cooling with liquid nitrogen may be necessary for long-wavelength detectors.

Table 8.3 Common near-IR and mid-infrared detector materials

Detector type	Spectral response range (μm)
Si	0.2–1.1
InGaAs	0.5–2.6
InAs	1–3
InSb	1–6
InAsSb	5, 8, 11
HgCdTe (MCT)	2–16
HgCdTe (MCT)	2–16
PbS	1–2.9
PbSe	1.5–4.8

8.6 Near-IR and Mid-IR Gas Spectroscopy: Future Prospects

The last decade has seen major advances in the availability, performance and versatility of compact mid-IR laser diodes, particularly in the form of quantum well, inter-band cascade and quantum cascade laser sources, reaching the performance levels of near-IR sources. There have also been significant developments in the demonstration of frequency comb sources and dual comb spectroscopy in the mid-IR, based on either down-conversion from femtosecond near-IR fibre lasers or in direct, compact form with quantum cascade laser sources. Mid-IR detectors too can perform as well as their near-IR counterparts, with appropriate cooling when necessary, while photoacoustic spectroscopy in the mid-IR makes use of the same acoustic detection. Unfortunately, however, fibres and fibre components for the mid-IR cannot currently equal the performance and range of that available for the near-IR spectral region. In terms of future developments, silicon has many attractive features as a platform for the development of compact, integrated silicon-chip devices for both near-IR and mid-IR gas spectroscopy, including a wide-transparency window over 1.2–8 μm, high refractive index for tight confinement in ring resonators giving high power density in non-linear interactions and the relatively high-power fractions possible in the external evanescent field. Several examples of the integration of key elements on a silicon chip have already been demonstrated, including evanescent-wave gas cells in the near-IR, frequency combs, heterogeneous integration of QCLs and ICBs and hybrid integration of suspended AlGaAs waveguides [87]. The capability of mid-IR gas sensing is growing rapidly, but the near-IR region still holds the advantages of lower component costs, much greater flexibility by the use of standard fibre optics and a high level of maturity. Also, for applications which rely on natural back-scattered light, such as those discussed in Chapter 5, Section 5.6.1, for remote detection of gas leaks with hand-held instruments or unmanned aerial vehicles, the near-IR has the advantage of much stronger back-scatter as a result of the shorter wavelengths employed. For the foreseeable future then, near-IR gas sensing will have an important role to play in those applications where a high level of sensitivity is not the primary issue, but other factors such as cost and versatility are more important.

References

1. F. K. Tittel, R. Lewicki, R. Lascola and S. McWhorter, Emerging infrared laser absorption spectroscopic techniques for gas analysis, in *Trace Analysis of Specialty and Electronic Gases*, W. M. Geiger and M. W. Raynor, Eds., Hoboken, New Jersey, John Wiley & Sons, Inc., ch. 4, 71–109, 2013.
2. J. Hodgkinson and R. P. Tatam, Optical gas sensing: a review, *Meas. Sci. Technol.*, 24, (1), 1–59, 2013.
3. C. S. Goldenstein, R. M. Spearrin, J. B. Jeffries and R. K. Hanson, Infrared laser-absorption sensing for combustion gases, *Prog. Energy Combust. Sci.*, 60, 132–176, 2016.
4. N. Picqué and T. W. Hänsch, Mid-IR spectroscopic sensing, *Opt. Photonics News*, 26–33, 2019.

5. M. Ebrahim-Zadeh and I. T. Sorokina, Eds., *Mid-Infrared Coherent Sources and Applications*, Dordrecht, The Netherlands, Springer, 2008.
6. Eblana Photonics Ltd. Specialty laser diodes. 2019. [Online]. Available: www.eblanaphotonics.com/optical-sensing.php (accessed April 2020)
7. Nanosystems & Technologies GmbH. Distributed feedback lasers. 2019. [Online]. Available: https://nanoplus.com/en/products/distributed-feedback-lasers/ (accessed April 2020)
8. Sacher Lasertechnik Group. Laser diodes. 2019. [Online]. Available: www.sacher-laser.com/home/laser-diodes/distributed_feedback_laser/ (accessed April 2020)
9. A. Spott, E. J. Stanton, A. Torres, et al., Interband cascade laser on silicon, *Optica*, 5, (8), 996–1005, 2018.
10. F. Capasso, C. Gmachl, D. L. Sivco and A. Y. Cho, Quantum cascade lasers, *Phys. Today*, 55, (5), 34–40, 2002.
11. J. Faist, *Quantum Cascade Lasers*, Oxford, UK, Oxford University Press, 2013.
12. R. Pecharromán-Gallego. An overview on quantum cascade lasers: origins and development. 2017. [Online]. Available: www.intechopen.com/books/quantum-cascade-lasers/an-overview-on-quantum-cascade-lasers-origins-and-development (accessed April 2020)
13. M. Rose, A history of the laser: a trip though the light fantastic, *Photonics Spectra*, 53, (6), 39–49, 2019.
14. M. S. Vitiello, G. Scalari, B. Williams and P. De Natale, Quantum cascade lasers: 20 years of challenges, *Opt. Express*, 23, (4), 5167–5182, 2015.
15. Thorlabs, Inc. Quantum cascade lasers. 2019. [Online]. Available: www.thorlabs.com/newgrouppage9.cfm?objectgroup_ID=6932 (accessed April 2020)
16. Hamamatsu Photonics K. K. Quantum cascade lasers. 2019. [Online]. Available: www.hamamatsu.com/eu/en/product/lasers/semiconductor-lasers/qcls/index.html (accessed April 2020)
17. J. Wallace, Commercial quantum-cascade laser technology matures, *Laser Focus World*, 53, (7), 23–25, 2017.
18. L. Bizet, R. Vallon, B. Parvitte, et al., Multi–gas sensing with quantum cascade laser array in the mid-infrared region, *Appl. Phys. B*, 123, (145), 1–6, 2017.
19. P. M. Hundt, B. Tuzson, O. Aseev, et al., Multi-species trace gas sensing with dual-wavelength QCLs, *Appl. Phys. B*, 124, (108), 1–9, 2018.
20. Sacher Lasertechnik Group. Tunable external cavity quantum cascade lasers. 2019. [Online]. Available: www.sacher-laser.com/home/scientific-lasers/quantum_cascade_laser/quantum_cascade_lasers/tunable_external_cavity_quantum_cascade_laser.html (accessed April 2020)
21. DRS Daylight Solutions. About mid-IR quantum cascade lasers. 2017. [Online]. Available: www.daylightsolutions.com/home/technology/about-mid-ir-quantum-cascade-lasers/ (accessed April 2020)
22. R. Ghorbani and F. M. Schmidt, Real–time breath gas analysis of CO and CO_2 using an EC–QCL, *Appl. Phys. B*, 123, (144), 1–11, 2017.
23. R. Heinrich, A. Popescu, A. Hangauer, R. Strzoda, S. Höfling, High performance direct absorption spectroscopy of pure and binary mixture hydrocarbon gases in the 6–11 μm range, *Appl. Phys. B*, 123, (223), 1–9, 2017.
24. M. Tacke, New developments and applications of tunable IR lead salt lasers, *Infrared Phys. Technol.*, 36, (1), 447–463, 1995.
25. L. Dong, B. Samson, Mid-infrared fibre lasers in *Fiber Lasers: Basics, Technology and Applications*, Boca Raton FL, CRC Press, Taylor & Francis Group, ch. 14, 255–268, 2017.
26. S. D. Jackson, Towards high-power mid-infrared emission from a fibre laser, *Nat. Photonics*, 6, 423–431, 2012.

27. Keopsys Group. CW thulium fiber laser. 2016. [Online]. Available: www.keopsys.com/products-services-lasers-amplifiers/continuous-thulium-fiber-laser/ (accessed April 2020)
28. Thorlabs, Inc. Mid-IR laser sources: 2.7μm. 2019. [Online]. Available: www.thorlabs.com/newgrouppage9.cfm?objectgroup_ID=10061 (accessed April 2020)
29. Y. Cui, W. Huang, Z. Wang, et al., 4.3μm fiber laser in CO_2-filled hollow-core silica fibers, *Optica*, 6, (8), 951–954, 2019.
30. Eksma Optics. Nonlinear and laser crystals. 2019. [Online]. Available: http://eksmaoptics.com/nonlinear-and-laser-crystals/ (accessed April 2020)
31. I. Armstrong, W. Johnstone, K. Duffin, et al., Detection of CH_4 in the mid-IR using difference frequency generation with tunable diode laser spectroscopy, *IEEE J. Lightwave Technol.*, 28, (10), 1435–1442, 2010.
32. C. Fischer and M. W. Sigrist, Trace-gas sensing in the 3.3-μm region using a diode-based difference-frequency laser photoacoustic system, *Appl. Phys. B*, 75, 305–310, 2002.
33. M. Giguère, V. N. Dang and J. Salhany, Fibre lasers: mid-IR laser source is widely tuneable for standoff explosives detection, *Laser Focus World*, 51, (4), 59–61, 2015.
34. Genia Photonics. Picosecond tunable mid-IR laser. 2019. [Online]. Available: www.geniaphotonics.com/products (accessed April 2020)
35. Toptica Photonics. FemtoFiber dichro midIR. 2019. [Online]. Available: www.toptica.com/products/psfs-fiber-lasers/femtofiber-dichro/femtofiber-dichro-midir/ (accessed April 2020)
36. APE-Berlin. Mid-IR laser Carmina. 2019. [Online]. Available: www.ape-berlin.de/en/ssnom-afm-ir-with-carmina/ (accessed April 2020)
37. Toptica Photonics. TOPO widely tunable high-power CW OPO laser system for mid-IR spectroscopy and applications. 2019. [Online]. Available: www.toptica.com/products/tunable-diode-lasers/frequency-converted-lasers/topo/ (accessed April 2020)
38. APE-Berlin. OPO – optical parametric oscillator. 2019. [Online]. Available: www.ape-berlin.de/en/opo-optical-parametric-oscillator/ (accessed April 2020)
39. G. von Basum, D. Halmer, P. Hering, et al., Parts per trillion sensitivity for ethane in air with an optical parametric oscillator cavity leak-out spectrometer, *Opt. Lett.*, 29, (8), 797–799, 2004.
40. I. Coddington, N. R. Newbury and W. C. Swann, Dual-comb spectroscopy, *Optica*, 3, (4), 414–426, 2016.
41. N. Picqué and T.W. Hänsch. Frequency comb spectroscopy, *Nat. Photonics*, 13, 146–157, 2019.
42. A Ruehi, Advances in Yb:fiber frequency comb technology, *Opt. Photonics News*, 23, (5), 31–41, 2012.
43. F. Adler, K. C. Cossel, M. J. Thorpe, et al., Phase-stabilized, 1.5 W frequency comb at 2.8–4.8μm, *Optics Lett.*, 34, (9), 1330–1332, 2009.
44. Z. Zhang, T. Gardiner and D. T. Reid, Mid-infrared dual-comb spectroscopy with an optical parametric oscillator, *Opt. Lett.*, 38, (16), 3148–3150, 2013.
45. Y. Jin, S. M. Cristescu, F. J. M. Harren and J. Mandon, Femtosecond optical parametric oscillators toward real-time dual-comb spectroscopy, *Appl. Phys. B*, 119, 65–74, 2015.
46. Menlo Systems GmbH. Mid-IR optical frequency comb. 2019. [Online]. Available: www.menlosystems.com/products/optical-frequency-combs/mid-ir-comb/ (accessed April 2020)
47. E. Baumann, F. R. Giorgetta, W. C. Swann, et al., Spectroscopy of the methane ν_3 band with an accurate mid infrared coherent dual-comb spectrometer, *Phys. Rev. A*, 84, (062513), 1–9, 2011.
48. G. Ycas, F. R. Giorgetta, E. Baumann, et al., High-coherence mid-infrared dual-comb spectroscopy spanning 2.6 to 5.2μm, *Nat. Photonics*, 12, 202–208, 2018.

49. G. Ycas, F. R. Giorgetta, K. C. Cossel, et al., Mid-infrared dual-comb spectroscopy of volatile organic compounds across long open-air paths, *Optica*, 6, 2, 165–168, 2019.
50. F. Cappelli, G. Villares, S. Riedi and J. Faist, Intrinsic linewidth of quantum cascade laser frequency combs, *Optica*, 2, (10), 836–840, 2015.
51. G. Villares, J. Wolf, D. Kazakov, et al., On-chip dual-comb based on quantum cascade laser frequency combs, *Appl. Phys. Lett.*, 107, (251104), 1–6, 2015.
52. Q. Y. Lu, S. Manna, D. H. Wu, S. Slivken and M. Razeghia, Shortwave quantum cascade laser frequency comb for multi-heterodyne spectroscopy, *Appl. Phys. Lett.*, 112, (141104) 1–6, 2018.
53. J. Hillbrand, A. M. Andrews, H. Detz, G. Strasser and B. Schwarz, Coherent injection locking of quantum cascade laser frequency combs, *Nat. Photonics*, 13, 101–104, 2019.
54. G. Scalari, J. Faist and N. Picqué, On-chip mid-infrared and THz frequency combs for spectroscopy, *Appl. Phys. Lett.* 114, (150401), 1–5, 2019.
55. IRsweep. The IRis-cor –turnkey mid-IR dual comb source. 2019. [Online]. Available: https://irsweep.com/products/iris-core/ (accessed April 2020)
56. A. Hugi, A. -M. Lyon, M. Mangold, et al., Mid-Infrared spectrometer featuring μ-second time resolution based on dual-comb quantum cascade laser frequency combs in *CLEO Conference Photonic Instrumentation & Techniques for Metrology & Industrial Process*, San Jose, CA, 2017.
57. A. G. Griffith, R. K. W. Lau, J. Cardenas, et al., Silicon-chip mid-infrared frequency comb generation, *Nat. Commun.*, 6, (6299), 1–5, 2015.
58. B. Stern, X. Ji, Y. Okawachi, A. L. Gaeta and M. Lipson, Battery-operated integrated frequency comb generator, *Nature*, 562, 401–405, 2018.
59. M. Yu, Y. Okawachi, A. G. Griffith, M. Lipson and A. L. Gaeta, Mode-locked mid-infrared frequency combs in a silicon microresonator, *Optica*, 3, (8), 854–860, 2016.
60. M. Yu, Y. Okawachi, A. G. Griffith, et al., Silicon-chip-based mid-infrared dual-comb spectroscopy, *Nat. Commun.*, 9, (1869), 1–6, 2018.
61. K. Tanaka, K. Akishima, M. Sekita, K. Tonokura and M. Konno, Measurement of ethylene in combustion exhaust using a 3.3μm distributed feedback interband cascade laser with wavelength modulation spectroscopy, *Appl. Phys. B*, 123, (219), 1–8, 2017.
62. L. M. Golston, L. Tao, C. Brosy, et al., Lightweight mid-infrared methane sensor for unmanned aerial systems, *Appl. Phys. B*, 123, (170), 1–9, 2017.
63. A. Upadhyay, D. Wilson, M. Lengden, et al., Calibration-free WMS using a cw-DFB-QCL, a VCSEL, and an edge-emitting DFB laser with in-situ real-time laser parameter characterization, *IEEE Photonics J.*, 9, (2), 6801217, 1–18, 2017.
64. S. Welzel, R. Engeln and J. Röpcke, Quantum cascade laser based chemical sensing using optically resonant cavities, in *Cavity-Enhanced Spectroscopy and Sensing* (Springer Series in Optical Sciences, vol. 179), G. Gagliardi and H.-P. Loock, Eds., New York, Springer, ch. 3, 93–142, 2014.
65. J. Morville, D. Romanini and E. Kerstel, Cavity enhanced absorption spectroscopy with optical feedback in *Cavity-Enhanced Spectroscopy and Sensing* (Springer Series in Optical Sciences, vol. 179), G. Gagliardi and H.-P. Loock, Eds., New York, Springer, ch. 5, 163–209, 2014.
66. I. Ventrillard, P. Gorrotxategi-Carbajo and D. Romanini, Part per trillion nitric oxide measurement by optical feedback cavity−enhanced absorption spectroscopy in the mid−infrared, *Appl. Phys. B*, 123, (180), 1–8, 2017.
67. J. H. van Helden, N. Lang, U. Macherius, H. Zimmermann, and J Ropcke, Sensitive trace gas detection with cavity enhanced absorption spectroscopy using a continuous wave external-cavity quantum cascade laser, *Appl. Phys. Lett.*, 103, (131114), 1–4, 2013.

68. A. Foltynowicz, P. Maslowski, A. J. Fleisher, B. J. Bjork and J. Ye, Cavity-enhanced optical frequency comb spectroscopy in the mid-infrared application to trace gas detection of hydrogen peroxide, *Appl. Phys. B.*, 110, (2), 163–175, 2013.
69. A. Khodabakhsh, L. Rutkowski, J. Morville and A. Foltynowicz, Mid-infrared continuous-filtering Vernier spectroscopy using a doubly resonant optical parametric oscillator, *Appl. Phys. B*, 123, (210), 1–12, 2017.
70. H. Tai, H. Tanaka and T. Yoshino, Fiber-optic evanescent-wave methane-gas sensor using optical absorption for the 3.392-μm line of a He-Ne laser, *Opt. Lett.*, 12, (6), 437–439, 1987.
71. L. Tombez, E. J. Zhang, J. S. Orcutt, S. Kamlapurkar and W. M. J. Green, Methane absorption spectroscopy on a silicon photonic chip, *Optica*, 4, (11), 1322–1325, 2017.
72. W. M. J. Green, E. J. Zhang, C. Xiong, et al., Silicon photonic gas sensing, in *Optical Fiber Communication Conference (OFC) 2019*, San Diego, CA, USA, OSA Technical Digest, paper M2J.5, 2019.
73. M. A. Butt, S. N. Khonina and N. L. Kazanskiy, Silicon on silicon dioxide slot waveguide evanescent field gas absorption sensor, *J. Mod. Opt.*, 65, (2), 174–178, 2018.
74. F. K. Tittel, Mid-IR semiconductor lasers enable sensors for trace-gas-sensing applications, *Photonics Spectra*, 48, (6), 52–56, 2014.
75. I. Sadiek, T. Mikkonen, M. Vainio, J. Toivonen and A. Foltynowicz, Optical frequency comb photoacoustic spectroscopy, *Phys. Chem. Chem. Phys.*, 20, 27849–27855, 2018.
76. R. Paschotta. Mid-infrared fibers. 2008. [Online]. Available: www.rp-photonics.com/mid_infrared_fibers.html (accessed April 2020)
77. Fiberlabs Inc. ZBLAN Fluoride glass fibers & cables. 2019. [Online]. Available: www.fiberlabs.com/fiber_index/ (accessed April 2020)
78. Thorlabs Inc. Mid-infrared optical fiber. 2019. [Online]. Available: www.thorlabs.com/newgrouppage9.cfm?objectgroup_id=7062#ad-image-0 (accessed April 2020)
79. H. Pitman, Gravity-free optical fibre manufacturing breaks earthly limitations, *Laser Focus World*, 55, (1), 93–95, 2019.
80. H. T. Munasinghe, A. Winterstein-Beckmann, C. Schiele, et al., Lead-germanate glasses and fibers: a practical alternative to tellurite for nonlinear fiber applications, *Opt. Mat. Exp.*, 3, (9), 1488–1503, 2013.
81. J. Knight, D. Hand and F. Yu, Hollow-core optical fibres offer advantages at any wavelength, *Photonics Spectra*, 53, (4), 53–57, 2019.
82. A. Rogalski, *Infrared and Terahertz Detectors*, Boca Raton FL, CRC Press, Taylor & Francis Group, 3rd edn., 2019.
83. A. Rogalski, Graphene-based materials in the infrared and terahertz detector families: a tutorial, *Adv. Opt. Photonics*, 11, (2), 314–379, 2019.
84. Hamamatsu Photonics K. K. Infrared detectors. 2019. [Online]. Available: www.hamamatsu.com/eu/en/product/optical-sensors/infrared-detector/index.html (accessed April 2020)
85. Thorlabs Inc. Detectors, 2019. [Online]. Available: www.thorlabs.com/navigation.cfm?guide_id=36 (accessed April 2020)
86. Vigo System S.A. IR detectors, 2019. [Online]. Available: https://vigo.com.pl/en/products-vigo/ (accessed April 2020)
87. J. Chiles, N. Nader, E. J. Stanton, et al., Multifunctional integrated photonics in the mid-infrared with suspended AlGaAs on silicon, *Optica*, 6, (9), 1246–1254, 2019.

Index